Theater of the Mind

Theater of the Mind

IMAGINATION, AESTHETICS, AND
AMERICAN RADIO DRAMA

Neil Verma

The University of Chicago Press   CHICAGO & LONDON

NEIL VERMA is a Harper Fellow in the Society of Fellows in the Liberal Arts and a Collegiate Assistant Professor in the Humanities at the University of Chicago.

The University of Chicago Press, Chicago 60637
The University of Chicago Press, Ltd., London
© 2012 by The University of Chicago
All rights reserved. Published 2012.

21 20 19 18 17 16 15 14 13 12   1 2 3 4 5

ISBN-13: 978-0-226-85350-5 (cloth)
ISBN-13: 978-0-226-85351-2 (paper)
ISBN-10: 0-226-85350-0 (cloth)
ISBN-10: 0-226-85351-9 (paper)

Library of Congress Cataloging-in-Publication Data

Verma, Neil.
Theater of the mind : imagination, aesthetics, and American radio drama / Neil Verma.
  p. cm.
Includes bibliographical references and index.
ISBN-13: 978-0-226-85350-5 (hardcover : alkaline paper)
ISBN-13: 978-0-226-85351-2 (paperback : alkaline paper)
ISBN-10: 0-226-85350-0 (hardcover : alkaline paper)
ISBN-10: 0-226-85351-9 (paperback : alkaline paper)
1. Radio plays, American—History and criticism—20th century.
2. Radio broadcasting—United States—History—20th century.   I. Title.
PN1991.3.U6V47 2012
812'.02209—dc23
2011036722

CONTENTS

Acknowledgments vii

Introduction: What Is the "Theater of the Mind"? 1

PART 1 Radio Aesthetics in the Late Depression, 1937–1945

1 Dramas of Space and Time 17
2 Producing Perspective in Radio 33
3 Intimate and Kaleidosonic Styles 57
4 Norman Corwin's People's Radio 73

PART 2 Communication and Interiority in 1940s Radio, 1941–1950

5 Honeymoon Shocker 91
6 Dramas of Susceptibility and Transmission 115
7 Eavesdropper, Ventriloquist, Signalman 135

PART 3 Radio and the Postwar Mood, 1945–1955

8   Later Than You Think? 163
9   Just the Facts 181
10 In Trials 203

Coda: Instruction and Excavation 225

Guide to Radio Programs 231

Notes 241

Index 277

ACKNOWLEDGMENTS

This book is the sum of the generosity and dedication of many supporters who believed in me and in this project. During my early research, I was mentored by a group of advisers that would be the envy of any junior scholar. Tom Gunning was a gracious collaborator and a thoughtful critic who made it possible for me to do my best work. It was a privilege to have his intellect informing my process. Like all Tom's students, I constantly sense his enthusiasm and insight enriching every level of my work, and my debt to him only grows. I also owe a great deal to Bill Veeder, whose direction helped this project mature and whose detailed criticism kept it on the right course. I am lucky to have him as a mentor and grateful to have him as a friend. Bill Brown drew on his extraordinary breadth of expertise to help me shape many aspects of this research, and I feel especially thankful to him for his advice on its underpinnings and structure. Jim Sparrow taught me to think like a critical historian, introduced me to essential literature, and went out of his way to help me to evolve my ideas at every stage from conceptualization to publication. Jim has championed my work since the beginning, and I truly appreciate that.

I got my start on this project during a year I spent at Chicago Public Radio, an opportunity made possible by Jim Chandler and the Franke Institute for the Humanities. The experience was so rewarding thanks to Alison Cuddy, Delia Lloyd, Becky Vlamis, Steve Waranauskas, and especially Josh Andrews and Gretchen Helfrich. I am grateful to the Division of the Humanities at the University of Chicago for the award of a Mellon Dissertation-Year Fellowship, which carried forward my research tremendously. For their support, suggestions, and inspiration, I'd like to recognize Chicago faculty members Shadi Bartsch, David Bevington, John W. Boyer, Margot Browning, Thomas Christensen, Bert Cohler, Philippe Desan, Chris Faraone, Bob Kendrick, Jim Lastra, Armando Maggi, Joe Masco, William Mazzarella, W. J. T. Mitchell, Deborah Nelson, Barbara Stafford, Yuri Tsivian, David E. Wellbery, Rebecca Zorach, and the late Miriam Bratu Hansen. I am grateful to the staff of the Committee on the History of Culture, the Departments of English, History, and Cinema and Media Studies, the Franke Institute, the College, and the Society of Fellows in the Liberal Arts. My thanks also go to all of my students, who inspire me every day.

As I developed the manuscript, I got help from many exceptional scholars. I am grateful to Michele Hilmes and Jonathan Sterne, both of whom showed outstanding generosity of spirit in helping me to revise my work, challenging me intellectually, and prompting me to make the adjustments needed to complete this project. My thanks to Katie Chenoweth, Andrew Dilts, Judith Goldman, Dina Gusejnova, Markus Hardtmann, Leigh Claire La Berge, Spencer Leonard, Megan Luke, Ben McKean, Tim Michael, Lauren Silvers, Audrey Wasser, and all the rest of Chicago's Harper-Schmidt Fellows, for their friendship and collegiality. My ideas continue to be expanded by my fellow instructors in the Media Aesthetics sequence of Chicago's Humanities Core, including Tim Campbell, Hillary Chute, Xinyu Dong, Berthold Hoeckner, Reggie Jackson, Heather Keenleyside, Benjamin Morgan, John Muse, Larry Rothfield, Lisa Ruddick, and Jennifer Wild. For their kind encouragement from near and far, my thanks to A-J Aronstein, Scott Balcerzak, JoAnn Baum, Benjamin Blattberg, Jay Beck, Chris Buccafusco, Angela Bush, Myles Chilton, Neil Chudgar, Cheral Cotton, Nick Cull, Doron Galili, Mollie Godfrey, John Griswold, Elissa Guralnick, Michael Harnichar, Adam Hart, Anna Friedman Herlihy, Mark James, Alysha Jones, Andrew Johnston, Amanda Keeler, Sarah Keller, Elizabeth Kessler, Julia Klein, Jennefa Krupinski, Sara Beth Levavy, Salinda Lewis, Daniel Morgan, Tom Perrin, Christina Peterson, Chris Piatt, Alexander Russo, Shayna Silverstein, Michael Stamm, Timothy Stewart-Winter, Jennifer Stoever-Ackerman, Jeff Sypeck, Bethany Thomas, Tom Thuerer, Caity Tully, Julie Turnock, Shawn VanCour, Simeon Veldstra, Jessica Westphal, and the late Doug MacDonald.

I have presented some of the material in this book in the following workshops and conferences at the University of Chicago: the Mass Culture Workshop (2006 and 2009), the History of Culture Workshop (2006), the "Elements of Style" Conference (2007), the History of Culture Symposium (2007), the

Interdisciplinary Approaches to Political History Workshop (2008), and "Contradiction," the Weissbourd Conference for the Society of Fellows in the Liberal Arts (2011). I have also presented some of this work at conferences elsewhere: "The Frankfurt School Reconsidered," York University, Toronto (2007); "Justifying War," the University of Kent, Canterbury (2008); "Back Down to the Crossroads," the American Studies Association Annual Meeting, Albuquerque (2008); and "Media Citizenship," the Society for Cinema and Media Studies Annual Meeting, New Orleans (2011). My thanks go to the organizers of these events, and to participants for their feedback. A version of chapter 5 was published in the *Journal of American Studies*, and I thank the reviewers and editor Susan Castillo for their constructive criticism during that phase.

This book was completed only thanks to the encouragement and dedication of the staff at the University of Chicago Press. It has been a great pleasure to work with Robert Devens, who took a welcome interest in my project, asked challenging questions, found ideal readers, and designed a process to help me transform and polish its early incarnation. Russell Damian, Anne Goldberg, and Ruth Goring have been a terrific help in bringing this project to completion. For their hard work creating illustrations, I want to thank Amber Joliat and Kathy Moretton, as well as my dear friends Ruth Silver and James Hetmanek, who really came through for me, as they always do. My thanks also go to those who helped me out with images, including Kirsten Reoch of the Park Avenue Armory, Bob Kosovsky at the New York Public Library for the Performing Arts at Lincoln Center, Jeremy Megraw at the Billy Rose Theater Collection, the US Army Center for Military History, Geoff Swindells of Northwestern University Library, the Tee and Charles Addams Foundation, the Bill Mauldin Estate, the Thomson Company, and the Cartoon Bank at Condé Nast. A special thanks to Carlisle Rex-Waller, who did such a thorough and professional job editing the final manuscript.

I am so grateful to my family members, who have shown such faith in me over the many years I've been working on this project: Caryl Verma, Harish Verma, Arun Verma, Sonia Dinnick, Wilf Dinnick, Delia Enright, and our newest addition, my lovely daughter, Margaret Rose Verma. My family is very close in spirit, and it is only through the support and devotion of each one of these remarkable people that this project came to be. I love you all so much. Most of all, I thank my wife, Maureen McKinney, who has been an indispensable critic, ally, and advocate in the many years that it took to write this book. Her sacrifice, faith, and encouragement made this project possible. Maureen is my best friend, and I am blessed each day by her generosity, wit, and love. I can hardly begin to thank her for the happiness of our life together.

Finally, a special thanks to Norman Corwin. It was an honor to speak with him now and then over the years as I researched and wrote this book. Norman was the last greatest living bridge to the radio age, and he remained to the end of his days a passionate believer in what he liked to call "the unadorned human voice." I dedicate this book to him and to that belief.

INTRODUCTION

What Is the "Theater of the Mind"?

In a 1956 article in the *Los Angeles Times*, entertainment reporter Walter Ames wrote that although television was fast becoming the most common form of day-to-day entertainment in the United States, broadcasting executives were discovering that a few marquee radio programs still drew audiences of eight million or more.[1] This persistent popularity was partly attributed to the low cost and portability of radio receivers. As one industry leader put it, "You can't get television in a canoe." But CBS executive Guy della-Cioppa had a different idea, telling Ames that as networks curtailed their soap operas and adventures, he began to receive letters from listeners requesting new dramas. In response, CBS commissioned scripts about crusading district attorneys and rugged Civil War heroes, even briefly reviving *The Columbia Workshop*, a highly inventive anthology program that had been off the air since 1947. What accounted for the sudden appeal of a genre that had only recently faded away? A onetime radio dramatist himself, della-Cioppa speculated that television viewers might have reached a point of saturation at which they started to yearn for something that would involve more of their own "imagination." He continued: "As a little boy in Tampa once said while watching a television story, 'You know, mamma, I like stories better on radio 'cause the pictures up here [pointing to his head] are better.'"

Della-Cioppa's enthusiasm proved wishful. In the American broadcasting system, the network's true customer is the sponsor, not the listener, and in the end few new radio dramas attracted the kind of advertising dollars needed to justify national distribution in the late 1950s. Nevertheless, by linking radio to "imagination" and "pictures in the mind," della-Cioppa's anecdote exactly captures how Americans thought about radio plays during the twentieth century. In 1956, there was less nostalgia for the content of old programs than there was for the process listeners used to unravel them. That affection still resonates today, justifying every revival of old-fashioned radio in the intervening decades. Indeed, in one form or another, the story of the Florida boy remains one of the most venerable clichés in informal conjecture about medium-specificity the world over. To hear echoes of it, tune into an "old-time radio" show on terrestrial stations or satellite services; listen to dramas on NPR, BBC, or any public radio network from South Africa to Canada; search hundreds of fan websites devoted to collecting classic broadcasting; watch a film depicting the golden age of radio like Woody Allen's *Radio Days* (1987), Ken Burns's *Empire of the Air* (1991), or Eric Simonson's *On a Note of Triumph* (2005); or read one of many nostalgia books on radio culture. Do any of these things and you will eventually encounter an expression that recapitulates the apocryphal boy's gesture toward his temple. Wherever old radio plays resound, we utter the same phrase every time: "Radio, the Theater of the Mind."

It is a saying so consistently associated with radio drama that it often serves as a synonym for the genre, if not the entire medium, and it has become a part of how writers approach what was perhaps the most prolific form of narrative fiction for two decades of the twentieth century. Yet despite its currency, the coinage of the phrase is unclear. The *New York Times* first ran the words "theater of the mind" not referring to radio at all, but as the title of a 1949 television show about neurotics.[2] Quotation anthologies contradict each other, with one nominating crooner Smilin' Jack Smith as the first to use the phrase and another linking it to radio curator Ken Mueller.[3] Some historians point to poet Stephen Vincent Benét, but others associate the term with actor Joseph Julian.[4] Actually, in his memoir, Julian describes radio not as a theater of the mind but as a theater *in* the mind.[5] Usage is unaffected by the substitution, which suggests that the preposition *of* specifies a location, hence the boy pointing to his head, the area of the body in which, as W. J. T. Mitchell has noted, all minds and mental images are housed.[6] Perhaps "theater in the mind" is a superior way to describe what the phrase is after, since internalization is the principle that governs the saying, which names one medium (radio) by its capacity to nest a second medium (theater or pictures) in a third (mind or imagination). That suggests an irony to our common notion of radio as a "mass" medium. The voices of Amos 'n' Andy, Jack Benny, and the Shadow lived in American collective experience only by existing in tens of millions of mutually isolated theaters at once. Maybe all narratives ultimately take place "in the imagination" in this introverted manner—David Hume wrote of the mind as a "stage" on which perceptions mingle and strike postures like theatrical players—but fans and theorists of radio drama successfully particularize its credentials around that scenario.[7] Radio is not *a* theater of the mind; it is *the* theater of the mind.

This idea merits more critical thought than it tends to receive. After all, can we dismiss the relation of the psyche and medium in the "theater of the mind" as little more than a Mcluhanesque conceit? Is the historical value of the phrase exhausted by the idea that radio makes pictures inside the imagination? And how do you "draw" those pictures, anyway? Ascertaining the source of the phrase "theater of the mind" is less important than finding a better way to study the receptive habit that stands behind it. With that idea as its starting point, this book is an aesthetic and cultural history of classic American radio drama about the ways in which writers and directors "worked in the mind," treating this evolving and amazingly undertheorized process as one of the defining activities in quotidian American life in the mid-twentieth century. Like any structure, I argue, the theater of the mind had to be built. By focusing on a few important stages in that assembly, I will show that the insinuation of congress between mind and medium through radio stories arose because the aims of the networks and the ideas of key broadcasters intersected with a particular set of political, technical, and cultural developments. During the golden age of the genre, radio dramatists confronted the caprices of their medium, invented ways to guide listeners in stories, and also spoke to upheavals precipitated by hardship and war, three simultaneous errands that involved a suite of overdetermined questions about relationships, suggestion, and interiority. At stake in solving elementary problems of representation in radio was a pragmatic understanding of the process by which the mind acts on signals received from the mass media, so a genre believed to make pictures in the "imagination" defined a vernacular sense of that very psychic faculty. The matter of the mental image, as Mitchell points out, is a battleground for theories of the mind.[8] In short, in what follows, I argue that as American broadcasters built a theater *in* the mind, radio drama necessarily became a theater *about* the mind, in an era in which that concept was a site of extraordinary contest. Placing this insight at the center of this book, I hope to rethink a number of classic broadcasts, call attention to some that have escaped critical notice, and transform the theater of the mind from an idiom into a heuristic, thereby more profitably conceptualizing a dramaturgy whose complexities are at least equal to the enchantment that we ascribe to them.

STUDYING RADIO PLAYS

To get at the experience that we call the theater of the mind and to fit this medium into its milieu, this book draws on manuals, reports, commentaries, and trade publications, but the center of gravity of my research rests in the audio recordings themselves. To study old radio tracks in an organized way, the first challenge is abundance. In 1937, by one reckoning, American radio put five million shows on the air.[9] Although only a fraction of this material has survived, it remains quite difficult to decide which genres, airtimes, and programs to select in order to arrive at a fair account. Existing indexes are unhelpful for this purpose. The best way to know whether or not a play is important is to listen to it, but with tens of thousands of recordings available, there is almost no limit to the time such research might require. Measures of audience size can help, but criteria based on sheer popularity constitute a bad metric for a history of style, and since hundreds of thousands of listeners were definitely out there for even the most poorly rated network shows, it seems trivial to quibble over how large an audience is required to merit consideration. Mindful of these issues, I selected about seventy programs to study in depth (4 to 650 broadcasts each, spread across available broadcast seasons), along with more than ninety secondary shows (1–3 broadcasts each), totaling approximately six thousand unique broadcasts.[10] I included plays in hour-long, half-hour, and fifteen-minute formats and drew from programs that aired throughout the years covered by this book. I studied anthology programs, sagas, and serials from each of the main networks and included shows affiliated with most of the ad firms involved in brokering the sale of network time. I chose very popular programs such as *The Lux Radio Theater*, but I also addressed experiments such as *Lights Out!* My research ranged from daytime to evening and from comedy to docudrama, soap operas, and westerns, but this book focuses on experimental playhouses, crime shows, and thrillers that aired in the evening because I believe that the set of conventions that we call the "theater of the mind" primarily developed in them, and I will make that case over the course of these chapters. My sample is unusually capacious relative to other work on the history of radio drama, but it cannot help but remain imperfect. It is not my design to obviate counternarratives on this subject but to invite them.

To describe these broadcasts, I use the terms "radio play" or "radio drama" in order to highlight the "theater" aspect of my triggering question. These terms have some historical justice. Radio drama hybridizes several trades and expressive registers, but as a rule the actors, writers, directors, musicians, engineers, programmers, spokespeople, and advertisers in the field tended to think of their work as "show business." Their work also satisfies many criteria that define dramas in the Western tradition.[11] The radio platform facilitates most of the elements outlined in Aristotle's theory of tragedy, even elevating plot, character, and diction, the components needed for purifying catharsis.[12] Indeed, the ancient Greeks may have practiced a primitive form of radio drama by wearing masks equipped with amplification devices and performing in auditoria that were acoustically designed so as to let voices reach distant seating. Nowadays we seek a more medium-specific idea about what a radio play is. Radio dramatists themselves often spoke of how the genre conjoined the three elements of voice, music, and sound effects. Here I opt for a minimal definition, construing the term "radio play" to denote a radio broadcast that has a scene in which a character performs an action. I emphasize "scene" so as to include re-creations of news events but to exclude reading aloud, in which a voice may speak in character but does not purport to occupy space other than the studio. A drama is a story but it also must conjure a place. As Roland Barthes noted in his reading of the legacy of the aesthetics of Diderot, the existence of a scene is not just a feature of theater but the very condition that allows us to conceive of theater.[13] My emphasis on "action" is meant to distinguish radio performance from the plastic and pictorial arts while also reconnecting the word "drama" back to its first denotation; as G. E. Lessing emphasized, unity of action across scene is among the most enduring of dramatic questions.[14] Of course, radio uses aural means to relate action, but that is true of many depictions on stage or screen. Czech critic Jindřich Honzl once defended voice drama on just these grounds. If a theater is no more (or less) than a system of signs, Honzl reasoned, then there is no difference between a voice with a visible body and one without.[15] The ghost of Hamlet's father has an identical ontological status as an apparition on stage and as a disembodied voice crying from under the stage, since both acts signify a character in the code of the performance. Under this logic, acts conveyed by movement are just as "dramatic" as those related by sound effects. Here I take Honzl's insight as inspiration, not dogma. The compositional machinery that evoked "scene" and "action" in radio is not quite so arbitrary as it seems in this account. Both terms contained representational conundrums and responded to heterogeneous and shifting audience expectations; both were in constant dispute throughout the golden age of radio, and that is just what makes them so rich in the context of this study.

This book covers the years from 1937 to 1955. My narrative begins not long after the FCC Act of 1934 solidified a business model in which large networks began to really dominate the staging of plays, and it concludes as that system unraveled, around the time that Guy della-Cioppa mistook nostalgia for his old métier as a sign of its continued life. This periodization is in part pragmatic, since many recordings have survived from these years. Dramas of the late 1930s to mid-1950s also feature enough continuity in creative personnel that we can often hear styles and approaches unfolding among a finite group of influential practitioners, thus giving my readings historical connectedness. But criticism must exceed biography. Like Raymond Williams, I see merit in studying drama primarily through "conventions"—evolving customs, assumptions, widely shared means to "win the consent" of the audience.[16] Conventions are the connective tissue between dramatists and listeners, between single plays and predominant sound, between a dramatic practice and the public it meets and creates. They bind the design of a studio to the voice of an actor, the use of a sound effect to a production note, and ultimately link drama to a social milieu. This book emphasizes how creators and listeners "came together" in conventions pioneered by dramatists such as Archibald MacLeish and Irving Reis in the late 1930s in part because of the distribution achieved in those years. The aesthetic system that we call the theater of the mind could not have been conceptualized without inventing a national audience to which radio could speak.[17] By sketching pictures that ranged from family tragedies to national ceremonies, dramatists of the 1930s played exploratory games with space and time while dealing with the crises of the era. That practice transformed during the war years, when a theater about drawing exterior places in the mind became one about interiority. With the rise in morale-building programs and psychological "shockers," 1940s dramatists drew the idea of influence into their practice in perhaps the most decisive stylistic development in the history of dramatic radio in the United States. After the war, programs such as *Dragnet* and *Suspense* found new ways to explore authority and inner life. In stories of undercover men and repressed housewives, radio dramatists "rethought" representational problems through the filter of postwar anxieties. In this way, dramatists gave descriptive force to the "theater of the mind" idea by making mass media and consciousness seem coextensive. That schema supported an enduring contemporary belief—the idea that questions about psychic life are always contiguous with questions about communication media, because classic radio stories present these two topics as coupled and mutually complicit.

I make that last point with caution. Here I am not positing that midcentury radio plays *pioneered* ideas about the coextensiveness of mind and medium. Several scholars have shown this idea to be not only older than the golden age of radio, but also a "media fantasy" associated with a wide variety of modern technologies.[18] Jeffrey Sconce, for one, has shown that a "shared representational strategy" links together telegraphs, radios, televisions, and computers across the modern imaginary.[19] In stories about each of these devices, flows of thought mingle with flows of energy and information, just as radio waves enter the mind and are transmuted into images in della-Cioppa's anecdote. In what follows, I promulgate no alternative "origin" for this fantasy but show how it became an important part of the craft of radio storytelling in the late 1930s and beyond, as a result of how dramatists conceived their work in a particular context. If radio truly became a theater about the mind, then it should come as no surprise that broadcasters aired many elaborations of Sconce's strategy, just as they would later air tales that use vulgarized Freudian accounts of the psyche. But the fact that radio appropriated such fantasies unevenly, plurally, and in a particular sequence is yet to be explained. You could think of the narrative below as the story of how dramatists used radio as a way to "do politics," then to "do consciousness," and finally to "do psychology." The preexistence of fantasies on all of these subjects has a limited ability to explain why things happened that way, and it tells us little about how aesthetic practices normalized fantasies and managed transitions from one to the next. Still, the fact that radio aired plays directly thematizing the interaction of electrical and psychic energies may have solidified that idea, making the relation between mind and medium seem appropriate to audiences of enormous size, thereby reinforcing this link with more fervor than perhaps any other single pop culture phenomenon. Ever since the golden age of the medium, we have tended to contemplate "the mind" through "the media" and vice versa, a predilection that is in part due to the legacy that this largely extinct genre had on American thought and folklore.[20]

THE AESTHETIC PERIMETER

In order to open up the history of radio plays, this book proposes interconnected frameworks for performing formal readings and digging into aesthetics, a relatively new priority in radio studies. In the past, books about classic radio have primarily concerned the design of networks, the role of government oversight, and the nature of listening communities.[21] Writers have tended to think about radio as an "industry" or a "technology" rather than as a performance platform, and so the logical approach has been to study its political economy using arguments associated with the "social construction" mode of critique, which stresses the social and political forces that shape technological systems over time.[22] With these approaches and others, scholars such as Susan Douglas, Robert McChesney, and Susan Smulyan have written bedrock scholarship that explains how the industry evolved from the Great War to the 1930s as corporations lobbied against high-powered stations while controlling small stations in order to privatize the airwaves, wire together networks through affiliations, and adopt advertising as a chief revenue stream. Defeating alternative models, business staged a de facto acquisition of the airwaves, creating a durable oligopoly. In this policy-centered critique, it hardly matters whether networks aired Chekhov or *Little Orphan Annie*. Since the literature is calibrated to show legerdemain behind how the airwaves became a private commodity, it is ill equipped to assess how content shaped meaning. Writers need not listen to many programs to relate the controversies of utmost concern. But this account is only one component in a variety of works on broadcasting that have been published recently by Michele Hilmes and others.[23] Influenced by social and cultural historians, these authors make audiences the centerpiece, treating broadcasts as hitherto ignored evidence of American perceptions about race, gender, consumerism, civil rights, and other issues.[24] Writers are beginning to treat broadcast recordings as seriously as federal records, surveys, and the press, changing the entire complexion of radio studies by asking how radio "talked about itself." To the scholarship on how and why people listened, we are now adding details on what people listened *to*, making broadcasts a prime source material for investigating the discursive construction of the medium.[25] This pursuit has led many writers to the methodological problem of how to read radio dramas in a way that ascertains how they produce images and beliefs in the context of a structured listening encounter, thereby drawing radio's cultural history into the uneven perimeter of its aesthetics.

Within that circle, writers find modest resources. While researchers have recently made many inroads at grasping the behind-the-scenes practices of programmers and the role of listeners in shaping narratives, students seeking methodical work on radio writers or deep critique of dramatic broadcasts turn up little, as humanistic writers have often limited their comments to appraisal, lauding an effective choice without explaining how it came to be thus. This situation has begun to change.[26] Critics such as Elissa Guralnick have provided models for close reading, theorist Andrew Crisell has published an influential semiotics-based approach, and scholars have rediscovered the work of Rudolf Arnheim, perhaps the only theorist to undertake a true "aesthetics" of radio. Still, the field has adopted no standard argot with which to perform routine interpretive tasks like describing scenes, explaining segues, or grappling with patterns in dialogue. Not long ago Rick Altman dryly noted that radio scholarship all but ignored the "actual sound" of radio.[27] Even now, it remains difficult to compare the styles of writers and directors, trace how techniques evolved, or argue over how plays reflect historical moments. Just as historians would benefit from extensive listening, critics might do more intensive listening. This book pursues both objectives, providing approaches to help explore portrayals of time, space, perspective, dialogue, narration, characterization, and other properties that require contextualized theorization in order to apprehend how radio managed the traffic between mind and medium that we have long intuited but lacked the vocabulary to unpack.

After all, with more than half a century separating the end of the golden age of the genre from today, why *is* there no major historical study of the dramatic conventions of radio? The problem may be a critical error perpetrated by the "theater of the mind" cliché itself. Radio has always seemed a poor candidate for theorization precisely because we associate it so strongly with the imagination. In fixing attention on the listener, we place directing, engineering, and writing beyond the ambit of research. For instance, writer Paul Fussell calls the 1940s a "special moment in the history of human sensibility," because radio honored the "creative imagination" by obliging us to fill in the "missing visual dimension"; radio is thus characterized as what Marshall McLuhan called a "cool medium," one that asks us to "fill in" perceived sensory gaps.[28] But the cool-medium model erroneously treats "imagination" as a force without structure or history. The model also impoverishes criticism by asking us to study radio by what it does *not* offer. We start out by supposing that every sensory datum must possess a visual in the first place, and then forever associate radio with metaphysical scarcity. That stance would be unacceptable in any parallel situation. The lack of visuals is not deemed to be the essence of recorded music, although the same situation applies; we routinely call radio a "blind medium" but never call photography a "deaf medium" or let its silence blind us to visual features.[29] Indeed, the Floridian boy in the anecdote above could have been describing reading novels rather than listening to radio, but it would be risible to argue that our ability to picture characters mentally makes literary criticism futile. Yet the rhetoric holds that, unlike every other means of representation, radio only fabricates lack. That is not so. Radio drama is a positive endeavor, evoking scenes through speech, reverb, filter, segue, and other devices directed at an imaginary allowing itself to be instructed. In belittling these efforts, we surrender to their spell too eagerly. Our reticence may stem from what media historian Alexander Russo has shown to be a cunning disavowal on the part of 1930s broadcasters: by plugging radio as a "blind medium" that uses the "imagination," dramatists reified their processes of production and at once enhanced and concealed their ability to draw pictures in the mind.[30] The "imagination" became a trap into which critique stepped. To escape that trap, we must resist obsessing over listener involvement and begin from what is undeniably *in* the broadcast, asking how it set parameters to that participation and seeking the repertoire of techniques with which dramatists made themselves invisible in a theater of their own construction. Paradoxically, to make a case for the "theater of the mind," you have to begin by dismantling it.

My approach to that task began as a series of ad hoc responses to the richness of the material that drew on formal stylistics, cultural history, and literary historicism and only later became systematized. This book uses close readings of plays and "distant reading" of programs, along with new kinds of illustration and old-fashioned shoptalk.[31] I occasionally argue at a microinterpretive level, positing that the acoustics of a play explicitly about democracy in crisis implicitly explore tensions in progressive thought, or that the syntax of a policeman's soliloquy conveys aspects of postwar social anxieties. I also chart dense constellations of traits evolving among several genres at once or across hundreds of broadcasts over decades. But the diverse local methods in this book are all guided by the global methodological premise that it is necessary to confront common narrative moves and aural details and provocations, as well as the overall form of a drama as revealed in directly perceivable qualities—or the "actual sound" of radio. I study both what characters say and how they say it, which is a fact of broadcasting that exists in the material, reflects experience, and belongs in the ambit of media history. As the boy in Guy della-Cioppa's story knew, the essence of radio is in what it stimulates us to do, and how. The chapters below therefore ask historically rooted questions about styles, norms, devices, and structures that prompt us to transform a story in the air into an inward-facing picture, thereby mediating inner and outer worlds. To express my pursuit, I employ the term "aesthetics" as it is used by critics who follow theorist Walter Benjamin in applying to the mass media the Greek notion of *aisthesis*, a theory concerned with sensory feeling and perception.[32] As Miriam Hansen argues, to study modern vernacular experience is to theorize the implication of technology in how modes of sensory life are at once produced and negotiated.[33] By virtue of existing exclusively in a mass media platform, radio drama could palpate preexisting habits of sensation and also reorganize them in a single gesture, inaugurating economies of innervation that remade the category of the sensible. Arnheim called radio a *Hörkunst*, a term that suggests both a "sound art" and also an "art of listening."[34] It is this double process of craft and reception, of instruction and response, that the framework of aesthetic theory, broadly construed, can access in a historical way. After all, if what we long for about the theater of the mind is less *The Shadow* and more the mode of "pictorial listening" that *The Shadow* invents, then an aesthetic history is inevitable, and it may provide one of the best chances to tell a precise story about American perception during the middle decades of the twentieth century.

So the question of how to "do aesthetic history" is the first methodological puzzle this book faces. The second is how to connect aesthetic history to other areas of critical and historical thought. The chapters below draw on and contribute to sound studies, drama criticism, narratology, and social, cultural, and intellectual history. In some areas, I treat plays as "ideology" in the traditional sense, arguing that radio broadcasts reproduce values rooted in the apparatus of commercial capitalism, perpetuating its mystifications; other readings treat that stance skeptically, emphasizing that dramas provide alternative options of knowledge and experience.[35] In the American context, radio dramas are always reinforcing the system of which they are appendages, but no status quo is perpetuated without leaving a remainder. Theodor Adorno once wrote that mass audiences could experience entertainment in doubled states of acceptance and disbelief at once, a "split consciousness" in which listeners held ideological mystification at critical remove as they also enjoyed amusement—no matter how powerful the "culture industry," its most passionate critic conceded, "society cannot have it all its own way."[36] That is just how many Americans listened to radio plays, in a state of incredulous credulity toward the inducements arrayed before them, and it would be unjust to oversimplify that encounter. Besides, whether it is due to ambition or haste, most of the plays here bear a plurality of messages—along with assumptions, ciphers, hunches, and accidents—and so assessment of them requires conjecture when it comes to particular properties, as well as agnosticism about the "ultimate" social meaning of the form itself.

I have a similar attitude when it comes to using recordings to learn about the past. I am committed to the idea that these broadcasts index a variety of aspects of social life, and I will argue that values and anxieties are encoded in particular norms. But that is only one part in a dynamic process. Performance theory has it right in positing that social performances inform drama, just as dramas provide patterns for what we do as social beings.[37] I think of radio drama as a voice echoed in an imperfect acoustic mirror, an utterance that begins inside and becomes a living voice that is later muffled, redirected, and transformed by encounters with various textured surfaces, before returning to its source of emanation. Dramatic radio rarely seems to be "saying things" and more often seems to be "thinking about things" in a restless, loose, and echoing manner. That process resembles what Raymond Williams calls "the structure of feeling."[38] For Williams, aesthetic history identifies the shifting affects that anticipate concepts and acts—a structure of feeling is "not feeling against thought, but thought as felt and feeling as thought"—and the study of it thus probes how shared moods and impulses are always being constructed, dissolved, and reformulated prior to any doctrine that might harness those impulses.[39] To put it another way, and to layer a third and last gloss on Guy dellaCioppa's old chestnut about how pictures are better in your head, I propose that in its heyday radio drama was not only a theater *in* the mind and one *about* the mind, but it was also a theater *for* the mind, a laboratory in which shared meanings—including ideas about mediation itself—were drawn in, examined, and refashioned before reemerging into public understanding in a sequence of events that was just as messy and recursive as it sounds. Perhaps that is what gives this kitschy material such critical allure. As Jonathan Sterne has noted, the study of the auditory past is limited to speculation by the fact that we can never quite know how any sound was experienced inwardly, since history is only exteriorities.[40] If that is right—and if old radio is called the "theater of the mind" because it perforated, instructed, and theorized interiority—then to study radio plays is to reach for something hanging halfway over the edge of critical availability.

THE LISTENING IMAGINATION

This book is structured in three parts, each describing aesthetic aspects of evening radio dramas during a particular period: 1937–45, 1941–50, and 1945–55. Each section contains its own story, as all three work together to account for the longer arc of how broadcasting practices evolved in concert with a series of historical developments. Early chapters tend to deal with granular issues of aesthetic composition such as amplitude, sound effects, and other aspects of "sound aesthetics" proper, while later chapters tend to concentrate on "larger" questions about character, plot, genre, and other issues linked to performance aesthetics more broadly. Each time frame is also presented as a period in which dramatists tended to produce work that closely corresponds to one of the three interpretations of "theater of the mind" that I have explored in this introduction—a theater in the mind, then a theater about the mind, and finally a theater for the mind. Of course, old radio is too sweeping in scale and too slapdash in execution for uniform development. Styles seldom make hairpin turns. The temporal overlap in my three sections reflects the fact that no one aesthetic campaign fully excludes the characteristics of the others. Still, it can be clarifying to work at the level of generalization, which is an inevitable feature of studying aesthetics over time, especially with so much material to sift through. So I have adopted a loose structural approach, if only to frame material across in a way that brings technique into focus and produces a classificatory array against which anomalies and complications might appear.

In part 1, I begin by contextualizing what I call a "drama of space and time" that became vibrant in the late 1930s and was rooted in techniques for the use of volume, acoustics, and sound effects to draw pictures in the mind. In four chapters, I read plays by a cohort of CBS broadcasters alongside themes that writers and other public figures used to discuss radio during the Depression, arguing that the exploratory quality of 1930s radio aesthetics is especially significant in an era in which radio's impact on society was conceptualized using spatial language. In this section, I concentrate on radio as practiced in the broadcast studio and provide terms to help read the phenomenon of perspective in programs such as *The Columbia Workshop* and *The Mercury Theater on the Air*. I also propose that there are two styles in this phase—the "intimate" and the "kaleidosonic"—that accounted for a double rhetoric of empathy and egalitarianism in many important performances of the Depression era, particularly those of playwright Norman Corwin. Part 2 argues that the entrance of the United States into World War II transformed the way that many Americans considered the medium, which now seemed less like a force binding together a nation and more like a weapon to impose belief. Dramatists began to conceptualize interiority, airing plays in which consciousness is tied to acts of communication. In three chapters, part 2 explains that in 1940s radio drama interiority is dominated by "signals" and "transmissions" in plays about people who are contiguous with vast communication systems. It was during the 1940s that American radio drama achieved its permanent place in our cultural memory by depicting what it also seemed to be enacting. Part 3 considers aesthetics during the late radio age in the 1950s, asking how crime serials, adventures, and science fiction genres reforged tools created for 1930s-era liberalism and wartime. In three chapters, I argue that as the industry adjusted to the advent of television and the Red Scare, dramatic narratives put the mind in new scenarios of risk. Dramas began to stage invasions and races against time, direct address became significant, and the mind was depicted as if only to exhibit its performance under duress. By the 1950s, the theater of the mind had reached a point of exhaustion, but it remained a means to elaborate fables about the purchase of media on inner life and to test an abundance of new models of how the psyche operates.

As the "golden age" of the medium passed and network content diminished, radio listening itself did not abate. In fact, programming formats flourished and evolved, as they still do. But radio *plays* ceased to be an everyday part of collective life in the United States. What dwindled was the habit of using the medium to make scenes expressively sufficient for "visual" internalization. That is not just an event in the economic roots of broadcasting, the side effect of the postwar shift of advertising to television and print, but also an event in the history of sensory experience, and this book endeavors to provide a replete way of understanding that development. I also try to pinpoint what endures from the radio age. Although radio is, as Lisa Gitelman puts it, "emphatically noninscriptive," its ongoing afterlife is made possible by an array of recording technologies, and so it still remains a factor in how we think about aural media.[41] When the theater of the mind faded to the background of the postwar soundscape, it became a notion worthy of great memorial among members of subsequent generations, who, like Guy della-Cioppa, projected onto the radio age fantasies of what their day-to-day mediated experience did not seem to provide. The fusion of mind to medium on which classic radio dramaturgy was predicated also persisted as a cardinal point of popular understanding, even as the innervating experience that empowered it became uncommon. By sketching pictures of the mind and sending them over the airwaves as invitations to the listening imagination—thereby summoning a new kind of imagination into existence—the evanescent radio play had succeeded in leaving a lasting mark.

1

Radio Aesthetics in the Late Depression, 1937–1945

CHAPTER 1

Dramas of Space and Time

On Sunday, April 11, 1937, the Columbia Broadcasting System produced one of the most openly polemical radio plays aired in the United States during the Great Depression, *The Columbia Workshop*'s "The Fall of the City." Broadcast from CBS flagship WABC New York to affiliates coast to coast, the program required four directors to choreograph several principal actors and scores of volunteer extras around an array of microphones transmitting live from the drill hall at the Seventh Regiment Armory at Park Avenue and Sixty-Sixth Street in Manhattan. One of the largest unobstructed interior spaces in New York City, the hall offered more than fifty-five thousand square feet of floor space with eleven elliptical wrought-iron arches under a barrel-vaulted roof supported by exposed arch ribs. The acoustic of the space could magnify piercing bursts of choral speech or convey unusually subtle effects such as the shifting of sandals on stone—audio so subdued that engineers called it "down in the mud" sound.[1] Both of these sounds were signature components in "The Fall of the City," a blank verse roman à clef about civic leaders and a restive mob in a huge square debating the prophesied arrival of a "Conqueror." CBS was openly venturing into public debates over isolationism, while aping the conventions of classical tragedy to court an urbane audience not ordinarily disposed to network fare. Such ambitions merited a sensation. The *New York Times* considered "Fall" a highlight of the broadcast year; *Newsweek* called it the first evidence that radio theater was "radically different" from other media; according to *Variety*, the show beguiled even the "hunky-dunky" lay auditor. Celebrating at the Stork Club after the show, CBS staff reportedly toasted a new era in radio literature.[2] "Fall" has remained iconic ever since. In 1939, CBS

FIGURE 1.1. *The Columbia Workshop* presents "The Fall of the City" at the Seventh Regiment Armory, April 11, 1937. Image courtesy of the New York Public Library, the Library for the Performing Arts, Billy Rose Theater Collection.

published it as "the first poetic work of permanent value" written for the air, and today a recording of it is one of just three anthology radio plays listed in the National Recording Registry of the Library of Congress.[3]

"Fall" was indeed a kind of "Armory Show" for radio, signaling the start of a period in which many programmers sought to modernize and invigorate broadcasting aesthetics and extend highbrow culture to the masses.[4] Scholars have long considered the play to be a cerebral anomaly amid hours of schlock, and several writers have pointed out the Popular Front politics of the men who mounted it and argued that the manifest story of "Fall" reflected geopolitical anxieties.[5] Yet little attention has been paid to its technical exigencies and the complexity that made the result estimable within the industry. Much remains unfathomed about why "Fall" required the acoustic that the Armory provided—what one review called the "outdoor effect" created when the sound of the mob was captured live and fed to speakers far at the back of the hall where the actors stood, making a hundred voices sound like thousands in layers that conveyed the depth needed to evoke "the City" in the mind of the listener.[6] But this grandiose setup was the whole point. Radio dramatists of the 1930s went out of their way to create dramatic situations in which time and space are featured as the most prominent design elements, depicting distended spaces, Byzantine sets, rapid segues, protean action, and distances that transform abruptly.

In part 1, I situate these "dramas of space and time" in Depression-era culture and I characterize them as a harbinger of the psychological radio that would emerge in wartime. In this chapter, I detail the context in which such dramaturgy developed on the major networks in the late 1930s, as dramatists became preoccupied with suggesting plastic settings in the listener's mind, encouraging listeners to explore imaginary space just for the sake of doing so. At the very same historical moment that radio culture itself seemed to be "homogenizing" the nation, broadcasters grappled with how to make the airwaves into a space that might be shaped. By building scenes in the mind, radio plays served as a way to come to grips with what proximal relationships were becoming in the media age. In chapter 2, I move into the studio, explaining how dramatists used amplitude, effects, and acoustics to create listener's perspectives that structure explorations of space. I also perform a close reading of "Fall," showing tensions in its artistic aims and political proposals, both of which hinge on "positioning" the listener. Chapter 3 takes a broader view, considering normative use of perspective in such programs as *The Shadow* and *The Mercury Theater on the Air*, highlighting two strategies that together account for the overall sound of the period, what I will call the "intimate" and the "kaleidosonic" styles. In the final chapter of part 1, I focus on the plays of writer Norman Corwin, who attempted to meld intimate and kaleidosonic designs into a rhetoric befitting the New Deal era's rituals of unity, drawing stirring pictures of America itself in the mind of the listener. Ironically, it is through Corwin's People's Radio aesthetic that the Depression-era's exploratory celebrations of exteriority transformed into tales of interiority, and dramas of space and time taking place in the mind truly became a theater about the mind.

A STAGE FOR THE WORD

It took a series of developments to bring "The Fall of the City" to the air, beginning with a premium that network programmers began to place on artistic brio in the late 1930s, along with a willingness on the part of broadcasters to push boundaries. The script for "Fall" came from Pulitzer Prize–winning poet Archibald MacLeish, who was soon at the forefront of an emerging group of high-profile writers at CBS, a place cemented by his 1938 "Air Raid," which inspired an admired set of plays on the titular theme.[7] A lecturer at Princeton and an editor of *Fortune*, MacLeish used his modest prominence to campaign for *The Workshop* in the press and canvassed friends like Ernest Hemingway for scripts. In this period, it was common for intellectuals to endorse the medium as an artistic venue alongside reform groups lobbying for pedagogical mandates and public programming; MacLeish was willing to work within the corporate framework of the networks.[8] Plugging "Fall" in the *New York Times*, MacLeish decried not the low commercialism of network radio, but mulish literary disregard of it: "Every poet with dramatic learning," he emphasized, "should have been storming the studios for years."[9] For MacLeish, verse literature was uniquely disposed to radio: "The word dresses the stage. The word brings on the actors. The word supplies their look, their clothes, their gestures." The comment alludes to a question at the heart of modern dramaturgy. As Eric Bentley once argued, by following Edward Gordon Craig to elevate stage management, lights and scenography, many modernists of the early twentieth century pursued "theatricalism" at the expense of "theater," diverting performance from the verbalization that anchors its realism.[10] Unfettered by the histrionics of that overvisual theater, MacLeish's "stage for the word" promised to return theater to its writers and provide a virtually boundless public forum to boot.

Such rhetoric reflected a new trend. As an expressive activity, radio drama is as old as commercial licensing. The first American radio plays were pioneered in local stations like WGY Schenectady in the early 1920s, and the format slowly spread across the country.[11] In the subsequent decade, dramatic practice consolidated as part of a large-scale concerted campaign for quality content on the national networks. In 1930 NBC produced a dramatic adaptation show called *Great Plays* that soon fizzled, but seven years later the network revived the show to become part of a suite of content that included middlebrow programs such as Arturo Toscanini's *Symphony Orchestra* and Clifton Fadiman's *Information Please!*[12] Meanwhile, rival network CBS hired reviewers such as Alexander Woollcott while luring young dramatists such as Irwin Shaw and Arthur Miller, who launched their careers on *Dick Tracy* and *Cavalcade of America*. Programmers invited work from New York's Theater Guild and Actor's Repertory Theater—one historian calls the 1930s American stage a veritable "feeder program" for radio—and in the summer of 1937, CBS and NBC aired rival Shakespeare festivals with the likes of John Barrymore and Tallulah Bankhead.[13] That year also saw plays on air by such authors as Molière, Tolstoy, Eugene O'Neill, and George Bernard Shaw; by 1938 adaptations of such authors as Henrik Ibsen ran even on the commercial *Lux Radio Theater*.[14] Federal Theater writers from the Works Progress Administration also produced up to three thousand broadcasts a year, as networks promoted their work.[15] On December 8, 1938, Zora Neale Hurston read prose about African American folktales on CBS's *The American School of the Air* and was interviewed for NBC's *Meet the Author* the next day.[16] Around this time, major newspapers established radio beats, trade publications began to run reviews of broadcasts, and more than 350 high schools and colleges across the nation began to teach radio drama, including Harvard and Brown. When poets Norman Rosten and Edna St. Vincent Millay followed MacLeish to CBS, prestige trickled down to the hacks. In 1938, the largest radio advertising firm required its writers to relinquish authorship claims, but four years later *Variety* ran columns listing the scribes of popular dramas.[17] The storming of the airwaves had begun, if not quite according to MacLeish's plan. But the poet's voice in broadcasting would amplify as Pearl Harbor approached and he recruited dramatists into the war effort during his subsequent career as Librarian of Congress and as an adviser to Franklin Roosevelt's Office of Facts and Figures and the State Department.

With "Fall," MacLeish cast himself as an outsider raiding the airwaves, but the majority of *The Workshop*'s staff came from the industry. Irving Reis, the head of the program, had begun his career as a sound engineer, and he first pitched the idea for an experimental weekly anthology with bandleader Artie Shaw in 1936. Their timing was fortuitous. That year was an unusually aggressive period of competition between NBC and CBS, as each network wooed the stations of the other into its fold, often employing as bait network-funded "sustaining" content to fill programming gaps between sponsored programs.[18] This was just one of several factors driving the push for "culture" at the networks. Around this time, some advertising firms took production away from actual stations—in 1936, Blackett-Sample-Hummert centralized their radio portfolio of some thirty accounts at what *Variety* called a "radio factory" on Long Island—so CBS and NBC began to expand staff contracts, maintaining a pool of performers who hung around cafés near flagship stations waiting for a gig.[19] That drew the attention of unions such as Actor's Equity and the Chicago Federation of Musicians, both of which began a series of attempts to unionize various staff as professional artists. Fear of radical reform of the industry also encouraged highbrow content as a way to curry favor for the network oligopoly that had been only recently solidified in the FCC Act of 1934. As Robert McChesney has shown, commercial broadcasters publicized all efforts at "high-grade cultural programming" as part of educational offerings intended to fulfill a "public interest" requirement enshrined in communications law.[20] Perhaps Columbia bestowed greater largesse on upmarket fare because it was publicly traded, unlike NBC or the regional networks; so CBS chief Bill Paley had to quell the anxieties of investors worried that the FCC might withhold licenses and undermine the basis of commercial radio. With this in mind, programmer Paul Kesten strategized to brand Columbia the "Tiffany Network" by hiring a shop censor, curbing "bodily function" ads, and limiting ads to just 10 percent of evening airtime.[21] Programs such as *The Workshop* also emerged along with similar shows that did not hawk the soap, yeast, and Jell-o on which commercial broadcasting had grown fat, publicizing instead something more valuable—the *idea* that the public interest was nourished by letting radio continue to gorge itself.[22]

The policy had impact. In a 1938 special issue on radio, *Fortune* likened the CBS strategy to graceful fencing.[23] Reis's *Workshop* certainly exhibited agility, promising in its inaugural broadcast both experimental drama and reports on the use of radio in shipping and "electrical surgery." The *Washington Post* published a feature that depicted Reis's musicians "performing the unconscious" on cellos and sound men roaming New York "holding up a microphone with a rapt expression and a vacant smile."[24] Reis explained his initiative: "We don't know what a microphone can do—nobody knows yet. We're going to put them in queer places, submit them to unusual vibrations. . . . We want to break every rule of broadcasting, jump the track routine and explore the mysterious maze of electrical phenomena." For a broadcaster who cut his teeth at a mixing board, the naïveté was a little affected, but the swagger proved genuine. Over the years, programs included interviews of some of the last Civil War veterans, lectures on tremolo vocal work, compositions by Igor Stravinsky and John Cage, plays by Dorothy Parker and William Saroyan, and adaptations ranging from Keats and Shakespeare to Dickens and T. S. Eliot. *The Workshop* received some seven thousand unsolicited scripts a year in its heyday, and Reis earned stature among his peers. One reporter described a moment in the midst of a *Workshop* show when staff noticed that the production was lagging by twenty-five seconds and was unlikely to conclude on schedule.[25] Rather than slashing lines according to custom, Reis called master control asking to borrow time from the subsequent program, a brazen request, even for accident-prone live radio.

Although Reis is the person most fêted for what historian Michael Denning considers to be a "Renaissance in aesthetic innovation" at CBS, his tenure proved short.[26] By 1940 he was on contract to Paramount Pictures, directing such films as *The Bachelor and the Bobby Soxer* (1947) before his death in 1953. But Reis's achievements relied heavily on a circle of talented staff. One writer has credited the showy swordplay of CBS not to Reis at all, but to talent recruiter William Bennett Lewis.[27] *The Workshop* also became important because of the centralized architecture of the industry, which enabled a small group of dramatists to develop conventions that would in time influence so much of American broadcasting. The shape of that system can be traced at least as far back as the 1927 Radio Act, after whose passage local stations with low to medium power linked up to national networks in order to capitalize on national audiences but avoid antitrust law.[28] "Wired networks' many local units, rationally joined together, not only gave the appearance of competition and diversity, but fit well with familiar business practices," writes historian Susan Smulyan, "The network system resembled other national distribution systems set up in the 1920's, such as dealerships, franchises, and chains."[29] It represented a massive centralization of content. Networks owned a few dozen stations outright as their affiliates dotted the landscape, so one way or another most of the shows that reached listeners by the late 1930s had the fingerprints of NBC, CBS, or Mutual. Regional networks persisted—the Yankee network in New England, webs owned by tycoons like William Randolph Hearst and General Electric's Sam Insull—but the expense of renting lines to achieve national delivery hindered the startup of true competitors.[30] In 1935, NBC alone paid $2.4 million to rent lines from AT&T, and the next year, stations affiliated with the big networks out-grossed independents five to one—*Variety*'s radio editor cracked, "Business is so good the tablecloths aren't big enough."[31]

Centralization and oligopoly serendipitously drew drama innovators together in a way that transformed the craft of broadcasting. When *The Workshop* debuted, CBS and NBC had a combined payroll nearing $80 million, but they employed only some fourteen hundred full-time staff in New York and Los Angeles together.[32] In fact, four more CBS directors helped to bring "The Fall of the City" to the air. Brewster Morgan, who supervised MacLeish's script, also produced Fred MacMurray's *Hollywood Hotel*, the blood-and-thunder show *Lights Out!*, and various Shakespeare specials. Earle McGill, who mixed sound for "Fall," wrote a textbook on radio directing while serving as president of the Radio Director's Guild and producing educational programs. Director William N. Robson, who handled "Fall's" mob scenes, first achieved fame for a 1935 script about a prison break at San Quentin. The play aired just hours after the actual break. When news broke that the fugitives had been apprehended, Robson rewrote the final scene and rushed it to the actors in the studio in mid-broadcast.³³ A flashy figure who commanded great respect from his peers, Robson helmed *The Workshop* after Reis left, directed war programs such as *The Man behind the Gun*, and earned one of several Peabody Awards for *An Open Letter on Race Hatred* following the 1943 Detroit race riots. "I remember one spoke in awe to Bill Robson," recalled actress Rosemary Rice, "like you would to Orson Welles."³⁴ Welles himself was the lead actor in "Fall of the City." The twenty-one-year-old had just embarked on his radio career as an actor for *The March of Time* and *The Shadow*, while preparing his directorial debut in a seven-part adaptation of Victor Hugo's *Les Misérables* to air in the summer of 1937. Welles's Mercury Theater troupe came to CBS airwaves one year later.

Boasting thousands of program credits between them, these broadcasters devised marquee content, reaching dozens of affiliates and millions of listeners in a way that had a disproportionate impact on the overall sound of the medium, if only because most of the actors, writers, engineers, and musicians who worked in radio drama during its flourishing decades were at one time or another professionally linked to those who aired MacLeish's play in 1937. Many *Workshop* staff earned little more than the contract minimum of $18.50 per show, while Jell-o paid Jack Benny's troupe $6,400 a week.³⁵ But credits on *The Workshop* opened doors to a burgeoning field. In 1938, *Variety* noted a rise in sixty-minute anthology playhouses, and an FCC study revealed that although music accounted for 53 percent of all American airtime, drama was nearly twice as likely to attract sponsors.³⁶ By 1940, plays would account for nine thousand hours of network air, and three years later they would bring in more revenue than any other format.³⁷

Beginning more than fifteen years after the first station licenses and ten since NBC pioneered the network model, *The Workshop* hardly inaugurated the golden age of radio, but it galvanized professionals at a moment when their collaboration could provide a lasting pedigree and establish interconnected production norms. *Workshop* drama also distinguished itself by concentrating on techniques unique to the medium. The 1928 show *Amos 'n' Andy* has been called "radio's *Birth of a Nation*," and Cecil B. DeMille's *Lux Radio Theater* brought melodrama to evening schedules in 1934; but each was coded by another format (the minstrel show and Hollywood feature), and few considered them to be stylistically novel.³⁸ During years in which *Lux* drew forty million listeners per week, a dearth of "radio showmanship" was lamented by personalities including Eddie Cantor and Mary Pickford in the pages of *Printer's Ink*.³⁹ Even sympathetic entertainment critics conceded that a true "radio technique" remained undiscovered. By contrast, Reis put radio through its paces, training a cohort of broadcasters who would lead innovation well into the television age.

TEXTURING SPACE

With innovation at a premium, radio was ready to trace pictures in the mind in an entirely new way, but that alone still does not explain the exigencies of staging "The Fall of the City." Why should spatial gimmicks emerge as foremost among the criteria used to gauge directorial pizzazz? What made complicated settings smack of "showmanship" in radio plays? Part of the answer lies in how the medium was broadly conceptualized as a cultural form destined to overcome traditional forms of distance. In *Being and Time* (1927), Martin Heidegger declared that wireless expanded "the everyday environment."[40] Because Being (*Dasein*) has a tendency to closeness, the philosopher wrote, this expansion brings about a "de-severence" between Being and the world. Heidegger used this as an example of the conquest of "remoteness" in modern life. Readers could find a similar assertion in John Dos Passos's *U.S.A.* (1930–36), which noted that broadcasting killed "the old Euclidean god," something tantamount to a "philosophical" revolution; meanwhile, futurist writer F. T. Marinetti credited radio with the "immensification" of space into the afterlife.[41] According to theorist Douglas Kahn, such associations were shared among the avant-garde of the 1910s and '20s, a generation for which radio was "coterminous with a range of mythological, theological, and literary instances where communication at a distance produced compensatory and exaggerated relationships among objects and bodies."[42]

That opinion is surprisingly similar to that of industry spokesmen, who often used metaphors in which venues seem to combine in radio-space. Highlighting his public service bona fides, CBS vice president Paul Kesten touted his network as "a concert hall, a herald of news, a field of sport, a hall of learning, a carnival of music, of laughter," arguing that the medium democratized culture previously only enjoyed by elites.[43] NBC boss David Sarnoff, who helped coin the term "mass communication," predicted a day when "the oldest and newest civilizations will throb together at the same intellectual appeal, and the same artistic emotions."[44] Theorist Rudolf Arnheim argued that centrifugal radio dissolved "the boundaries between countries but also between provinces and classes of society, while critic Lewis Mumford suggested that the greatest "social effect" of radio had been "the restoration of direct contact between the leader and the group."[45] For many commentators, Roosevelt's Fireside Chats were believed to renew democracy by circumventing intervening institutions to bring appeals directly to the populace. Radio culture seemed to dismiss fusty old boundaries that had outlived their usefulness, and all people were enjoined to inhabit radio's commodious plenum as equals, just as they were welcomed at Franklin Roosevelt's hearth.

Along with this optimism, there also appeared melancholy. In the 1930s, researchers began to notice that American habits were becoming detached from locale, as evidenced by the atrophy of municipal singing societies and local dialects. According to Robert and Helen Lynds' studies of small-town Indiana, radio put localities into wider America, but folkways dwindled. "Like the movies and the national press services in the local newspapers," they argued, radio "[carried] people away from localism and [gave] them direct access to the more popular stereotypes in the national life."[46] The Lynds concluded that the medium made the nation more homogeneous. A similar conclusion was reached by the President's Research Committee on Social Trends, whose 1931 report cataloged more than 150 aspects of diminished localism, including "homogeneity of peoples because of like stimuli," the fading of pronounced "regional differences in culture," and "favoring of widely spread languages."[47] In *The Psychology of Radio* (1935), the first scholarly work to serve as a touchstone for the broadcasting industry, Hadley Cantril and Gordon Allport attributed homogeneity to the network model itself: "When a million or more people hear the same subject matter, the same arguments and appeals, the same music and humor, when their attention is held in the same way and at the same time to the same stimuli, it is psychologically inevitable that they should acquire in some degree common interests, common tastes and common attitudes."[48] At the end of the 1930s, writer E. B. White expressed the view that Americans tended to give "over-meaning" to their radio sets, which were treated less like objects, and more like "a pervading and somewhat godlike presence" in their lives.[49]

This discourse endures today. "Homogenization," "centralization," and similar terms feature prominently in writing on radio, as authors explain how listening formed generational identity, developed national uniformity, and bound together social classes.[50] Writers approach classic broadcasting with social theories that employ spatially coded terminology—what the Lynds called the "space-binding" effect of radio has become an "imagined community" following Benedict Anderson, or a kind of "public sphere" following Jürgen Habermas.[51] Several authors have discovered ambiguities in these developments. Bruce Lenthall has argued that many listeners actually used the mass media to personalize impersonalizing modern experiences, while Paul Young suggests that radio's homogenizing power represented a threat to Anglo-American culture, and Alexander Russo has shown that when you consider the endurance of local content, it seems that the degree of uniformity at hand has been exaggerated. Indeed, as Bill Kirkpatrick has explained, the mass media age had as much a part in reconfiguring concepts of "the local" as it did "the national."[52] Besides, the space-binding idea is a perennial feature of the discourse of media fantasy. Steve Wurtzler has argued that radio's vaunted ability to "eclipse space" was "but an intensification of the earlier promises of mail service, telegraphy, telephony, even canals and rail service, or celebrations of the cinema's ability to serve as a 'universal language.'"[53]

Here I am more interested in the prominence of the fantasy of space-binding than in whether or not it stands up to scrutiny, but it is important to note that for people who worked in radio drama, centralization seemed to be less about democracy or language, and more about the industry's move to produce most of its dramas in New York and Hollywood. As late as 1934, aspiring dramatists found work at stations in Cincinnati, Buffalo, St. Louis, and Portland. But after opening sophisticated new West Coast studios, CBS and NBC nabobs moved a great deal of drama production to California. In 1935, the *Chicago Tribune*'s critic Larry Wolters boasted that up to forty national shows were networked out of the Windy City, but by 1937 he was bragging about how many Chicago talents had been wooed to the West Coast.[54] Ad firms required centralization and consolidation to give sponsors marquee names for products, during a time when just seventeen advertising accounts bankrolled 45 percent of all network business and NBC was virtually an appendage of consumer product goliath Standard Brands.[55] It is not surprising that the idea of "centralization" took hold after CBS opened Columbia Square, its new $1.5 million headquarters at Sunset and Gower in Hollywood, near the studio lot of major shareholder Paramount Pictures. NBC moved in not far away.

And so by 1939, anyone wanting a career in radio production had to seek it at a flagship station. Radio reviews of the late 1930s reflect rapidly diminishing regional production in every category. Local "liars clubs" were swept off the air by national programs sponsored by some of the few brands to increase their share of cupboard space during the Depression—Chase and Sanborn Coffee (*Eddie Cantor*), Jell-o (*Jack Benny*), Campbell's Soup (*Burns and Allen*).[56] The mid-1930s craze of local amateur hours vanished in the face of celebrity-driven variety shows such as *Major Bowes*, *Kate Smith*, and *Hollywood Hotel*. Programs on insular issues by parochial societies gave way to *America's Town Meeting of the Air* and *The University of Chicago Round Table*. Regional thrillers like *Maryland Mysteries* ceded to gothic institutions like *Inner Sanctum Mysteries*. And unique local content diminished. *Scarettes* on WGAL

Lancaster, traffic accidents dramatized on WJAY Cleveland, and *Around the World in a Giant Amphibian* from KGW Portland left the air, along with dozens of "story-time" programs.[57] Standard Brands and standard fare even demanded standard time. By the 1930s, sponsored shows led the way in regularizing broadcast schedules, as commercialization promoted what historian Shawn VanCour calls a new "stopwatch aesthetic" that ensured each program would proceed seamlessly into the next.[58] To rationalize sponsorship sales, the thirteen-week "season" was born, and by the mid-1930s, sixty-minute anthologies and thirty-minute serials replaced many five- and fifteen-minute time slots; with the third-shift of wartime, many American radios received twenty-four-hour content for the first time.[59] Just as speech now lacked localization, chronology ceased to be studded with unique individuated events but was smoothed into a single calculable flow marked by clear cycles and predictable punctuations.[60] Radio integrated with the rhythms of everyday life. It also constructed those rhythms, becoming a permanent part of the modern understanding of orderly interval.

Despite scholarly interest in this homogenous space-time and the programs that embody it, we lack an account of its ramifications for the aesthetic agendas shaping the radio play, a form of expression that was changing too. At the moment that radio homogenized the nation, its narratives began wrestling with time and space in every available way. Although time was standardized, Walter Winchell's breakneck newscasts packed data in rapidly, as if to overcome the limits of his fifteen-minute *Jergen's Journal*, while Orson Welles's "first-person" narration skewed our point of view on action so as to provide expressive angles on events that challenge models of flat uniform space. VanCour has argued that one feature of 1920s radio drama was the establishment of narration for setting scenes.[61] Dramatists thought they were writing for a "blind man" and used speech to convey anything that would be silently visual in the theater. By the late 1930s, that preference had changed. Radio listeners became accustomed to recognizing purely aural rules that signified movement around the world of a drama. For instance, around this time it became common for narrators to cut in and out suddenly, while "scenes" faded in at the beginning and out at the end. Listeners learned that the longer the pause between scenes, the greater the distance traversed in the world of the fiction.[62] Many plays were designed under the rationale of "realistically" depicting space, while others used dramatic license. Consider Norman Corwin's "They Fly through the Air with the Greatest of Ease," which features two principal locations for a key bombing sequence that are hard to reconcile dramatically—the ground and the interior of a bomber. As Corwin's production notes explain, in reality the surface explosions would probably not be audible in the bomber itself, as altitude and velocity would result in a significant interval between ground concussions and any report in the cabin. Yet the notes suggest for directors to illustrate the explosions in both areas synchronously, since radio can foreshorten. "The brevity of the interval will horrify professional bombers," Corwin confesses, "but five seconds of airtime is the equivalent of twenty seconds anywhere else."[63] That "realism" was a concern is testament to the sway that such relations held over storytelling; that dramatic license was taken suggests that writers did not think of their dramatic canvas as flat or uniform. Space had become the servant of dramatic effect, a development that radically altered scenes and the techniques with which dramatists conveyed them.

Around this time, many long-running programs altered their formats. In 1934 *Amos 'n' Andy* was a fifteen-minute sketch serial, usually with two or three speaking characters, and it was common for much of the broadcast to consist of dialogue taking place at just one or two locations. But by 1943, the program was a thirty-minute episodic show, with twelve to fifteen scenes all over Harlem and beyond, each one framed by orchestral segues and fitted with up to six characters, bringing emphasis to action and movement.[64] The same change came to *Easy Aces* and *Mr. Keen, Tracer of Lost Persons*. Soap opera directors adapted well to listeners expecting spatial texture because their work was always defined by pace. Classic soaps took up to eight hours to play out a story arc, and writers also had to ensure that each twelve-minute segment had some kind of climax—trying to find a genre that faces similar challenges, one director analogized soaps to Icelandic epic poetry.[65] If whole broadcasts seemed to have more dimension, so did introductions, which often start out with the feeling of going somewhere. *The First Nighter* pioneered a device for this in the early 1930s. When the program opened, we took a cab to "The Little Theater off Times Square," moved through the lobby crowd, and slid into our seat next to Mr. First Nighter; ten years later, *Inner Sanctum Mysteries* opened its famous creaking door on Sunday evening, as our host Raymond offered us a chair on the other side of an imaginary room. Such characters greeted us with warm phrases like "Good evening" and "Shall we get into the taxi, mustn't be late," the sort of utterances that expect a response from an entity in the diegesis capable of making it.[66] Similar spatial texture can be found in the representation of microaction. During the later 1930s, footsteps became the most common of all sound effects, as directors believed that they helped listeners to imagine blocking. Engineers kept several sets of "walking shoes" for on-air use, using cornstarch, stone, and plywood to walk characters across snow, asphalt, or floorboard, textures that convey proximities between mobile characters.[67]

"What a temptation to put footsteps into a radio play!" wrote Earle McGill in a 1940 manual on directing. "If one were to believe the radio producers, this is the most uncarpeted nation in the world."[68]

In the context of these developments in spatial texture, it is easier to fathom Irving Reis's project to glean sounds from the nooks and crannies of Manhattan, which literalized a proclivity shaping the sound of the genre. Even his correlation of broadcast drama with "radio shipping" proves more astute than expected. The radio play had begun to evolve from an expressive activity driven by narration to one driven by scene. *The Columbia Workshop* was devoted to tricks that explore space and time. Its educational lectures demonstrated for the audience the sense of proximity and "sound focus" provided by new velocity microphones and illustrated the difference between shortwave and wired networks, explaining why programs originating in New York reached Australia slightly before they reached Chicago.[69] In exigent dramas, meanwhile, *The Workshop* offered impressions of locale. Leopold Proser's "Broadway Evening," for instance, blended stylized dramatic scenes staged in the studio with live exterior audio. To provide the latter, several engineers set up mikes along Broadway, beaming sound to the station's mixing board to evoke a surrealistic impression of billboards, barkers, brawls, and burlesques that approach and recede at a walker's pace. We experience these cacophonies beside a hypothetical couple strolling the Great White Way in a structured succession of overheard scenes that links streets, cafés, theaters, and subways into an expression of pure exterior space. *Workshop* sound design also distinguished physiques. For an adaptation of *Gulliver's Travels*, for instance, Irving Reis used four separate studios, each calibrated to "color" specific sonic frequencies. This resulted in a countertenor for Gulliver and a bass for the giants, which successfully sketched out a differentiation between their anatomical proportions. In other dramas, segues became vital elements. Vic Knight's "Cartwheel" told the yarn of a dying man's search for a valuable coin he had admired as a youth, a fifty-year story with thirty-four characters told in twenty-three scenes over fifteen minutes, statistics that the program announcer hastened to point out to listeners after the show.

The public took notice and tradespeople took notes. For NBC programmer Max Wylie, "Cartwheel" grappled with *the* problem in scripting: "Radio has no movement," he insists in a 1939 writing manual. "Like the novel it is forever committed to be one-dimensional. Whatever action occurs is the result of illusion."[70] The challenge was how to invent rules for those illusions. By conveying settings and motion, 1930s radio taught listeners to decode audio in sophisticated ways and generated expectations underlying many future innovations. Even news formats felt the impact. When Archibald MacLeish famously told Edward R. Murrow that his Blitz broadcasts "burned the city of London in our houses," he was crediting a series of choices that conveyed a sense of locale.[71] Consider Murrow's broadcast of September 21, 1940:

> I'm standing on a rooftop looking out over London. . . . For reasons of national as well as personal security I'm unable to tell you the exact location from which I am speaking. . . . Streets fan out in all directions from here. . . . Off to my left far away in the distance I can see just that fiery, red angry snap of antiaircraft bursts against the steel-blue sky. . . . More searchlights spin up over on my right—there they are! . . . The plane is still very high and it's quite clear that he's not coming in for his bombing run. . . . Out of one window there waves something that looks like a white bed sheet, a curtain swinging free in this night breeze. It looks as if it were being shaken by a ghost. . . . The searchlights straightaway miles in front of me, are still scratching that sky. There's a three quarter moon riding high. There was one burst of shellfire almost straight in the Little Dipper.[72]

Murrow refuses to divulge his location, yet he triangulates his listener in imaginary exterior space by stretching out a sense of streets, explosions, searchlights, ghosts, and stars. He suggests trajectories just like any other exploratory broadcaster, and it was after all a design choice to mike London from its rooftops, thus inextricably tethering us to Murrow as real air raids approach grippingly from a deep aural distance, the perception of which gave the scene its gravitas.[73] Not unlike Reis's cityscapes, Murrow's broadcasts used practices that suggest that the dimensions are plastic features to be organized for dramatic effect. To put it another way, although we have no D. W. Griffith to single out, it is not much of an exaggeration to say that what parallel editing did for film grammar in the 1910s and '20s, the drama of space and time did for radio grammar in the 1930s.[74] And just as Griffith's games of time and suspense both celebrated and overcame the syncopation of mass industry in his era, 1930s radio plays expressed a sense of flat national plenum precipitated by network culture, while at the same time undermining that idea with literalization, trickery, and play. This synchronization between public debate and aesthetic practice suggests that radio dramaturgy had become articulate in its own conditions of possibility. To explain how this development influenced the way that productions like "The Fall of the City" were conceived, we turn to how directors actually used their studios to shape spaces for their audiences to imagine and made the single most important aesthetic choice for any tale of exploration: point of view.

CHAPTER 2

Producing Perspective in Radio

To create plays that evoked spatial and temporal structures in the mind, 1930s dramatists required a set of sonorous marks that could inform auditors where they "were" and signify movement from one scene to the next. Perhaps the most expedient of these devices was music, a few bars of which could, in the words of one CBS executive, "take the place of scenery, lighting and costumes" or, as an NBC writer put it, "span continents or centuries in a few seconds . . . like a magic carpet."[1] A hint of stride piano evokes Harlem in the 1910s; "La Marseillaise" suggests France—in this way dramatists neatly incorporated into the play a network of preexisting associations and understandings among listeners in an audience, simplifying the task of representation by deploying fertile seeds of information. Sound effects could also be used as scenography. Sound engineer Walter Pierson credited illustrative sounds with creating "the environment of an act" by coding scenes and vicinities, thereby circumscribing the range of acts that might happen in them.[2] Opening scenes in radio often used what theorist R. Murray Schafer calls "keynotes"—background sounds linked with certain locales—that draw on cultural memory to counterfeit cliché sets, such as foghorns that indicate ports (and suggest adventure) or cricket chirps that denote darkened prairies (and establish the conditions for romance).[3] Keynotes can be constant, like the sound of the sea, or they can return punctually, like the faraway barking of a dog, helping to create rhythm or subterranean coherence. Critic W. J. T. Mitchell has suggested that effects are a form of ekphrasis, "verbalizing" an unavailable picture without using words, but relying instead on properties of iconicity, metonymy and "customary contiguity."[4] Radio director Himan Brown once put it this way: "When I did

Bulldog Drummond, what did I need? I needed London, and what is London? London is foghorns, and Big Ben, and so on."⁵

Keynotes establish places, but to create kinetic dramas of space and time—to convey depths into which sound might propagate to illustrate action and convey the sense of a journey projected through an imaginable space that obeys rules similar to those of other spaces we know—directors also had to move listeners around within those places. To accomplish that, sound had to produce an auditor's vantage point in the scene, a place we "hear from." This device changed the way meaning could "happen" in a radio drama by inventing a point at which sound is brought to a stop in the drama as it exits toward the listener, what Roland Barthes called a "threshold of ramification."⁶ Radio director Earle McGill described this "sound perspective" as nothing less than "the rational basis upon which radio drama as we now know it stands."⁷ McGill's approach to structuring listener perception was to shape an illusion to which a mind might wholly subscribe, and in this sense his craft was moving in the opposite direction of modernist and avant-garde theater. While prominent dramatists like Bertolt Brecht proposed to attack the conventions of the bourgeois stage by strategically tearing down the "fourth wall" of the traditional proscenium, radio dramatists erected one, as a way to focus listener attention on actions of dramatic importance and make the production have the feel of a stage.⁸

In this chapter, I investigate how Earl McGill's "rational basis" developed, focusing on the perspectival qualities in the conventional use of acoustics, sound effects, and amplitude.⁹ These conventions gave directors and engineers tools to turn their static "pictures in the mind" into veritable "motion pictures," providing otherwise simplistic broadcasts real depth and complexity. This is especially true of Archibald MacLeish's "The Fall of the City," whose ideological dynamics depend to a large extent on a basic discrepancy between "perspective" according to the words and "perspective" according to the sound. Toward the end of the chapter, I will explicate this discrepancy in detail and argue that in the dramas of space and time, theory of sound perspective can help grasp palpable tensions. Because it is a radio play and not something else, "Fall" presents us with two simultaneous sketches for the listener's imagination. In the tension between these two configurations, there is a double alignment that represents conflicting aspects of New Deal ideology that fail to map onto one another. In this way, the play not only shows how crucial sound perspective can be for critical engagement with radio aesthetics, but also demonstrates how those aesthetics worked as a site for expressing and negotiating cultural and political tensions during the Depression.

POSITIONING THE AUDIENCE

Terminology to describe the listener's "vantage point" is awkward. In the past, writers have used the film studies term "point of audition" to denote the listener's site vis-à-vis sonorous bodies in the fiction.[10] But as film theorist Michel Chion has observed, scholars use this formula almost entirely because of its resemblance to "point of view," a correspondence that makes it simple to parse sound film sequences in which listening contrasts visuals in interesting ways, as when two senses seem to offer contrasting or synchronous information, which enables writers to use film aesthetics as a framework to study unity itself.[11] That task can be misleading in radio studies, producing an unhelpful tincture of incompleteness before we even begin. Besides, the compound "point of audition" is too cumbersome for extensive use. Instead, here I will use "audioposition" to indicate the place for the listener that is created by coding foregrounds and backgrounds. This term also seeks to make the idea available as a verb, stressing that it is always fabricated. Listeners do not just "have" a point of audition; they are "positioned" by audio composition and components of dialogue. The concept is handy to analyze 1930s radio, which tended to audioposition with great care, even upending source texts to do so. Consider Welles's adaptation of *Les Misérables*, which features a scene in which hero Jean Valjean is smuggled from a nunnery in a coffin to undergo a mock burial. The acoustic ambiance of this sequence is designed so that we hear events as if we are positioned inside the coffin, even though Victor Hugo's text keeps the reader very much on the outside as Valjean emerges. By changing audioposition, Welles swapped reader surprise for listener suspense, as was his wont. He made an identical adjustment to a similar passage in Alexandre Dumas's *The Count of Monte Cristo*, positioning us with the hero as he is thrown in a bag to the sea by soldiers who believe him dead. If we were to enumerate the aesthetics of Wellesian radio, we might start with his penchant for positioning the listener in proximity to sham resurrections.

While some writers used audioposition for emotional effects, others used it to solve mechanical problems. Perhaps no writer had more of these problems to deal with than Carleton E. Morse, writer of *I Love a Mystery* and *Adventures by Morse*. These 1940s shows aired stories in ten-episode cycles, usually set in closed environments, such as islands, graveyards, oilfields, and secret cities in British Columbia, Chile, or Indochina. In these tales, multiweek scenes could feature such exotic set pieces as a tangle of vines hanging from the apse of a pitch-dark jungle temple. Environments like this could not be readily summoned by keynotes because they do not rest on preexisting listener memory, nor could listeners be expediently prepared to imagine their stage business: "Cobra King Strikes Back" climaxed with seven characters climbing rope ladders above a bottomless pit beneath a lost Cambodian wat hidden behind waterfalls while pursued by werewolf priests. To complicate matters, Morse also broke conventions about the number of voices that a scene could feasibly bear. Welles expressed the prevailing attitude tersely: "Of radio script shows there are many kinds; commercial and sustaining, good, bad and indefensible, and among these there is only one that you will listen to: *the kind you can follow.*"[12] Among radio directors, the worry was that listeners might have difficulty telling which character was speaking at a given time. That is why radio frequently uses hackneyed accents to mark one character from another, avoids names with assonant likeness (few radio plays have both a "Tom" and "John"), and limits itself to three or four voices that typically bear the brunt of a scene, so that listeners at home do not lose track.[13] Disobeying these conventions, Morse's "Tropics Don't Call It Murder" used up to seven voices at once, "The Thing That Cries in the Night" occasionally had eight, while "The Pirate Loot of Skull Island" could feature as many as thirteen speaking characters at a time. Not only could listeners lose track, so could performers, many of whom played more than one character at a time. In her memoir, *I Love a Mystery* actress Mercedes McCambridge joked that she once had to settle an argument between two characters she was playing.[14]

Under these circumstances, it proved challenging to differentiate characters, illustrate settings, and ensure a coherent story arc. Morse took on this challenge by moving audioposition through an ingenious routine. In most of his stories, sets contain closed pockets in which action can take place among one group of characters just beyond the earshot of others, such as tents in camps, bedrooms in hallways, or the front and rear of single-file treks through jungle. Thus the titular "Temple of Vampires" is corrugated with staircases, chambers, and ledges on a number of levels. Characters often become trapped in one of these pockets, as some heroes go exploring, others are lured off by succubi, and search parties head out in pursuit. Listener audioposition then segues from one group to another in a "parallel editing" pattern that can last for several episodes. Because audiopositions exclude one another, producing discrete areas of earshot, one group of characters regularly misses out on story developments involving another. To recap events from previous weeks, Morse's drama reunites parties every few episodes to let them verbally relate experiences that took place in their respective audiopositions, keeping us abreast of developments as they do each other and turning exposition into dialogue.[15] These reunions motivate more detailed description of the setting, which in turn makes it easier to set up scenes later on. The partition/reunion device is often plainly artificial. In "Pirate Loot," before splitting up for patrol, heroes Jack, Doc, and Reggie design elaborate sleeping arrangements for the cast of characters on the pretense of testing a theory about the mystery. When they reunite three episodes later, the reason for the sequestering is forgotten, but recaps of the actions performed separately in the interim clarify the environs and propel the tale. In this way, Morse conveyed tortuous material by scattering highly mobile audiopositions around the world of the drama, a system that made his convoluted adventures vivid and memorable. Decades after it aired, McCambridge still met fans with detailed recollections of "Temple of Vampires."[16]

For critical listeners, audioposition can help decode dramas that are dimensionally ornate. Once we know the characteristics of "our" audioposition in the fiction, that is, additional shells of sound can be introduced (interiors and exteriors, foregrounds and backgrounds) by qualifying audio relative to the color and volume of sound that we already know to be proximal. It is no coincidence that dramas of space and time became possible just a few years after sound came to be measured in decibels, which made it easy to control amplitudes numerically in a way that was conducive to establishing precise spatial tiers.[17] The very simplicity of that work makes audioposition at once so easy to perceive and difficult to theorize. As critic Rick Altman once put it, "Constantly delayed, dampened, reinforced, overlapped and recombined, sound provides us with much of the information we need to understand its origins and its itineraries."[18] In classic radio, the apprehension of those origins and itineraries is a prerequisite for any listener trying to decode narrative effects ranging from rudimentary intelligibility to complex moods. Directors often used amplitude operators to indicate "near" and "far," illustrating aerial dogfights on *Hop Harrigan* and evoking the sense that the cops in *Gangbusters* are closing in on the hideout. Reverberation was also used to situate voices in a vivid world.[19] If Dick Tracy's voice is the loudest of all, we know that we are near him; reverb tells us that we are in some kind of enclosure; a filtered scream suggests that someone is in danger beyond a wall. Introduce approaching footsteps, and you have already built a scene full of conflict and possibilities that are sketched almost entirely through volume and ambiance. But to complete the illusion, the listener must indulge in a fallacy of unprompted perception, feeling that the "mental picture" is his or her own invention.

From the 1940s into the 1950s, *Suspense*'s William Robson, Norman Macdonnell, and Elliot Lewis—all of whom worked at *The Workshop*—coaxed listeners to become invested in dramatic outcomes by first establishing audioposition alongside a character in distress and then introducing menacing sounds from the outer edge of earshot, placing listeners alongside Kirk Douglas hiding from an enraged butcher stalking the aisles of a darkened grocery store, beside Keenan Wynn running into the jungle pursued by dogs, or huddling with Richard Widmark on an atomic test atoll overnight, listening to a dying animal shriek beyond the blasted hill. Anyone stalking these men also stalks "us," a phenomenon that makes the illusion of space much bolder. Listeners believe that the scene indeed has a spatial depth because variation in the volume of sounds compels us to strain to hear more deeply into it. I would argue that the very existence of a prominent audioposition in the drama alters our comportment as listeners, prompting us to put aside the appreciative mode in which we listen to song and take up the raptness with which we undertake echolocation.[20] Conventions of alignment also reflect changing social values. In 1940s-era noir love triangles on *Suspense*, we are usually positioned with a man who is being lured into murdering his lover's husband, but by the 1950s we are more often aligned with a cuckolded husband who overhears the plot through a wall or telephone line. Whatever conclusion there is to draw about this change, a hypothesis can only be formulated by attending to how our audioposition is contrived. G. E. Lessing once argued that when we experience fear in drama, it is the result of the similarity of our position to that of the sufferer, the sense that what happens to the sufferer might befall us; fear is thus "compassion referred back to ourselves."[21] In the *Suspense* scenarios, compassion produces fear, which produces compassion, in a recursive loop that blends the experience of a listener into that of an imagined character.

DUMMIES, JEWELS, SPITFIRES

Producing rapport between characters and listeners is a concrete task that takes place in the studio, where physical blocking, microphones, and other sound equipment craft the "place" where auditors imagine themselves to be, a process that can be likened to decisions about how to frame a shot in a film studio or select the appropriate depth of field for a photograph. Consider Irving Reis's "The Finger of God," a play that used a parabolic microphone, which converges sound as a concave mirror focuses light. When Reis aimed this apparatus at moving actors and focused it on certain exchanges between them, stage blocking generated a unique volume level and degree of reverberation for each voice that together construed its place relative to others, adding depth to scenes while situating us before character traffic and "turning us toward" certain voices at particular times.[22] The microphone ceased being an amplifier and became a surrogate theatergoer; by mimicking line of sight, the microphone transferred a visual orientation of perception into audioposition. It was a technique especially suited to radio. By the late 1930s, most films equalized vocal amplitude and color, so that sounds were almost always clear, close, and nonreverberant.[23] After a decade of competing paradigms of filmic audioposition, as James Lastra has shown, the loose formal system of film sound began to prioritize narrative over all other concerns, which in most cases meant foregrounding "essential" sounds over noise, breaking scale relationships between image and sound, decoupling the mike from the camera, and abandoning unitary point of audition.[24] As a rule, late 1930s feature film voices seemed both everywhere and nowhere in the scene area, so that if we close our eyes, we tend to learn relatively little about the camera's point of view or the scenography. "Finger of God" tried to use acoustics and volume to convey precisely these sorts of details and using some of the same equipment employed in film. Indeed, radio engineers were keenly aware that they faced a fundamentally different problem from their peers in film. Projection screens have height, width, and time, but their depth is an illusion; while radio has only depth and time, and height and width are illusions. As a result, the challenge in radio is to create sound in four dimensions and translate it into a two-dimensional signal that can again suggest a quadrilateral world, one that may not be identical to the studio that formed it.

While these depth effects could be achieved at a mixing board, interviews and production notes suggest that 1930s directors preferred to block actors. "Radio acting is a physical thing for me," recalled Himan Brown. "The actors don't stand in front of a microphone like automatons. They relate. They touch. They move;" Irving Reis evinced a similar preference, instructing his actors to avoid "standing like a circle of tailor's dummies," or even to ignore the microphone completely, a practice that gave "a very definite impression of perspective and movement."[25] Prior to the time and space cycle, the "dummy" scenario had been common. In 1936, actor Lionel Barrymore used a similar metaphor, noting that watching a radio performance could be as dull as "watching a suit being made."[26] By then such an arrangement hardly began to exploit equipment on hand, and it seemed pedantic to those aspiring to artistry. Without creative mike technique, tiresome exposition was the only way to position listeners before action in a meaningful way, a troublesome snag for drama on the move.

Around this time, critics and directors published several books about miking and studio layout. Rudolf Arnheim's *Radio* (1936) devotes some fifty pages to the rhetoric of mike technique, pointing out that radio cannot convey

FIGURE 2.1. "Here is your chance!... The window is open!... Go ahead, why don't you?... You don't dare push me out!..." A 1941 *New Yorker* cartoon by Charles Addams, poking fun at the discrepancy between the world of the studio and that of the setting it evokes. Note that our audioposition is about to segue from an apartment to the ground outside, both of which are produced inside a single space. © Charles Addams. With permission of the Tee and Charles Addams Foundation.

direction, only distance.[27] According to Arnheim, directors ought to resist the mike's natural "spiritual and atmospheric nearness" and use distance as expressive content, establishing what he called sound "vectors."[28] Similar ideas appear in handbooks. In Erik Barnouw's *Radio Writing* (1939), the CBS writer describes drama as a "trio for three singers"—sound effects, music, and speech—all of which move in unison throughout a broadcast, and any given moment in a play has one of these elements "emerging" relative to the others, with prominence or diminution a function of relative volume, "radio's spotlight."[29] Barnouw's ideas help to explain verbiage then appearing in radio scripts, much of which dealt with directing attention—"establish," "register," "sneak out"—and his model suggests that the framework of a deep stage with mobile elements was replacing the sartorial circle. Engineers used more lively jargon, calling actors who inched closer to their mikes over the course of a program "creepers," far-off sound events "out in the alley," and programs with no definite aural scheme "clambakes."[30] Production notes also reveal fussiness about adjusting sound levels and mixing equipment to suggest dynamic relationships. In Norman Corwin's notes for "The Plot to Overthrow Christmas," for instance, the author explains that at a certain juncture listener audioposition descends to the underworld, a task for which he advocates using a slow oscillating effect: "It's a long way to Hell and you don't get there in a flash."[31]

The growing significance of these arrangements is also evident in how broadcast studios for drama evolved physically. As historian Emily Thompson has shown, 1930s music studios were increasingly clad in dampening surfaces in order to even sound pickup and standardize listening. Such environments sought minimal acoustic profiles that were free of reverberation, where voices could become "clear and direct, efficiently stripped of all aspects unnecessary for communication."[32] As a result, recordings of song no longer registered acoustic details about the unique characteristics of the location in which the singing took place. But this was not the case with drama studios, which were built differently. Typical studios had "dead" walls hung with monk's cloth to nix reverb, but drama studios always had at least one "live" end with hard walls that reflected sound for scenes taking place in imaginary enclosures.[33] Drama studios offered carbon and ribbon microphones that had omnidirectional, bidirectional, and cardioid responses, all of which were mounted, angled, or filtered to put only select areas, voices, and frequencies "in beam" at given moments.[34] Director Earle McGill remarked that rehearsing the actors was a negligible part of radio technique, noting that actual vocal qualities were only of "incidental interest," since the real art lay in "rehearsing the microphones."[35] That rehearsal came with its own props: "sound tents" were used to create heterogeneous pickup environments within shared studios; moveable "fins" appeared in the studio walls, each of which had a different surface that could color sound according to the needs of a play; studios added echo chambers, sound caves, and effects stations into their architecture.[36] Norman Corwin recalled frequently employing "dead booths"—studios within studios that could be erected and collapsed around microphones silently by the actors in mid-broadcast, in order to produce the effect of a theatrical "scene change."[37] In the seventeenth century, Francis Bacon had dreamed of "sound-houses" for studying artificial echoes, re-creating sounds and extending the powers of the ear.[38] For a few seasons in the late 1930s, CBS broadcast studios seemed out to fulfill that utopian fantasy.

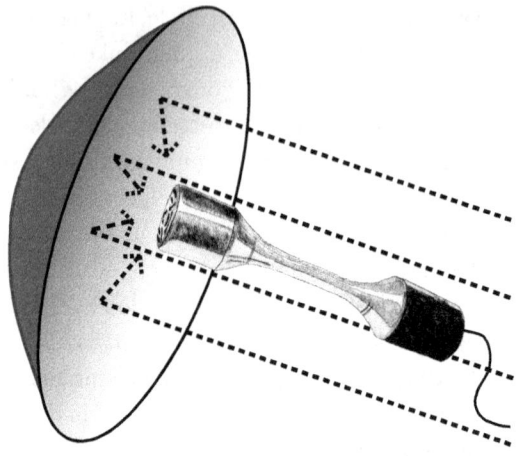

FIGURE 2.2. A "parabolic" microphone.

By the end of the 1930s, the drama studio was far more acoustically complex than its musical neighbor, and when the audience was positioned adeptly, this complexity yielded stylizations. Amplitude was used creatively. In William N. Robson's "Incident in the Pacific," for instance, a submarine is evoked when commands are relayed by a series of actors standing at incremental distances from the mike, as if they are seamen stationed at a series of bulkheads stretching away from our audioposition to the back of the boat. The scheme gives the drama a sense of closeness that empowers its suspense. Other plays use filters to conduct words through putative objects and substances, as in *Suspense*'s "The Burning Court," in which a dead detective speaks "through" a poisonous liquid that a woman is about to give her husband. With sound color, new conventions could also arise. When a carbon microphone was used in a play otherwise miked by much more sensitive ribbon mikes, listeners knew that thinner voices came from a diegetic telephone or phonograph record. Meanwhile, the use of high ambient resonance identified voices connected to ghosts, invisible people, alter egos, and God. Few programs would attempt complex segues forward or back in time using audio alone until they had such tools and facilities, and even popular programs such as *The Lux Radio Theater* failed to develop many scenographic innovations because they usually aired from theatrical rooms.[39] Staff could not easily control a stagey acoustic, so *Lux* did not vary its sound qualities, and its plays often focus on banter rather than blocking.

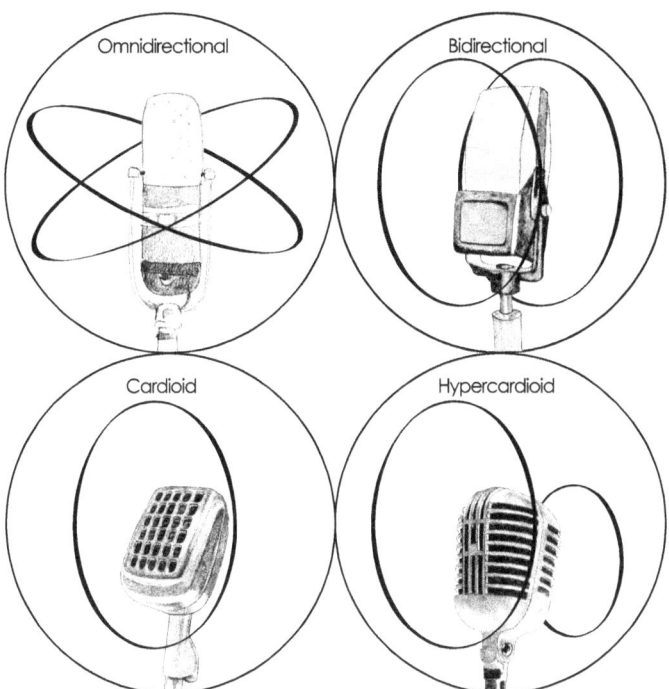

FIGURE 2.3. Common microphone polar patterns. As the name implies, a "cardioid" polarity stretches forward in a "heart-shaped" zone of pickup. These response patterns had as much to do with usage as they did with design. The omnidirectional microphone, for instance, could be tilted sideways to behave more like a cardioid microphone.

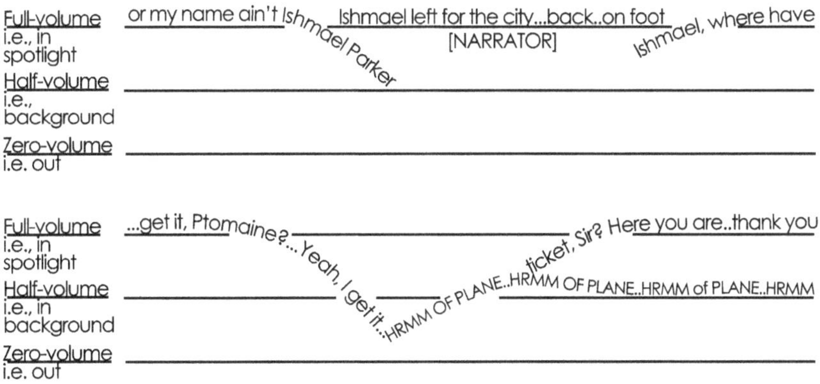

FIGURE 2.4. Volume as "radio's spotlight." Based on an illustration in Erik Barnouw's *Handbook of Radio Writing* (Boston: Little, Brown, 1939).

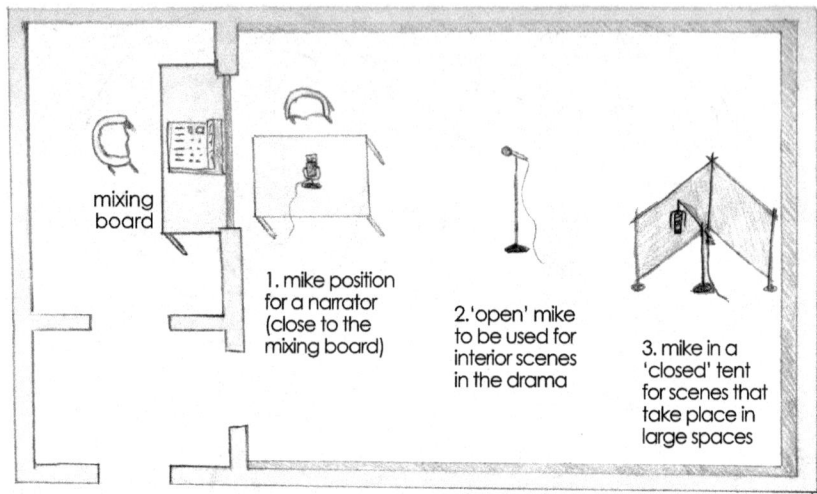

FIGURE 2.5. A simple layout for a radio drama.

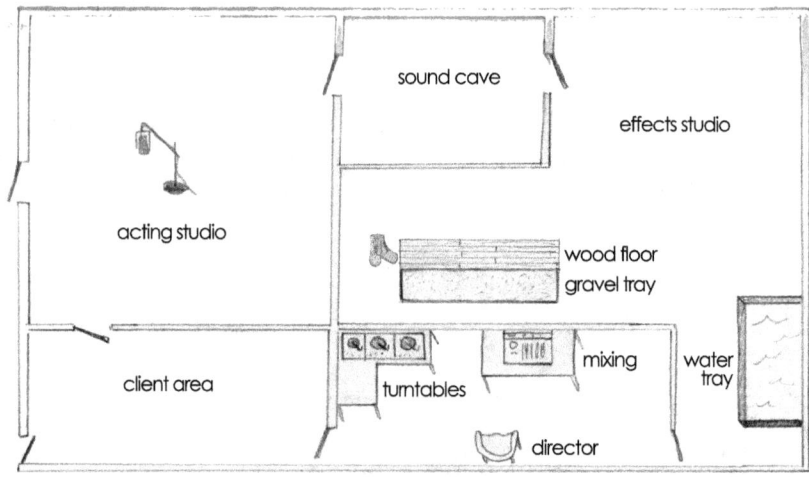

FIGURE 2.6. Studio layout for *The Lone Ranger*. The "sound cave" could be used to illustrate action that takes place far from our audioposition. The floorboard and the sand and water trays provided a range of walking surfaces to suggest settings. Based on a sketch reproduced in Robert L. Mott, *Sound Effects: Radio, TV, and Film* (Boston: Focal Press, 1990).

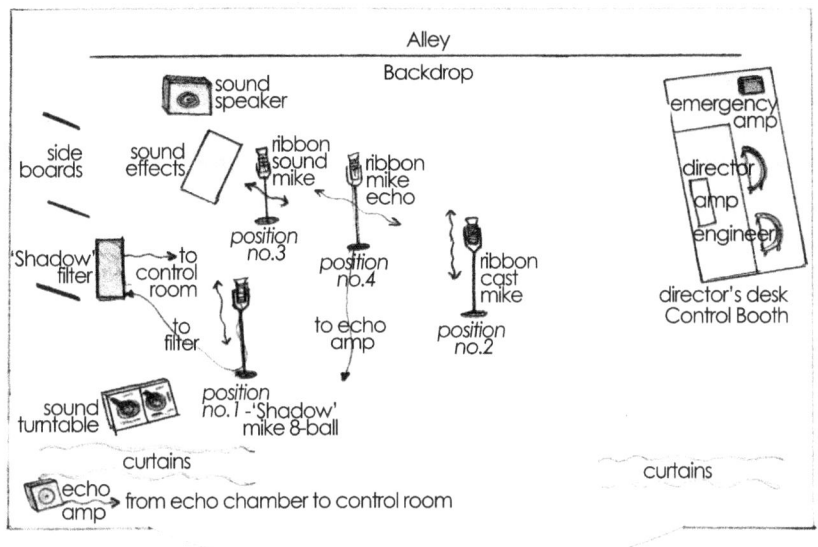

FIGURE 2.7. Studio layout for *The Shadow*. Based on a sketch reproduced in Earle McGill, *Radio Directing* (New York: McGraw-Hill, 1940).

As trade concepts and studio geography evolved, so did the practices involved in creating sound effects. Rudimentary slapsticks and gongs came to the medium from the tradition of vaudeville trap drummers such as Ora Nichols, the founder of the sound-effects department at NBC, who almost singlehandedly developed the first shared vocabularies for representing action in radio aurally.[40] By 1937, according to the memoirs of sound-effects artist Robert Mott, Nichols had accumulated as many as a thousand sound schemes, from ingenious manual effects to complex machines, many of which recall devices found in the music of the avant-garde.[41] One device featured ten motors, a series of compressed air tanks, and several whistles. Five feet high and two deep, the machine took ten months to build and could reproduce sounds ranging from a bird chirp to five hundred gunshots per minute. In the 1930s, such one-of-a-kind devices began to diminish as effects standardized. Engineers swapped real guns for the Foster gun machine, which could produce anything from a ricochet to machine-gun fire. Scarcer sounds soon became available on disks. In 1936, *Variety* profiled Tom Valentino, who traveled the globe to record the sound of New Jersey Holsteins and gurgling South American brooks, which were subsequently sold at two dollars a disk, a "universally marketable commodity for international radio."[42] A recording of the love call of the hippopotamus at the Cincinnati Zoo soon began to stand in for virtually all hippopotami aired anywhere, as effects became a lingua franca that was assimilated wholesale, almost unconsciously, by mass media societies the world over, a development that concretizes historian Michele Hilmes's idea that one of radio's chief accomplishments was to distribute a system of meanings.[43]

But what should the relationship between an object and its sound be? It is an ancient question. The Roman poet Lucretius imagined that sound was a corporeal substance that impinges upon the ear as particles emanating from a source, suggesting that the characteristics of sounds resemble their origin in a direct and unimpeachable manner. "When the snarling barbarous trumpet brays loud and deep with raucous reverberating boom," he wrote, "the elements that invade the ears are not of the same form as those that enter when swans from Helicon's mazy mountain vales uplift the mournful strains of their melting melody."[44] In the mass media era, this poetic speculation became a matter of some real professional debate. How *does* one decouple a sound from the object that makes it? For many, this act of alienation at once produced and also disturbed models of authenticity. As Jonathan Sterne has argued, the very idea of "faithfully" reproducing the sound of an original is itself an artifact of the culture of sound reproduction.[45] With the appearance of disks and sound machines, many sound artists suddenly began to believe that manually improvised "spot" effects somehow seemed "closer" to "the real thing." To re-create a real jewelry heist on one episode of *Gangbusters*, for instance, producer Phillips H. Lord hired the very stolen diamonds themselves (all $29,000 worth) from the Los Angeles jeweler in question after they had been recovered.[46] To catch decapitated heads for the guillotine sequence in his adaptation of Dickens's *A Tale of Two Cities*, Orson Welles reportedly tried a dozen different woven baskets before settling on the right one.[47]

Such fuss waned in the canned sound era. By the end of the 1930s, the new gauge of creativity lay not in bringing beasts or firearms into the studio, but in making a panther's cry by decelerating a record of a woman's scream, or a depth charge by speeding up the sound of a cement mixer. In the 1940s, one record of waterfalls was popular because it could be adjusted to evoke rifle shots, jets, or even an atom bomb, a sound rendered according to the engineers' imaginary associations with it, and not on any real-world knowledge.[48] Most radio engineers recognized that effects often involve appealing to qualities that may not be made by the actual object being imitated, because some of the psychic associations that we have with those objects belong to no particular sensory faculty. On the other hand, the use of disks could also provide counterfeit sounds that provided scrupulous verisimilitude whenever manual materials fell short. Ora Nichols's sound-effects engines could mimic airplanes, but by wartime not just any plane would do. As Robert Mott recalled, "If the script called for a dogfight between a Spitfire and a Messerschmitt 109, you'd jolly well better use the recordings of a Spitfire and a Messerschmitt 109!"[49]

Effect standardization also allowed sound to construe audioposition, expanding options to make events seem located in a certain place and heard from a certain direction. That is, once a pistol shot was uniform across many broadcasts, listeners could over time acquire what composer Otto Laske calls "sonological competence," an aptitude that makes references increasingly legible.[50] As a result of familiarity, subtle alterations of the sound emerge into prominence. Miking and coloration can thus more easily indicate what kind of space the shot takes place in, how far away it is, or if it is resounding in a character's memory. Engineers now wrote in a known language, a development that diminished the need for dialogue to confirm the nature and source of sound events.[51] Standardized effects within a TV series could also give programs their own unique spatial codes. In the early TV soap opera *The Edge of Night*, for example, a different buzzer was used for each of characters' homes, which helped notify listeners when they were positioned before Eric's house or Adele's. On the radio, disks also allowed sound effects to be offered in sequences—a car crash was in fact several tracks delivered in a "phrase" of skid, pause, impact, crunching metal, breaking glass, and silence.[52] The crash could move over its duration either as indicated by the volume of each element or as a full phrase, the car careening either away from or toward us. The intro of *I Love a Mystery* was such a sequence, while *Calling All Cars* began "inside" a moving police car picking up a bulletin, after which the car started its siren and sped away, leaving "us" behind. In this way, effects provided the operators of exploratory radio, such as "toward," "away," and "with": listeners rode beside cowboy Tom Mix in his intro, but they could not keep up with the Lone Ranger, whose program was one of many prewar shows about adventurers who were devilishly hard to catch, including *Speed Gibson*, *The Air Adventures of Jimmy Allen*, and *Captain Midnight*.

By 1937, thanks to these new practices, an agenda to arrange meaningful audioposition affected all levels of production. Whole aesthetic projects hinged on the result, sometimes with political effects, since audioposition suggests alignments. Producers were perfectly aware of this and figured it into their program design. For instance, because producers of *The University of Chicago Round Table* current affairs program opposed argument for argument's sake, they chose an omnidirectional microphone around which three professors were seated equidistantly, and the result sounded more like a friendly conversation about the issues of the week than a polarized debate. Miking and vocal technique also helped to create the "interlocutory protocols" that produced the feeling that Franklin Roosevelt was "speaking to you."[53] Engineers testified that while avoiding monotony with "vivifying use of emphasis, pause and reiteration," the president's voice never seemed to peak; speechwriter Robert Sherwood recalled that Roosevelt paid attention to pace, rarely exceeding his allotted time by more than a fraction of a second; one professional imitator of FDR likened his tone to bel canto.[54] Architectural acoustics also accomplished qualitative rhetorical work. FDR's Pearl Harbor speech showcased the large-room ambiance of the House of Representatives in which it was delivered, the president spoke the words so slowly and evenly that the first two sentences took nearly a minute. By contrast, just a few hours before Roosevelt addressed the nation, Walter Winchell had spoken to "Mr. and Mrs. America and all ships at sea" with a closer mike and delivered twelve separate news items on Pearl Harbor in under two minutes, speaking almost two hundred words per minute.[55] What Winchell said in this broadcast is one critical level of his historical meaning, but we misunderstand its total effect if we neglect the quality of his voice, which rhetorically positioned his listeners on the frantic streets of Broadway, a world of flash and panic, the polar opposite of the reassuring comfort of sagacious corridors in the FDR speech. Just as audioposition provided mobility, coloration and delivery put listeners "in their place" and could change that place rapidly over the course of a program, between programs, and week to week.

The work of Reis's cohort of broadcasters was thus a "drama of space and time" not just because it often thematized exploration, but because it used equipment to seat auditors before worlds that it chose deliberately and switched often, turning scenes from mere static pictures in the mind into worlds of quasi-cinematic pans, cuts, and zooms. Contrast this textured aural landscape with today's terrestrial radio, in which almost everything happens flush against the microphone. The sound of radio's golden age was very different. As audioposition moved, one's living room was the front of the Spanish Civil War one moment, Broadway the next, then Brobdingnag, and then a music hall. On a phenomenal level, programs destabilized the place of listening by sketching, erasing, and redrawing backdrops for the listener's mind as a way of fashioning vicinities between voices to implicate them in one another. While theaters dismantled fourth walls, radio erected them; while cinema flattened amplitude, radio serrated it; while music studios abrogated reverb, radio put it to work. By exploring time and space, radio became amenable to craft subjectivity because it could tacitly quarter addressees in whatever ground would be most rhetorically powerful.

AUDIOPOSITION IN "THE FALL OF THE CITY"

I think that listening for audioposition is a justifiable basis on which to begin to access and critique the experience of radio in the late 1930s, when a great deal of radio drama was crafted to encourage listeners to perceive position dynamics at the forefront of the aesthetic encounter. By studying the property, we get a little closer to the listening habits of the period as they were at once produced and practiced. Listening for audioposition can also be an aid when it comes to unpacking sophisticated productions such as "The Fall of the City." In radio we cannot do "close reading" or shot-by-shot analysis because we hear several layers of sound simultaneously, which makes it hard to put them into small units that can be subjected to the kind of granular analysis that we associate with readings that are grounded and rigorous.[56] By listening for audioposition, a critical listener has such a granule, one that corresponds directly to an aesthetic choice made on the part of a broadcaster, whether intentionally, absentmindedly, or (more often) as a matter of custom and convention. While listeners may disagree about where audioposition rests in a particular scene, they at least have a common framework in which such disagreements can take place, a basis on which to force diaphanous "pictures in the mind" into a tangible format where they can take shape.

In the case of "Fall," listening for position matters because it reveals important conceptual frictions that shape the political rhetoric of the drama. In this broadcast, director Irving Reis and writer Archibald MacLeish each sought to vet new techniques, but their approaches were not entirely cooperative. As noted in chapter 1, MacLeish conceptually focused on how radio could be a "stage for the word," while Reis was far more invested in showing "what a mike could do" in the acoustic environment of the Armory hall in which the play was staged. With this divergence in mind, I would like to suggest that the design of "The Fall of the City" is a duet for two schemes that often harmonize but ultimately describe two mutually incompatible audiopositions. As a result, there is a tension in this play between verbal information and acoustic information, a fissure threatening to undermine the stability of the enterprise. This discrepancy is covert enough not to disrupt clarity, yet obvious enough to enrich the piece by exposing the subterranean issue of whether the word or the mike could better dress the stage and summon the actors. In this conflict lies a struggle between two models of radio dramaturgy within the theater of space and time, as well as a conflict between corresponding sensibilities in New Deal political thought. In its totality, I argue, the broadcast is about a political rhetoric going in both technocratic and populist directions at once, just like midcentury radio itself.

The broadcast begins with an expository dialogue, a controversial convention. Reis advised that writers cease using spoken material to convey stage blocking and scenography in order to curtail what he derisively called "the 'here comes so-and-so' school of exposition."[57] Nonetheless, MacLeish's script opens with a "Studio Announcer," who offers what seems at first to be expository commentary concerning a prophecy recently issued by a woman returned from death in the City. Because his voice is in the "noplace" at the outset of the play, he seems to be an external host setting up ensuing events, and he has all the aural markers that tell us this—he is closely miked, alone, and seems to occupy a small enclosure, one designed perhaps to be not so different in ambiance from the living rooms of his listeners. Such a host was standard in many serials that had to recap previous episodes ("Last time we met our hero . . ."), and was also common in playhouses that set a scene verbally ("Our curtain rises on . . ."). MacLeish actually favored the convention, deeming the announcer the most "useful dramatic character since the Greek chorus" and bestowing him with just the mediating duties associated with the latter.[58] But for an announcer, the tone of the dialogue in the opening moments of "Fall" is extremely atypical:

> In a time like ours seemings and portents signify,
> Ours is a generation when dogs howl and the
> Skin crawls on the skull with its beast's foreboding.
> All men now alive with us have feared.
> We have smelled the wind in the street that changes weather.
> We have seen the familiar room grow unfamiliar:
> The order of numbers alter: the expectation
> Cheat the expectant eye. The appearance defaults with us.
> Here in this city, the wall of the time cracks.

This is hardly here-comes-so-and-so. MacLeish accomplishes stage-dressing work at a discursive level beyond that of denotation. The atmosphere associated with its caesurae and argot—style rather than content—appeals to cultural memory to cover the play with a mood of fatalism. "The City" deserves its definite article, because stylized vocalization has prepared us for an allegorical world filled with folk eating from maize leaves before a temple and awaiting a prophecy.

This stylization of dialogue has the secondary effect of calling into question the "noplace" associated with an announcer's speech. Like a chorus, the Studio Announcer speaks as one personally involved in diegetic outcomes, held fast on the inside of the narrative as well as at its exterior. In other words, as the speech continues, it becomes clear that the speaker is standing on the very stage that he is building. Schiller once called the classical Greek chorus a "living wall" between domains that preserves poetic freedom on one side and contemplation on the other; in modern times, the use of a chorus could enable poets to bring forth poetry from vulgar modern life, rebuild the palaces of antiquity, and bring the problem of representation back to the essence of the human.[59] MacLeish certainly does the latter, but he also puts unusual pressure on the living wall, to the point where the Announcer's "studio" falls into the fiction, dragging listeners alongside. This effect primes the auditor to draw parallels between the play and the real world, an effect fully triggered when the correspondent on the scene of the central square (Orson Welles) speaks in direct address. In this way, the Announcer's true dramatic function has been to transform us into listener-characters "in" the drama. He occupies a place inside the fiction in order to create somewhere for us to be in it as well—as literary critic Gérard Genette once put it, "To an intradiegetic narrator always corresponds an intradiegetic narratee."[60]

MacLeish's "word" has thus far created sophisticated effects merely by using intonation, meter, and diction, leading up to the first transition of the play, as the Announcer "takes us to the scene" using a choppy segue in which we suddenly "appear" in a locus overstructured by its boundaries by way of an emphatic gesture, as if the voice itself were a vehicle. This recodes the space-time of the play, matching the interior of the drama with its exterior by means of a common "now," priming listeners to be positioned "on the scene," where the murmurs of a crowd act as a keynote and the ambiance opens up into a wider arena as indicated by the lag between an utterance and its second resounding. After offering a new manner in which to imagine space, the play first erodes "location" and then offers a very strong sense of time and place, with the effect of "bringing us into the real" as the drama in fact brings us into the fiction. "The word" has coded space, decoded it, and recoded it once more, at no time requiring unusual acoustic color. The auditors are drawn into a ground inside the story for them to occupy as a single unity with others. By the time that we meet the city crowd, we ourselves are a second multitude very much in the fiction, crammed into Welles's vicinity and the recipients of his direct address. When Welles says, "I wish you could see this," he intends the pronoun "you" as a fictional second-person plural in which we imagine ourselves included. Indeed, most of his self-references are in the first-person plural, even though he tends to refer to the crowd in his vicinity as "they." This usage insists on a fantasy of collocation; we are at once listeners and quasi-characters, listening to a projection of our own presence.

Later on in the broadcast, "the word" cooperates with "the microphone" when it comes to revealing political warnings, sketching out space, and illustrating action. For instance, after the Oracle predicts that a Conqueror will come and "the city of masterless men will take a master," pacifist Priests and Ministers are drowned out by Welles's comments far more often than the orators who argue to resist, an orchestration that reveals the writer's sympathies through the relative volume of coincident speech. Meanwhile, Welles verbally confirms key silent actions, such as the distribution of food by the appeasement faction, and he also identifies otherwise ambiguous effects, including the orgiastic dance of the crowd, which is richly illustrated by music, drumming, and laughter, but also clarified by narration. A similar harmony of word and sound occurs when the Conqueror's approach is aurally connected to a concert of speech rhythms. Welles's crescendo of commentary counterpoints erratic far-off drum beats, as well as the approaching metronomic steps of the metal monster, each step louder and closer than the last. Finally, the crowd is oddly muted in its frenetic supplication before the empty shell of the Conqueror, so that Welles can tell us that the great villain is no more than a hollow cavity and then intone the alert at the core of the play: "The people invent their oppressors: they wish to believe in them."

Since words and sound cooperate so well, it is quite surprising that, on close inspection, our position on the plaza as described by dialogue is in fact incompatible with our position as suggested by the sound. In the text, Welles explains that he is on "a kind of terrace" situated "near" the Ministers on a raised platform, which is "well off to the eastward edge" of a plaza framed by hills and streets leading into the distance. There is a crowd of ten thousand, "maybe more," down below, awaiting the Dead Woman, who draws near from a tomb that is evidently much more distant. All that we know from the dialogue is that it is "off to the right somewhere." Welles's elevation above the crowd is unclear, but it is likely that his terrace is more than midway between the ground and the Ministers, since from his vantage "all is faces" and he can see men hoisting children on their shoulders and patterns in crowd behavior—"a serpent of people, a current of people coiling and curling through people." The Priests in their "high pyramid" are probably on the southern edge nearby; when a bare-breasted girl heads toward the pyramid, she is "coming" rather than "going" toward its steps. Dialogue suggests that Welles has mike in hand, as he repeatedly asks us to "listen" to a specific voice as if he were pointing the microphone toward a source. According to what we know, there are no other live microphones in the space and no indication that Welles moves. Indeed, a static gaze is suggested by Welles's penchant for noting the circling of hawks above (as if to cordon him) and also by the time it takes for him to discern the sources of commotions. Welles is thus neither panoptic nor panauditory, but "off-center" and physically aligned with elites—when a messenger approaches the Ministers, Welles says he is coming to "us." This position serves a critical function in the climax of the play. Thanks to this elevation, Welles can see into the empty armor of the colossal Conqueror ascending the pyramid and witness the mob's imprudent submission to him.

But audioposition does not match this blocking at all. According to the dialogue, we are "with" Welles at all moments, and each featured speaker occupies a significantly different place on the scene relative to us. The plaza is presumably large enough for distances between the orators to result in a different acoustic profile for Minister, Priest, and Oracle. In a "natural" scene, each voice would feature a unique color formed by elevation, adjacent surfaces, and the power and distance of vocal projection relative to us. The Ministers at their balcony just above would certainly be the loudest of all characters in the City. In the performance, however, each public speaker sounds equidistant because they are equally amped and acoustically identical, as if at an equivalent radius to our center. Moreover, the people "down below" are much louder than the orators, even before they begin climbing the platform in the finale, and they seem considerably closer to our audioposition than MacLeish's dialogue suggests. If the audio matched the script, the voices of citizens would also feature reverb as they ascended to us, yet they have no such color, suggesting that these voices are projected laterally rather than vertically, which explains why they sound muffled by adjacent soft bodies and clothing, precisely how they would if Welles were in their midst. If MacLeish's "word" built space through diction, meter, and segue, Reis did so with ambiance, acoustics, and volume, offering nothing to contradict the stations of most of MacLeish's dramatis personae, but changing the listener's place among them. To match MacLeish's script more precisely, Reis could have added reverb to crowd speakers, and directed the Priest and Dead Woman to deliver lines off-mike. Instead, Reis chose to create a simpler aural scheme that had the result of wrapping Welles

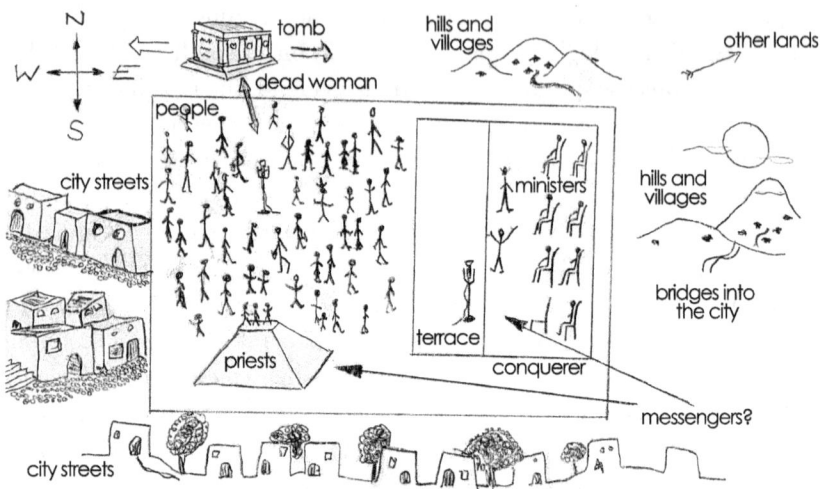

FIGURE 2.8. A diagram of the principal setting of "The Fall of the City," April 11, 1937, according to dialogue and sound. The entrance of the Conqueror is hard to ascertain. He seems to enter the City from a bridge, then passes through streets full of defenders. Note that the standing microphone seems to be in two places at the same time: on the terrace and at the center of the crowd.

in two envelopes of humanity—one depicted by the crowd in the Armory, and one drawn in from the beyond to occupy a proximity at which Welles spoke. One interesting feature of this situation is how easy it is for us to take both Reis and MacLeish seriously at once, despite their incompatibility. Perhaps that is a clue to what makes a theater in the mind unlike every theater outside of it. The "picture" of "The Fall of the City" would be paradoxical on paper, stage, or screen, but it resolves effortlessly in the only medium whose relation to spatial laws never exceeds emulation: the imagination. "The Fall of the City" works best in the mind; it is also trapped there.

Critic Béla Balázs once wrote that the "form-problem" in radio is that it cannot give us a sense of stage.[61] But when it comes to drama, the opposite is true. Radio's problem is that it cannot *help* but make a stage, and in "The Fall of the City" the conflict of word and sound manifests itself as a struggle to govern that inevitable propensity. MacLeish positions us with the correspondent, near elites, and gives us reason to worry that "the masses" will endanger their own liberty, contrasting the crowd under the influence of traditional orators with "us," the "second crowd" in the story being addressed by the perceptive correspondent. According to this reading, the play shows people being misled by politicians but enlightened by radio announcers. Only with an objective, enlightened commentator standing physically between the orators and the people can the truth be known. But this reading does not account for the sound design, in which Reis has focused more on the crowd than on the announcer and pushed our position off the terrace and down among the masses. Reis's correspondent is both "of" and "with" the people. If his audioposition from their midst discerns the mendacity of rhetoricians, then there is no reason to suspect that the diegetic masses cannot do so as well. More importantly, rather than drawing distinctions between the foolish mob and the well-informed listener-as-character, Reis brings these groups into sonic coalition, as we all share a single imaginary audioposition.

So two conflicting configurations are at work. On the one hand, the "stage for the word" offers pessimistic liberalism by featuring a mob that submits to demagoguery with unlimited obeisance, as "we" listen to the correspondent, a civic-minded opinion-leader providing sage technocratic guidance. On the other hand, the theater of acoustic design offers a populist schema that does not distinguish between "us" and the crowd that listens to the orators, a Popular Front parable that makes the trial of the people our own. MacLeish's correspondent is a hero because he guides the listener away from the folly of the people in the square, while Reis's is a hero because he brings us together with them. Given that the play is expressly concerned with the politics of communication to begin with, perhaps the best way to read the broadcast is as a meditation on contradictions lurking in how New Deal culture processed the idea of addressing multitudes. In its formal clash, the play offers us a version of what James Carey has identified as a polarity between concepts of mass communication as an elitist "science of society" and mass communication as a populist "science in society."[62] Should media leaders like the correspondent guide citizens to the truths that they cannot see for themselves, or should they work among the people to improve communication between groups? Embodying the notion that this is an insoluble choice, "The Fall of the City" oscillates between the two positions without synthesizing them in a dramatic plenum that uniquely allows for such equivocation. Both schemes are copresent and mutually confronting, as the technocratic word and the populist sound give an interplay that lends richness to the performance.

This reading suggests that radio was much more than just "the way" that New Deal policies reached regular people. Radio's exploratory dramaturgical conventions were also sites in which key problems about the place of the media in Depression-era thought were negotiated. In "The Fall of the City," radio was engaging the contradictions precipitated by its own existence at the level of the senses, precisely the juncture where, as Miriam Hansen has argued, media at once mass produce experience and also provide an aesthetic horizon for its negotiation.[63] And this interplay only emerges because my reading considers where we are according to what we hear, on the pictures in our heads as "drawn" by dialogue, held up against those sketched by acoustics and volume. Audioposition makes the duel between word and sound possible; neither agenda can muster conceptual ends without positioning us in one way or the other. In few earlier cases does audioposition matter so much to dramatic aims. On *Amos 'n' Andy*, it was far more important to decode Andy's malapropisms than to ascertain his vicinities, but every ambition of "The Fall of the City" is inextricable from its aesthetics, so without studying this broadcast in sound form and "reading" for audioposition, the deep work of the program is unknowable. One reward of such analysis is that we can use it to discover common conventions in audioposition that exhibit how feeling was structured prior to and within everyday life. In the next chapter, I present just such a hypothesis about Depression-era radio drama, arguing that two normative schemes of audioposition emerged in this era, each of which is dimensional in nature and was created to help tell stories that responded to the social and political upheavals of the decade. In fact, one of these styles is defined by a property on which the two schemes of MacLeish and Reis each scrupulously insist, our "intimate" proximity to the reporter on the scene.

CHAPTER 3

Intimate and Kaleidosonic Styles

Alive with the bustling traffic of unruly mobs and untrustworthy ministers, "The Fall of the City" suggests the degree to which political friction lies restless beneath the audioposition choices made in many scripts and studios of 1930s radio. But there is something even more valuable about an examination of these choices than the access that it gives to hitherto unnoticed complexities in one play or another. Using this kind of reading, we can isolate strategies used to structure listener alignment, systematizing how directors of the era drew images in the mind that responded to and shaped the public to which they spoke. One broadcast does not make a trend, but the conventionalization of the style of a play can have a greater reach than a single broadcast in that it points toward a normativity among the aesthetic instincts of members of a listening community. So in systematic stylistic observation, it becomes possible to make new types of claims about what radio plays say about American understanding. This is the rationale behind the argument of this chapter, which considers such programs as *The Columbia Workshop*, *The Shadow*, *The Mercury Theater on the Air*, and *The March of Time* and suggests that late 1930s directors developed two audioposition formulas that account for the overall sound of the period, what I call the intimate and the kaleidosonic styles. I will argue that each of these styles represents one way that prominent broadcasters solved representational and narrative problems, and that each also embodies an aspect of the political rhetoric of the period. It was only when the balance of these styles was upset that radio aesthetics ended its love affair with space and time and dramatists embraced stronger models that proved

Dorothy Thompson's 1934 prediction that what the airplane was to international warfare, radio was to international propaganda.[1]

THE INTIMATE LURE

I have argued that for the polemical speeches of "The Fall of the City," Irving Reis and Archibald MacLeish manipulated audioposition to expose charged political questions. Other *Workshop* productions evince a similar preference for form to fit content. Consider William March's isolationist "Nine Prisoners," which concerns the execution of Germans during the Great War. The narrative is related after the fact in nine direct-address confessions by the soldiers ordered to carry it out, so as to fragment the sequence of the narrative just as its events have alienated the men from one another. The play's argument is prosecuted by the manner in which it is related. *Workshop* directors likewise synced emotional effects with audioposition choices. Brewster Morgan's version of Edwin Granberry's "A Trip to Czardis" is typical. As a poor family journeys from a Florida backwoods to the town in which the father works, the auditor is positioned near two young boys. Jim, the elder, regales young Daniel with tales of balloons and lemonade at the Czardis midway, to which Jim expects their father may treat them. For much of the play, it is as if we sit between Jim and Daniel in the buggy on its way to town, enraptured by their colloquy and hearing of wondrous landmarks that approach and recede from us in the aural background. Because our position stays with the boys for most of the play, while adults remain taciturn at the outer edge of earshot, listeners only gradually come to realize that the family is making the trip to visit the boys' father prior to his execution. For cathartic effect, Morgan wants us to apprehend this at the same instant as Jim. After sharing childhood with both sons, the play yanks us back into maturity, and the auditor and Jim now share an understanding of which the third member of the triad is excluded. We witness Jim transform from Daniel's peer to his steward, as the execution reorganizes family roles. This is the source of the play's poignancy, and it would have been shortchanged by a different audioposition or excessive segue. It would have surely been possible to focus on why the father was convicted, or to follow him to his fate, but instead the narrowness of the foreground creates a sense of solitude. The town approaches and later recedes; the father fades away from the son. In this way, tight audioposition lures us toward a grim act of official violence through a place of quiet melancholy, in which we are intimately caught.

In several shows, *The Workshop* mingled similar "intimacy" schemes with appraisals of the Depression emergency, as writers picked characters such as the boys in "Czardis" to carry listener position through the fiction. According to convention, we engage social problems staged from oblique angles: in "The Story in Dogtown Common," we hear about trials on the seas by way of the lamentations of mariner's wives awaiting their men; in "Bread on the Waters," we are aligned with waterfront urchins on Christmas Eve, meeting a procession of the socially displaced; in "Pepito Inherits the Earth," we follow a soldier meeting the victims of fascist atrocities before he dies helping his son escape Franco's armies over the Pyrenees. In these cases, we sympathize with a "near" character with whom we have positional intimacy so that we might empathize with "further-off" characters in distress and privation. This sympathy-empathy structure was common in 1930s documentary realism, which attempted to report on people of low social standing to render their lives vivid to those of higher standing.[2] Travel was a characteristic component of this artistic mode, as prominent writers such as Sherwood Anderson, Theodore Dreiser, and Gilbert Seldes published stories that exposed the inner feelings of "ordinary" people. The results were often grand celebrations of American life, yet morally circumspect. In *Let Us Now Praise Famous Men* (1940), James Agee expressed the trend as a hypocritical use of closeness "to pry intimately into the lives of an undefended and appallingly damaged group of human beings, ... for the purpose of parading the nakedness, disadvantage and humiliation of these lives before another group of human beings."[3] *The Workshop*'s structure pried "intimately" but seems less of a "parade of nakedness," since self-censorship precluded the most outrageous suffering fetishized by writers. In literature, it was the "artist" or the "journalist" who went out among the people, but in radio it was usually someone from the locale carrying us through it—often women, children, or another morally unimpeachable "earwitness." The upshot was that *The Workshop* had the sentiment of social documentary but avoided the consequence of subsidiary luridness.

While *The Workshop*'s political shows aligned us intimately with witnesses in order to involve a listener with the experience of social displacement, 1930s adventure programs instead positioned us intimately alongside primary heroes at the center of adventure, a choice that could create opportunities for the process of listening that affected the configuration of a narrative. Consider the usual plot sequence of *The Shadow*.[4] In the first act of a typical program, we tarry alongside Lamont Cranston and sidekick Margot Lane as they uncover a mystery and come across clues, until Lamont decides to transform into the Shadow to pursue a lead, just prior to a cutaway for a commercial break. As the next act opens, listeners seem to have lost Lamont and find themselves near the plotting villains at their lair. But soon the Shadow's trademark laughter filters into the scene, and the crooks curse themselves for having divulged their scheme within the invisible earshot we have been sharing with the hero all along. The compelling power of the program centers on this implicit homology between the main character and the listener, who exist as mutually invisible presences. In *The Shadow*, we shadow the Shadow. During the 1930s, the program also thematized audioposition, as Lamont frequently used his power to confuse villains' hearing, confounding their attempts to triangulate him in space. In tandem with invisibility, inaudibility proved handy in many enclosed drainage pipes, caverns, or vaults, where showdowns tended to occur.

Other programs depicted fruitful acts of echolocation. Carleton Morse's "The Thing That Cries in the Night," has our heroes baffled by a series of family murders in a mansion, until they find the source of a baby's cry that follows each attack. Meanwhile, in kidnapping stories on *This is Your FBI*, blindfolded victims frequently recall to Special Agent Jim Taylor the sounds at their detention site, details that lead the G-man to position the hideout in the world of the fiction. In such programs, we feel our way through an environment that becomes increasingly transparent. Over time, dramatists devised ciphered landscapes of sonorous bodies for us to decode while in a state of proximal intimacy with sleuths who help us cognitively transform sounds into maps. Noting these conventions, historian Kathleen Battles has written of a "dragnet effect" in 1930s crime programs, in which peace officers used their police radio systems to stage a ritual of inevitable apprehension in the course of which the understanding of radio as an "omnipresent force" was elided with the symbolism of ubiquitous police "presence" in modern society.[5]

Clearly, clever use of intimate listener audioposition could give plays emotional dimension or offer satisfying puzzles. But these options hardly exhaust the style. A case study in the form can be made using Orson Welles's *The Mercury Theater on the Air*.[6] Before he began the series, Welles favored a structure in which long passages of external narration set up a few sparse dramatic scenes. This is acutely evident in the 1937 adaptation of *Les Misérables*, in which Welles's prolix excurses eat up three minutes of airtime, an eternity in 1930s arrangements. After the first episode of the seven-part series, *Variety* quipped, "Maybe there will be someone left to listen to the finish six weeks from now, and maybe not."[7] On *Mercury*, Welles experimented. For his staging of *Dracula*, he preserved Bram Stoker's multinarrator system, one of the few interpretations to do so. In his adaptation of *Julius Caesar*, he added passages from Plutarch to Shakespeare to clarify action. Welles and producer John Houseman also hit upon a style based on first-person retrospective narration. In this common pattern, *Mercury* typically assigned an internal character the duty to relate the story to us ("It all began when . . ."). According to one interview, Welles considered this convention the best way for radio to achieve artistic independence without "clinging to a technique designed for the stage."[8] In the first-person singular genre for which Welles is known, we are stationed intimately near a single character as recipients of his or her direct address during present-tense narration (like the audience in "The Fall of the City") while we are invisible to—but physically alongside—his or her anterior self in the past (as in *The Shadow*) overhearing voices during the narrative's principal temporal span.

    This intimate formula established rapport and seemed less objectionable than external commentary, but it also presented problems. Because radio has no universal equivalent to the quotation mark, auditors could easily misinterpret whether a character was speaking to us in the present, subsequent to the main action, or to another character in the past during that action.[9] In writing, verb tense can clarify this issue. As narrative theorist Gérard Genette has pointed out, while you can tell a story without saying where it happens, it is impossible to do so without locating the story in time relative to the narrating act, since any story must be told in a past, present, or future tense.[10] For Welles, weighing down dialogue with "she said" and "I replied" was out of the question. To keep things legible, *Mercury* innovated a code of amplitude in which direct present-tense narration is denuded of effects and miked more closely than past-tense sound. The present is a state in which sound reveals nothing about location, while the past is a dynamic state in which sound is replete with detailed information. This set of conventions made exposition seamless, but at the same time it redoubled the gravity of the choice of which character would narrate. This figure "carries" the listener not just because the story tells us so, but also because on a visceral level he or she is "closer" to us than anyone else, as indicated by volume in narration. In a first-person radio play, anything close to us physically is also close to us temporally because the level of amplitude that codes distance also codes time. The aesthetic situation is that the sound quality of the narrating voice is unavoidably both "now" and "foreground." This voice always promises more magical intimacy than others, which is why it is so affecting when narrators turn out to be unreliable, dead, or insane. So in Welles's scheme, a character-narrator would have to be nominated, and because this nomination forms a profound intimacy between the listener and another being, the choice also implicitly decides whose place in the world of the story is surmised to produce effects that match dramatic aims. *Mercury* plays unravel to audioposition analysis exceptionally well. Consider "Hell on Ice," in which engine master Melville narrates the tale of the doomed steamship *Jeannette* by reading to us in the present from a log he kept of this misadventure. In segues back into the past, our audioposition remains "with" Melville, so the audience is close to the officers of the vessel but never one of them, always more aligned with the ship workers, who remain stalwartly devoted to the Captain as factionalism and mutiny break out among officers over the many years the *Jeanette* is frozen into the polar ice-pack. This scheme mimics a New Deal–style alignment of the people and leader against quibbling elites and involves us in the alliance to boot.

Welles, producer John Houseman, and adapter Howard Koch did not always take such a simplistic approach to the rule of miking a narrating character the closest. In Ellis St. Joseph's "A Passenger to Bali," Welles plays Reverend Walkes, a political agitator stuck on a ship wandering from port to port in East Asia. No nation will permit a man with Walkes's record of fomenting uprisings to disembark, much to the irritation of our narrator, Captain English, who has agreed to give Walkes passage on the basis of documents that turn out to be forged. In the play, Walkes's basso profundo is loud and sharp, so in spite of what we are told, he sounds twice as "close" to listeners as English does, which makes his presence as onerous to a conventional acoustic profile as his character proves to the ship. Walkes's entrance from a distance is marked as "unsafe" to sonic relationships long before his nefarious history is revealed. English cites "a physical revulsion the moment [he] laid eyes on" Walkes, and we feel the same the moment we lay ears on him because of his shrill proximity in an aural scheme that really ought to foreground English. Of course, the rule is broken for a reason. Our instinct about highly amplified characters is so strong that we cannot help but *feel* positional intimacy with Walkes, a design choice that gives dimension to "Passenger" by morally coupling us with English but sonically pairing us with both him and Walkes at once, something that is often spatially impossible. English becomes hostage to his passenger, and the sound becomes hostage to Walkes's paroxysms of laughter and penchant for yawping from deep fogs. Underneath its manifest events, the drama is about two voices that struggle to command the audio space, a tug of war in which our attention is the rope.

As the drama continues, we begin to wish that the Captain will yield to his temptation and throw Walkes overboard, simply to expunge the sound. Meanwhile, the vessel atrophies, seas slow, and the ship becomes unseaworthy under the pull of Walkes's gravity. Of course, in the end the ship will buckle and sink following a climax in which English confronts Walkes. Since this final confrontation coincides with a gathering storm, it is also a conflict between two voices struggling to top one another, a match that brings to a head the competition that the amplitude has been staging all along. English wins the contest, at last asserting command of the foreground. Yet the denouement is melancholic. As we row to shore in a lifeboat with English, we hear the begging calls of a chastened Walkes, left alone on the crippled ship. The scene moves us away slowly so that Walkes is brought into equilibrium with English, then drops out of the sonic world as our intimacy remains with the latter, healing the contusion of normative aural relationships. Yet the play becomes oddly hollow as Walkes's voice vanishes into the background, and we are wistful for our intimacy with the bombastic agitator. The play bends the moral sympathies of the listener around what acoustic theorist Barry Truax calls a "habituation syndrome"—we are like a city dweller irritated by urban noise who moves to the country only to learn he or she can no longer sleep without it.[11] In many dramas, we are passengers of other men, other women, and other creatures. "Passenger" vanishes just as we measure the allure of perilous intimacy against the surety of safe passage.

THE KALEIDOSONIC STYLE

The dramas cited in this book so far vary a great deal, but most conform to a signature audioposition design of the 1930s drama of space and time because in each play the listener seems to nestle intimately with one carefully selected character for the duration of the drama. "Broadway Evening" moves us through space beside ambling strangers, "Fall" befuddles our location, and "Passenger" misleads our alignment, but in each of these broadcasts, the listener always flanks a body "nearby." This "intimacy style" proved highly versatile. "Fourth of July" on *Calling All Cars* lets listeners "hear" the evening from a mobile Los Angeles radio police car, while "The Odyssey of Runyon Jones" on *26 by Corwin* brings us through the many domains of the afterlife alongside a young boy searching "cur-gatory" for his deceased dog. Pare Lorentz and William Robson's "Ecce Homo" attaches us to "Worker #7709" moving through industrial America from the typewriter factories of Syracuse to the irrigation dams of Missoula, gathering the voices of workers decrying parsimonious oligarchs. In "My Chicago," Arch Oboler gives us an encomium to the Second City through the ears of a wandering child, and with "Native Land," Carl Sandburg offers his own "City of the Big Shoulders" with a story of the poet gleaning an omnibus of sound bites. In a decade that journalist Murray Kempton called an "age of the counterfeit Whitmans," *Cavalcade of America*'s Robert Tallman even resurrected the poet himself, and affixed us beside Whitman's ghost as he searches for folk phrases across America in "I Sing a

New World."[12] Just a few years after scholars saw folk sayings vanish into the homogeneous radio age, intimate plays replicated these idioms in evocative paeans. As historian Daniel Czitrom has argued, this was among the central social purposes of the mass media in this period—to surround the promise of consumer goods with all available signs of the lost intimacy of a recent past that had become suddenly mythical.[13]

And as war came, the intimate model went international. Norman Corwin's *An American in England* series brought us to London alongside the author to hear plucky maidens working flak guns and the denizens of a small Norfolk village dousing lamps in case of an air raid.[14] The episodes of *Words at War* repeated the pattern: "Here Is Your War" has us follow Ernie Pyle across North Africa, which seems to be a miniature America in its GIs' accents. In "Assignment USA," we crisscross the United States on a train with journalist Selden Menefee, cataloging pernicious isolationism voiced across geography. With journalists, we circle the rim of the Eastern Front from the Aleutians to Greece in "Firm Hands, Silent People," "The Last Days of Sevastopol," and "Shortcut to Tokyo." "War Tide," "Paris Underground," and "White Brigade" align us with wily pensioners and undercover schoolgirls in occupied China and Paris. And in "Malta Spitfire," "Combined Operations," and "Dynamite Cargo," the servicemen of air, land, and sea are our vessels in several theaters. *Words at War* had reinvented the structure of "Czardis." We sympathize with one soldier so that we may hear another die in sacrifice, or with a reporter so as to hear a whole nation fall. Vicarious experience and empathy were now driving recruitment and Victory Bond sales.

The intimate position formula only became possible after broadcasters learned to build fore- and backgrounds in 1930s studios, using techniques to make sonorous bodies and events seem to approach and recede, creating deep spaces through which listeners travel. The formula had the stamp of showmanship. It felt close but was not quite "point of view" and could sustain both first- and third-person narration. It put listeners deep into churning seas but provided a pilot. It let auditors track a figure who keenly angled events, compressing bewildering material by narrowing the horizon of the soundscape to a single trajectory through it. The scheme formed a dyad of the listener and the magically proximal but mobile character, an allegiance that cast events in all sorts of lights, and one that could itself be put under strain. But no matter how stressed, the tether held fast, and sympathy with a traveler was almost always a convention of the orchestration, dictating how listeners used aural information to imagine events as they transpired. Gertrude Stein once wrote that the excitement of watching theater derives from the fact that members of the audience are always trying to reconcile the temporal lag between what they see happening on stage and what they feel about it.[15] Syncopation makes us long for a unification of rhythm. Intimate radio plays do something similar with space, providing a feeling of proximal copresence that constantly promises merger but withholds it.

Intimate audioposition became so widespread as a tactic that critics have tended to confuse its effects with those of the medium itself. Instead of considering intimacy to be the result of composition choices and volume levels, the radio device has the reputation of being innately intimate, putting the listener "in another man's shoes."[16] One problem with this shibboleth is that it ignores history. Carolyn Marvin has argued that the promise of "effortless intimacy" has been "perhaps the commonest of all prophetic themes about communication" since the nineteenth century.[17] Another problem is that it ignores the aesthetic components that created intimacy. When we hear what Dick Tracy hears from the spot that he hears it, yet hear Tracy's voice unmediated by bone and sinew, the listener is not in the detective's skin but cheek-by-jowl with him. We neither hear Tracy's thoughts nor experience sound as he does. And this feeling of proximal intimacy is not inherent to the triodes of the radio set but is the product of an aesthetic strategy. One voice cannot seem close unless others seem distant, and intimacy is thus no more than a discrepancy in volume and vocal color that conveys the impression that we are nearer to one character than another, a discrepancy that was especially available to listeners of the 1930s and '40s, whose programming made more frequent use of depth effects than radio does today. Intimacy is certainly an absorbing effect, but that is not to say that it is the only effect within radio's ken.

In fact, the best-known passage in 1930s radio, the first act of Welles's "The War of the Worlds," disobeys intimacy protocols entirely, instead opting for the second style that I want to suggest helped to form the Depression-era's overall dramatic sound. To introduce this second scheme, it is helpful to bear in mind some of the aesthetic peculiarities of this well-known passage of Welles's broadcast. After a perfunctory introductory setup, no figure in the drama emerges to offer us passage, no voice seems closer to us than any other for long, and we have no clear "perspective," let alone an ally providing it to us. Reports just seem to come in about an invasion that just seems to happen. Although dialogue frequently alludes to off-mike events, suggesting scenic depth, the acoustics and effects offer paltry illustration of them. In fact, scenes are so aurally shallow that according to what we hear the Martians almost never come physically close to a microphone, hovering in the background no matter where our audioposition rests. In intimate works, audioposition follows a character, but in "War of the Worlds" audioposition segues on its own accord from place to place.[18] And just as the play lacks scenic depth, it also offers little sense of gradual duration. Space seems flimsy and shapeless; events happen all at once. As a whole, "War" becomes knowable not through a terrestrial real-time journey at all—the kind that we usually take in intimate dramas—but by switching through a series of static fixed positions across its playing space, each going live one after the other. Using this type of segue, the action seems to culminate by aggregating. Listeners hear dozens of settings depicted or discussed: the Meridian Room in the Park Plaza Hotel, the Princeton Observatory, Grover's Mill in New Jersey, an artillery station, a bomber cabin, the roof of "the Broadcasting Building" in New York. From these we hear a series of performances and effects that resist coalescing into a unified scene: "Stardust," "La Cumparsita," a clock in the Trenton Observatory, "uneven cooling" of the Martian vessel, artillery fire, piano interludes, the sound of a microphone dropping. Meanwhile, we have to track many characters with little certainty who is most important or even still alive: Professors Pierson, Englehoffer, and Gray; band leaders Raymond Rochelo and Bobby Millet; correspondent Carl Phillips and farmer Mr. Wilmuth; Brigadier General Montgomery Smith, Captain Lansing, and the Secretary of the Interior; as well as a dozen announcers, operators, and soldiers. No voice speaks in all scenes, no place contains all effects, and no person frames the horizon of the fiction. If intimacy is like a tracking camera shot, "War" is a montage, a world that we teleport around instead of moving through.

Of course, the intimacy model will not help much with "War of the Worlds" because *Mercury* did not structure this half of the program as a first-person singular fiction at all, but instead drew upon a news format.[19] It is well known that the panic incited by the show arose owing to anxiety over world events whipped up by coverage of the Munich Crisis, and this could not be true if "War" did not bear some formal likeness to the conventions of news broadcasting.[20] While it has the up-to-the-minute feel of reporting on Munich, "War" actually bears much closer analogy to earlier 1930s programs—*News Comes to Life*, *Eye Witness*, *Front Page Drama*—"news dramatizations" with sometimes tenuous relationships to real events. These programs flourished in the period because documentary audio of the news was curtailed by a 1933 agreement between major newspaper chains and the networks.[21] Premiere among the dramatizations was NBC's *The March of Time*, Welles's first gig as a radio actor and a show whose format was one of the most legible to 1930s listeners. Called the most challenging show in broadcasting by *Variety*, *March of Time* offered a set of modular scenes depicting recent events that collectively spanned the world.²² According to a profile of the program in the *Washington Post*, director William Spier and his team began production by selecting eight stories in the newspapers that seemed to be of paramount interest.²³ Over the course of ten hours, these stories were checked, scripted, "placed in proper perspective to each other," and cast using actors who hung around Radio City, some of whom made careers impersonating perennial newsmakers like Chiang Kai-shek and Wallis Simpson. Eleanor Roosevelt was known to visit *March*'s control booth to watch Agnes Moorehead portray her.²⁴ After writing scores and running rehearsals, *March* went live with eight two-minute sketches, each with a lead, cumulative moment, and blackout. On October 5, 1934, for example, listeners heard General Hugh S. Johnson saying farewell to his staff at the National Recovery Administration, Queen Mary teaching young Princess Elizabeth the Highland Fling, Harvard president James Conant writing a letter to decline an endowment from the Nazis, French gendarmes nabbing a narcotics gang, Herbert Hoover debating Henry Wallace, an Irish marquis dying of a family curse, a horse winning a prize in "the longest shot of the year," and the sentencing of Lindbergh baby kidnapper Bruno Hauptmann.

Each story features a single person performing a single action in one scene, yet the result is worldmaking. Like "War of the Worlds," *March of Time* suggest that there are transmitters set out in a grid upon a map of the earth, a preexisting net of fixed audiopositions, each node of which can be switched on or off as if from a celestial vantage. Each story is entirely independent, but the characters in them exhibit an important similarity—although Windsor Castle, Harvard, and the New Jersey courthouse had slightly different acoustic signatures, Queen Mary, Conant, and Hauptmann are each precisely the same distance from "us." If the form of intimate sound ritualizes empathy, the design of *March* ritualizes equality. *March* had announcer C. Westbrook van Voorhis to introduce each module with a dateline and end it with his signature punch, "Time Marches On!," a tag whose delivery was likened to the force of a howitzer.²⁵ Thanks to this punctuation, the October 5 episode could collate oration, conversation, epistle, trial, debate, obituary, and colloquy while taking us to Washington, Scotland, France, Ireland, Massachusetts, and the Bronx, all without danger of becoming incoherent. The aesthetic also tended to flatten time, as no matter how the scenes are sequenced or dispersed, we get the feel of simultaneity.²⁶ Clarity was maintained because the broadcast routinely returned to van Voorhis, whom everyone called "the Voice of Time," a title that reflects how effectively the formula subsumed the process of rearranging the events and manipulating "the news" into just a few isolated scenes. Archibald MacLeish wrote of such announcers: "What he said was probably trite; it was also undeniable."[27] "The Voice of Time" is not a character in the ordinary sense. He exists at an unlimited physical distance from the events that he describes. If character-narrators in intimate plays are vessels, then "Time" is more like a safe harbor. Although he can interrupt scenes with commentary, "Time" can never be "in" the scene as Melville is in "Hell on Ice," nor involved in outcomes like the Studio Announcer in "Fall of the City," but he makes his world as legible as an intimate narrative, giving *March* an abstract but orderly sense of space and time.

Although the show clearly has a special style, one used for many other types of broadcasting, critics have struggled with how to describe it, either employing terms like "voice documentary," "cantata," "demophonics," and "telescoping montage" or likening it to painting on a large canvas or turning a kaleidoscope.[28] Here I modify the final term into the adjective "kaleidosonic," to describe the feeling of a shifting sonic world that is accessed through a central point that is itself static and removed from events. In the 1930s, the kaleidosonic style ranked with intimacy among the chief strategies used to form dramatic images in the minds of listeners. In intimate plays, our position follows alongside a character, whose place in the fictional world shapes content as we move in three spatial axes. Such plays offer scenic and emotional depth. Kaleidosonic plays leap from one mike to another, "objectively" arraying the world before us, with everything equidistant and accessed across just two dimensions. Such plays sound shallow but broad and highly public. The former type of play speaks to each of us as an individual and is obviously selective about its material; the latter speaks to us as a nation, denying that it selects voices of interest and disavowing the vast technical architecture that brings them to us. Writers tended to use the conventions of intimate radio structures for plays about places, while employing kaleidosonic structures for plays about events. These aptitudes were ideal for an era in which radio thematized space and time, but they also let radio encapsulate themes of empathy and collectivism that writers used to expose the crisis of the decade. Intimate sequences make us certain that we know people, but kaleidosonic sequences make us certain that we know "the people." To embody the New Deal era, radio would require both.

ISN'T THERE ANYONE ON THE AIR?

It is not surprising that kaleidosonic structures appealed not only to newscasters, but also to Popular Front writers, who used the egalitarian distances in *March of Time* to structure programs that station representatives of social groups at common distance relative to us—as Bruce Lenthall has observed, for many modernist radio writers, sound designs like this were a way to pluralize narrative perspectives in a socially inclusive manner.[29] That was the idea: "I never wrote down, thinking I was superior to my listener," Norman Corwin recalled, "and I never wrote up, thinking I was inferior to my listener."[30] Corwin's oeuvre is full of plays that use volumetric equivalence to achieve this directness, spanning the nation and binding it together in works such as "Between Americans" and "Psalm for a Dark Year," sound-collages of workers across the country. "Psalm" celebrates Thanksgiving Day, as did another kaleidosonic play, Stephen Vincent Benét's "A Time to Reap," which features farmers from across the country talking about crops that they have harvested. Benét's "Towards the Century of the Common Man" dispenses with the "Time" figure altogether, as does his "Listen to the People," whose script denies characters names, listing simply "Radical Voice," "Conservative's Voice," "Woman's Voice," and several others with no characterization at all. In kaleidosonic dramas, characters hide nothing, rarely struggle with internal forces, and hardly develop over time. These programs have such robust confidence in exteriority that inner life is moot. Indeed, didactic broadcasts like these are barely "plays" in the ordinary sense, yet they sound like a drama of space and time right away by segueing so energetically, offering a picture of an America that MacLeish himself championed, the nation composed of the ignored, the poor, sharecroppers, immigrants, African Americans, the oppressed, and the outlaw, all given equivalent amplitude in a conjured social space. "When an artist of the thirties wanted to present a fact in human terms," William Stott has written, "these are the kind of humans he chose."[31] Of course, nothing served network public relations aims better. As the airwaves grew ever more privatized, networks showcased voices drawn from all segments of society, which gave this highly exclusive medium the semblance of inclusively and made the nation seem like a republic of transmitters, while in reality the industry addressed a nation of receivers.

During wartime, the sense of community in Corwin's work would take on a global character. Corwin's "This Is War" (the first episode in a series of the same name conceived under the auspices of MacLeish's Office of Facts and Figures) depicts machinery and men from around the world churning as one. The whole planet seems to be a single factory of war production, whose shop floors and battle lines are each equidistant from us at home. In a series of modular scenes, the sound of airplanes is promised to allied nations on the other end of "the vast and bloody canvas" of the war against the "one big slavery stretching from pole to pole from dateline to dateline." As hortatory, the kaleidosonic style took on a lyricism to match its compass. Corwin's "Daybreak" uses the

	INTIMATE	**KALEIDOSONIC**
Sound	Slow fades Approaching and receding	Sudden segues Teleporting
Space	A few deep settings	Many shallow settings
Time	Long durations and chronologies	Rapid scenes in temporally independent modules
Dramatic emphasis	Places	Moments
Narrator	"Close" and "now" Usually an internal character Potentially affected by the action Potentially unreliable	"Close" and "now" Never an internal character Never affected by the action Always reliable
Tense	Present and past	Present
Characters	Known deeply Come and go Individuated	Known superficially Accumulate Part of a coalition
Listener	"Carried" through space by a character	Arrives suddenly at static points
Rhetoric	Makes you feel for another Persuasion	Makes you feel connected to others Patriotism
Political affect	Sympathy/empathy	Equality/pluralism

sun as a "Voice of Time" figure. In it, we follow solar audioposition, segueing from the voices of a serviceman in New York to a South American singer, lovers in Oklahoma and an army surgeon overseas working against hope. In 1945, Arch Oboler aired "Night," a similar production that cataloged evening scenes from a dime-a-dance hall to a late-night revival, all bound together by a nocturne. By 1944 Corwin had mastered the style, and one of his programs segued across the nation to glean the voices of union members, TVA workers, Great War vets, apple salesmen, housewives, and seventeen movie stars (from Rita Hayworth to Groucho Marx)—all endorsing Roosevelt's reelection. Browsing the script, Roosevelt reportedly exclaimed, "My God, can he do all this on one show?"[32] Undertaking the unique task of talking to all of America *about* all of America, kaleidosonic aesthetics are a specimen of the golden age of the point-to-mass mode of mediation, the aesthetic echo of the very concept of broadcast speech.

As extreme as these works could be, however, none of them approach the raucousness of "War of the Worlds." Here was kaleidosonic radio run amok. Welles and Houseman sabotage *March of Time*'s formula by dispensing with the "Voice of Time" figure and by jumbling scenes too swiftly for listeners to maintain order. That is why the show has so many "man behind the curtain" moments. Its rhythm lies not in *March*'s steadying increments, comforting us that time is indeed marching on, but rather in a series of interruptions as one format intrudes on another with a sudden cut, suggesting that the structure of time is devolving. It seems as if nobody is in charge of the broadcast—or the world. To compound this impression, "War" aggregates not only voices, but also program categories, offering material that would not be found in typical dramatized news of the period: weather reports, science updates, orchestral concerts, even switches to specialized radio bands, such as military and amateur channels, which are included as if only so that the story may annihilate them. The passage stutters through segue-interruption-segue until a collapse, and nothing characterizes the dialogue more than its many slipshod transitions: "Just a moment please," "The next voice you hear will be," "Late bulletin coming out of," "I'm speaking to you now from." Scenes end before the fruition of their action, and the forms of radio extant in 1938 seem to strip away. The coherence and order encapsulated by the *March of Time* breaks down into an aesthetic of perpetual interjection. The network is turned over to the army, three battlefield scenes occur suddenly and vanish, and the signal is crossed, returning to the network only as its New York hub is annihilated, along with the locus of "Time." As Jeffrey Sconce has noted, what makes the show so powerful is how scattered it feels, at once portraying military, media, and social catastrophes.[33] It is an apocalyptic form to convey an apocalyptic tale: radio itself disintegrates at the end of the act, with its memorable fade of a shortwave plea—"2X2L calling CQ. Isn't there anyone on the air? Isn't there anyone? 2X2L . . ."—followed by perhaps the most effective five seconds of silence ever broadcast. As the first act of "War" was fading, it is small wonder that some listeners found the claim that Martians were destroying broadcast towers plausible. After hearing everything from everywhere, the enlarging world fades into nothing. With neither anchoring berth nor intimate stranger, the only thing left to steady us is the radio set itself. The remarkable thing about the play is that it offers us *nowhere to be*. With no discernible proscenium, listeners assumed there was none. With a darkening stage and no seat before it, the edge of the fiction dissolves.[34] It is in the end a challenge to the very predictability that gives dramatic conventions their regularity, as well as to the sense of "scene" as a place that is not our own. The tale abandons us right where we really are, huddled beside a radiophonic life force dying before our very ears.

After an intermission, the second half of the program switches style entirely. For the remainder of the play, we move through deep space as part of an intimate dyad, sharing Pierson's audioposition as he wanders the ruins of New Jersey and Manhattan, a process that has the scenic and emotional depths that give clarity to events. This change is part of what makes "War of the Worlds" perhaps the definitive radio play of the 1930s. Its first act is a paradigmatic example of the kaleidosonic style, while its second is a paradigmatic example of the intimate style. Actually, the same sense of balance can be heard in Archibald MacLeish's two major radio plays, the first of which ("The Fall of the City") is intimate, and the second of which ("Air Raid") is kaleidosonic. But it would not be until after Pearl Harbor that Norman Corwin, radio's poet laureate, would attempt to synthesize the two styles fully, melding the empathic and egalitarian qualities of the 1930s sound into a true aesthetic for a radio drama "of the people." I turn next to how this attempt was diverted into an entirely different endeavor destined to drive the theater of the mind in coming decades, a drama of interiority in which Corwin and his talents eventually became unwelcome.

CHAPTER 4

Norman Corwin's People's Radio

I have argued that intimate and kaleidosonic audioposition schemes are the prevailing conventions structuring many innovative works in radio during the late Depression. These two aesthetic schemes were certainly ideal for an exploratory sound. Intimate plays take us around *places* (Broadway, "the City," ships, Cromer, Czardis, Chicago, America), listening to the testimony of voices in a succession; kaleidosonic plays tend to center on *events* (the news, Thanksgiving, the Depression, daybreak, a Martian attack), converging disparate voices all at once. Together the intimate and kaleidosonic styles offered a range of effects, either transporting us through environments alongside traveling characters or teleporting us around the earth. Both kinetic options are necessary for a drama of space and time, and so it is best to consider these styles not as discrete or competitive, but more like the poles of an aesthetic continuum along which it is possible to plot many important works of narrative radio from around 1937 through to the end of the war years. Of course, the continuum cannot account for everything, and it is debatable where a given play rests on it, but the model can help to organize a surprising amount of material that aired in the years during which the cycle of exploratory plays dominated broadcasting.

There is also a curiously political side to these two schemes. Although largely neglected by scholars who foreground inclusive "cultures of unity" in the New Deal imaginary, each of these strategies amounted to what Michael Denning has called an "aesthetic ideology," part of a repertoire of formal qualities that interlocked with populist and radical politics by creating an aesthetic to affirm and beautify solidarity in the public realm.[1] The balance between the intimate and the kaleidosonic was an area for structuring feeling—hypostatizing the political mood of a period, anchored in the pure aesthesis of listening, and always ramifying well beyond it. With this point in mind, this chapter foregrounds one version of this aesthetic ideology, Norman Corwin's hybrid of the intimate and kaleidosonic styles, perhaps the boldest aesthetic specimen of 1930s radio.[2] Corwin's "People's Radio" failed to become a common rhetoric after the war, but the attempt to make it one set the boundaries for the stories that would become predominant in the later 1940s, succeeding the dramaturgy of space and time.

GALLERY, ECHO, STOREHOUSE

One way to examine Corwin's aesthetic is by juxtaposing it with another artistic hybrid characteristic of the Depression era: documentary realist books published with accompanying photographs of the disenfranchised, such as James Agee and Walker Evans's *Let us Now Praise Famous Men* (1940), Erskine Caldwell and Margaret Bourke-White's *You Have Seen Their Faces* (1937), and Dorothea Lange and Paul S. Taylor's *An American Exodus* (1939). Like several of the plays cited in the chapter 3, these books took their readers among downtrodden Okies, sharecroppers, and miners across the nation. Caldwell's captions to Bourke-White's pictures are the kinds of phrases often found in the radio of Stephen Vincent Benét or Carl Sandburg–"There were plenty of people who couldn't get a living out of a farm long before the Government heard about it," "I used to be a peddler until peddling petered out."[3] The similarity suggests that we might say of radio what Alfred Kazin once memorably observed of Depression literature: "Here was America, all of it undoubtedly America—but America in a gallery of photographs, an echo of the people's talk, a storehouse of vivid single impressions."[4] Kazin's triad raises useful questions. Shallow, direct kaleidosonic radio resembles a gallery of photographs, while "echoes of talk" often illustrate the depths in intimate radio. But neither style can be described as a storehouse because radio cannot be aggregated like printed matter. As theorist Andrew Crisell has pointed out, because radio exists in time, its only pressing ontological obligation is that it must dissipate.[5]

It is not surprising that a frustration over pernicious impermanence is reflected in several plays. Many intimate dramas end like Corwin's *An American in England* series. After meeting feisty soldiers and civilians around the world over five weeks overseas, we are now alone with our traveler (Joseph Julian, as

Corwin), who returns home to New York and struggles to find a single phrase to sum up his encounters:

CORWIN: I decided after a tussle that you couldn't say it in one sentence, so I gave up trying. But just as I was turning in that night, I came across some old miscellany in the closet. And among the items was a quote from Benjamin Franklin, all framed and pretty and full of dust:
FRANKLIN: God grant that not only the love of liberty, but a thorough knowledge of the rights of man may pervade all the nations of the earth, so that a philosopher may set his foot anywhere on its surface and say "This is my country."
CORWIN: I dusted that off and searched for a hammer and some nails . . . [*sound of hammering*] "God grant that not only the love of Liberty, but a thorough knowledge of the rights of man . . ."

Because intimate structures deal in deep distances, we are always passing through. With presences unavailable, intimacy excavates an echo of Franklin's bon mot, a phrase that is only vivified by being reiterated, nailed on a wall that exists only in our imagination. Intimate dramas all end like this, with a memory of all of the sounds now lost to us, their echoes now immaterial. Kaleidosonic plays exhibit another kind of impermanence. Consider Corwin's "Unity Fair," a show structured around a carnival announcer drawing together voices to laud common people, "the ones who build the boats and man the trains." The play was inspired by the founding of the United Nations and the wave of "one-worldism" that crested in 1945. Voices at the celebratory fair identify themselves as they were before the war: "We were Chinese," "We were French," "We were Scandinavian." Then the voices come together at once: "We are people!" It is an earnest (if cloying) equivalent of Kazin's "gallery," complete with the two-dimensional superficiality that kaleidosonic radio shares with photography. But at the end of the broadcast, the people of the world climb aboard their new ship, the *USS Same Boat, Brother*. No matter how harmonious the concord, unity is on a vessel leaving us on shore. Like all kaleidosonic plays, the show offers us a world "all at once," but it is a great occasion always ending, about to disappear over the horizon. Intimate plays leave us nostalgic for what we've left behind; kaleidosonic plays leave us behind.

Together, the poles of exploratory radio provide "verbal" sense of place and "pictorial" sense of moment, but neither is able defeat the impermanence of the spoken act. So what about Kazin's storehouse, a repository of vivid impressions? A few 1930s shows grappled with the notion in social and historical dramas, a popular genre known for celebrating engineers, scientists, pioneers, and the "man in the street"—*Americans All Immigrants All*, *The Empire Builders*, *Freedom's People*, *The Romance of the Rancho*. *Cavalcade of America*, the leader of the genre, toyed with the storehouse idea by using analogy to characterize the American spirit. From 1935 to 1937, the show began with medleys of American song, combining folk music with light opera, from past and present, and all regions of the nation, a mingled overture that was a synecdoche for the whole enterprise. Afterward, *Cavalcade*'s external host framed two or three modular stories from different historical moments. Each story had to do with a wholesome form of labor or embodied a trait innate to American national character. "The Bridge of Builders" first depicts men building the Brooklyn Bridge in 1870 and then tells a second story of the 1930s building of the Golden Gate. The can-do spirit transcends locale and epoch, and the Golden Gate becomes but one in a smooth series of occasions stretching ever forward from a distant past. It is as if all builders always have been and always will be erecting the same bridge. "Tillers of the Soil" and "Showmanship" broadcast similar stories about farmers and actors. "The Will to Rebuild," meanwhile, depicts two tragic events "characteristic of thousands of other episodes in American history" that exhibit human resilience, one set in 1852 as Sacramento burns to the ground, and the second set in the contemporaneous Dust Bowl. In both cases, after a time of despair suffered individually, the community meets to plan rebuilding, at which point the dust begins to lift as if beaten back by an expression of unified will. The act of merger calls forth traditional know-how and allows indomitable Americanism to meet any challenge.

Of the first hundred episodes of *Cavalcade* from October 1935 to September 1937, almost two-thirds used this multigenerational quasi-kaleidosonic formula.[6] It is a format unimaginable outside of the Depression, and promoted that era's idea that toil in adversity could recuperate the nation and help those joining national culture from its margins.[7] In wartime, *Cavalcade* gave up its effort to portray these mergers and analogies, concentrating instead on individuals. When the program dramatized the early life of Mark Twain in 1936, it did so along with a sequence on Louisa May Alcott for a show about the "sense of remembered experience" in American writing. The focus was neither Twain nor Alcott, but their similarity to one another and the spirit of genius of which each was merely an instantiation. By contrast, when *Cavalcade* did the life of Twain again seventeen years later, it celebrated a writer of unique brilliance, saying nothing about his similarity to other writers or about America itself. The culture had changed, and so had the aesthetic means to speak to it.

But the change was not inevitable, as there were alternative means to portray the nation as a storehouse of vivid impressions. In December of 1941, Corwin wrote "We Hold These Truths," an hour-long special that fused both intimate and kaleidosonic radio styles and romanticized the storehouse idea best. The major work of the author's most productive year, "These Truths" aired on all US networks and abroad to reach an estimated sixty-three million listeners worldwide.[8] Corwin wrote much of the play on a train, and he received special permission to work in the Library of Congress after hours. The broadcast culminated a series of patriotic shows organized by essayist James Boyd that involved radio veterans Stephen Vincent Benét, Orson Welles, and Robert Sherwood, as well as writers such as Sherwood Anderson and William Saroyan.[9] Historian Michael Kammen has written that the series used modernist technique and unabashed nostalgia to conjoin populism and patriotism, hearing the past through the modern ear.[10]

"These Truths" is in many ways the capstone to the experimental age of radio, at least for the cycle of dimension-based dramaturgy initiated in "The Fall of the City." Corwin's play grappled with a symbolic past and a loaded present. Originally planned to celebrate the 150th anniversary of the Bill of Rights on Monday December 15, 1941, "These Truths" unexpectedly accrued new meaning after Pearl Harbor. On this double occasion, the program gathered talent to offer a show best described by its introduction. Delivered by Lionel Barrymore, the speech is saturated in New Deal rhetoric:

> My name is Barrymore. I am one of several actors gathered in a studio in California near shores that face an enemy across an ocean now "Pacific" in name only. We're here tonight to join 130 million fellow Americans in praise of a document that men have fought for—that men are *fighting* for—that men will *keep on fighting for* as long as freedom is a strong word falling sweet upon the ear. What we enact here tonight has been enacted many times before in living flesh and blood. The people we portray have walked the world. The drama is an ancient one, an endless one, the struggle for men's rights to live their lives out peacefully and profitably in a decent world. It may be that many of us people here are known to many of you people there. For with us, honored to have a part in this program of commemoration, are many whose names you may have heard. Such names as Edward Arnold, Walter Brennan, Bob Burns, Norman Corwin, Bernard Herrmann, Walter Huston, Marjorie Main, Edward G. Robinson, *Corporal* James Stewart—loaned to us for this occasion by the Army Air Corps—Rudy Vallee, and Orson Welles. In New York City waiting to join us is Dr. Leopold Stokowski and the Symphony Orchestra. In Washington,

the highest name in the land, the President of the United States, Commander in Chief of the Army and of the Navy, Mr. Roosevelt. But this is not a night of names, of personalities. Our names or any names are meaningless unless *your names* are added. Unless *you* join us. *You*, for whom the sacred rights were written, and to whom their keeping is entrusted. *You* the guardians of what has been bequeathed to you by millions like yourselves and by the toil of centuries as dark and menacing as this we live in. You, *the people of the Federated States*.

Few names come together so that more may join; the past becomes the present and future; commemoration turns into sacred perpetual stewardship. America is portrayed as a garrison of included people, born to receive, tend, and pass on their rights, a purpose with neither beginning nor end, neither center nor edge. The play is more than a drama of space and time; it's a drama of all space and all time.

A STRONG WORD FALLING SWEET UPON THE EAR

To evoke his utopia in the listener's mind, Corwin avails himself of both audioposition systems conventional to the era. As the dramatic portion of the broadcast begins, we are in Washington DC, positioned cheek-by-jowl with "the Tourist," whose intimate presence we sense in the sound of his footsteps and murmurs. Meanwhile, James Stewart addresses us in a "Voice of Time" role, guiding our audioposition and the Tourist through the streets. The city is depicted with traffic effects and Bernard Herrmann's loping score, which gives us the impression that we move about in deep space. The built environment itself begins to speak. Inscriptions on edifices fade in and out as we move closer and recede from the Lincoln Monument, National Archives, and Library of Congress, even the pedestrian walk signs. The city behaves like a recording mechanism, and we hear it speak in conjunction with narration that refers to Washington as a place where Americans "send their voices" from afar. Though our own voices may be fleeting, DC is a place where utterances congeal forever in durable stone; as we amble the city, we draw wisdom from these immortal voices without depleting their potent amplitude, forever stored and available. At the end of the scene, the Tourist reads the words of the Constitution under his breath. We are so "close" to him that we make out each whispered word. Stewart then guides us back in time "to one bright afternoon in Philadelphia." The Tourist's voice blends with those of founders reading the document, becoming one delegate, then a series of them, dissolving from anonymity into a coalition of famous men in 1787. In this way, an instant of bodily proximity becomes a moment of symbolic community as the intimate inverts into the kaleidosonic.

As the play proceeds, Stewart remains near us, but like the Voice of Time he is himself uninvolved in the world that we eavesdrop upon. He points men out and builds the Pennsylvania State House around our audioposition with expository commentary. The emphasis is not on any single Ben Franklin or Alexander Hamilton, but on their act of signing, and Stewart gathers their names just as Barrymore did those of the nation at the outset, in a list to be joined. Afterward, just as the singular place of Washington became the nation, the State House becomes all of eighteenth-century America as we shift into a kaleidosonic passage, segueing among farmers, clerks, "hackmen artisans," and "grease-grimed blacksmiths" in commons, parlors, and foundry rooms across the former colonies—shallow spaces marked both by homespun political talk and the keynote sounds of hammers or dishware. Two metal workers fear the establishment of a state religion. Bricklayers and war widows lament the lack of guaranteed rights. As the state-by-state constitutional ratification process proceeds, we move from state legislature to state legislature kaleidosonically until many voices at last coalesce at the First Congress. A place becomes a time, and then it becomes a place again. The pace of the play then grinds down as we wait for quorum at Federal Hall in New York City.

Suddenly, Stewart vanishes, and Orson Welles takes over. His oration grafts an unexpected kind of coalition onto the many we have already heard:

WELLES: Do you think 55 representatives of the American people sat in a hall in New York City, sat in a drafty hall and made up articles of freedom? . . . Oh no. They had much help. The many nameless, the unknown, from dust in quiet places, from broken bones deep in the earth, deep in *forgotten* earth, mixed with the empty clay, from bleeding mouths . . . from numberless and nameless agonies, the delegates from dungeons, they were there . . .
VOICE: I said that men were born equal . . .
WELLES: The delegates from ashes, they were there!
VOICE: The king did not approve . . .
WELLES: The gallows delegates, whose corpses lifted gently in the breeze, they too. Exiles, wanderers, Christians killed for being Christians, Jews killed for being Jews. . . . They made a quorum also. Must you know what they said; must you know what they argued? Listen! [Scream] *That* was an argument for an amendment. [Baby Cry] *That* was a speech in favor of an article of freedom [Another Scream] *That* praised the passage of a Bill of Rights.[11]

The voices of the dead pile atop one another in a kaleidosonic sequence that mimics the passage with bricklayers and widows. Welles suggests that present in New York also were such figures as Nero, Caligula, and Jesus. All history collapses into the moment. Once the Bill of Rights is signed, we cycle back through the coalitions and the state legislatures, as well as the common people of the eighteenth century at their forges and hearths. When we segue back to the present, Stewart asks, "Who better to ask about the Bill of Rights than the people?"

STEWART: Ladies and Gentlemen, an office clerk.
CLERK: Well, we know what freedom is now, we forgot for a while . . .
STEWART: Ladies and Gentlemen, an editor.
EDITOR: There have been attacks on the freedom of the press, but they've been broken every time, and today a man is free to start a paper . . .
STEWART: Ladies and Gentlemen, an autoworker.
WORKER: We got the right to organize . . .

And so forth. By toggling the intimate and kaleidosonic registers, Corwin accentuates both. We apprehend places like Washington through intimate proxies, afterward hearing static moments expressed through kaleidosonic sequences. And although the constituencies that compose America have changed since the 1790s, the way that the play arrays them does not. "The People" come into existence in sudden moments of shallow space that punctuate longer sequences of deeper space. They flash into our imaginary as the occupant and product of two aesthetic registers attempting to synthesize. Thanks to this unique oscillation, empathy effects and equality effects interlock, intimate strangers dissolve into pluralities, many places become one, and a series of constituent voices from the past transform into a list of names ever accumulating. Corwin's aesthetic literalized what critic György Lukács once called the "paradox" of modern drama, in that it made a dual aesthetic commitment to the "drama of individualism" and the "drama of milieu," thereby asking the degree to which the individual can believe in the pact of community.[12] Corwin's answer rings out in the affirmative. America is the ultimate drama of space and time, a picture in the listener's mind that includes his or her own face.

Corwin's play eventually drifts into free-form polemic as narrators trade off without cue, voices speak without context, and a broadcast that started out dedicated to putting us beside a corporeal traveler turns into a kind of dramatic essay outside of the world it describes. But as a whole, the People's Radio aesthetic of "These Truths" reaches further than perhaps any play toward

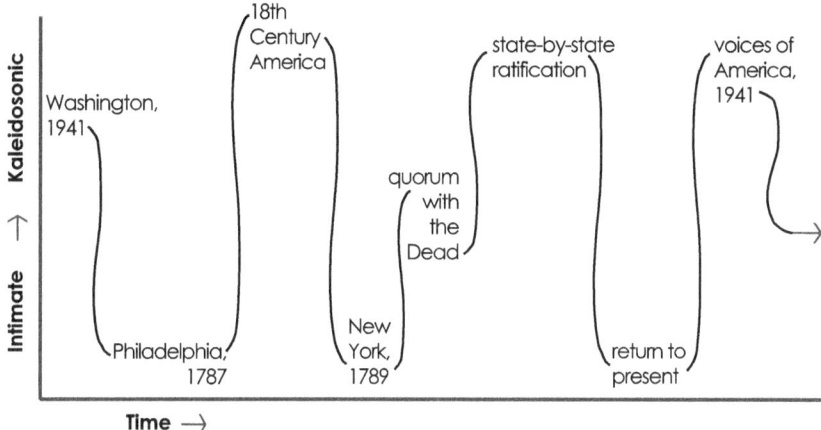

FIGURE 4.1. Norman Corwin's "People's Radio" aesthetic in "We Hold These Truths," broadcast on December 15, 1941.

Kazin's storehouse of vivid impression. *Variety* praised the program's "many-paced, many-shaded reading," while *Time* called it a "touchstone for future patriotic programs" and enshrined Corwin as radio's "wonder boy."[13] According to legend, James Stewart nearly burst into tears on air at the end of the broadcast. Corwin was now a household name, and his form of exploratory radio was no longer just a style, but a mystical vision of citizenship. In future works he would continue to search for the proper language to express his People's Radio, always with some hybrid of intimate and kaleidosonic schemes. While flirting with unctuousness, these plays would be great public and artistic successes, but none have the control of technique exhibited in "We Hold These Truths." Plays like "On a Note of Triumph" surely brooked far more polemic than "These Truths," but they are really more like essays about global citizenship and social justice than they are dramas of immortal nationhood.

They would also prove to be swan songs. By the later years of the war, venues to explore political ideology through aesthetics would slowly vanish. Since 1936, CBS had built eighteen programs to showcase talents such as Reis, Welles, and Corwin, but by 1943 all of them had been cut or sold to sponsors.[14] The *Nation*'s Lou Frankel observed that for some time radio had been like a printer, "putting words on the air as the printer sets them in type," a process that left room for an author to experiment; but by 1947 the industry was acting more like a publisher, exercising judgment on subject matter.[15] Poet Norman Rosten observed that by the 1940s commercialism precluded the possibility of radio art: "The word is fitted to the Product. The Product is God. The word is the interval between the announcements of God.... This is hardly a condition for literature."[16] Over dinner on a train from Pasadena to New York in 1948, CBS chief Bill Paley informed Corwin of the new status quo, describing aggressive plans for commercial radio and indicating that the company wanted their wonder boy in the soap opera business. "Where the hell indeed was I?" Corwin recalled,

> I was a radiowright—a new species. And according to some critics, best of the breed. But it was clear now that my outlet, the far flung and pervasive Columbia network, was no longer a Champs-Elysées down which I could walk, run or skip as I had done in the past.... I was on a train going East on no particular business, without portfolio, without a target.... Worn from the pace, sorry for myself, certain of the death of my medium, dull from Scotch, and unnecessary.[17]

Soon, loyalty programs swept through the industry, and Corwin's well-known radical sympathies turned his shower of wartime adulation into a purge.[18] In 1945 Corwin won the first Wendell Willkie "One World" Award and toured the globe, and in 1947 he became the first radio writer profiled in the *New Yorker*, but by 1948 the radio blacklist claimed him as perhaps its most eminent victim.[19] Where kaleidosonic aesthetics did not vanish it stagnated. A generation would pass before no less a figure than Glenn Gould would used a similar aesthetic in his *Solitude Series*, whose "contrapuntal radio" concept remains among the few clear relatives of Corwin's experiments.[20]

With Corwin's exile from mainstream radio, 1930s radio was finally over, and dramatic exploration dwindled after a moment of exacerbation. Indeed, during the war, the architects of the drama of space and time began to move on. Orson Welles and Irving Reis went into film. Earle McGill took on union duties, coordinating the Radio Director's Guild and the Writer's Guild. Stephen Benét died of a heart attack in 1943. Archibald MacLeish, Robert Sherwood, William Lewis, Brewster Morgan, and William N. Robson went to work for war agencies, where they still wrote and produced dramas but had a host of additional administrative duties. Arch Oboler and William Spier moved into suspense thrillers. Corwin's aesthetic died of commercialism, the Cold War, as well as inborn deficiencies—Benedict Anderson has argued that no nation can successfully "imagine itself coterminous with mankind" for long.[21] In 1943 critic Eric Bentley looked back on the "workshop" era and concluded in disappointment that it had been much more bombastic than popular.[22] And although intimate relations would persist, during the war radio gradually abjured the kaleidosonic style. Even *The March of Time* vanished, continuing only as newsreels after 1945. As early as 1941, the stylistic balance that defined Depression radio was already on its way to becoming permanently upset. Settings were no longer the mark of showmanship and segue patterns lost structuring importance, while the manner in which information flowed around the play began to have new significance. Complications of intimacy often gave way to complications of psychology, while rituals of empathy, equality, and unity were traded for trials of conscience, reversals of loyalty, and stories that recreated in dramatic form the influence of the mass media upon inner human life. If a core question of 1930s radio was the nation, a core conundrum of 1940s radio would be the psyche, and what emerged was a drama that no longer twisted time and space but warped communication and consciousness, giving radio its reputation as a theater of the mind. By the last years of the war, kaleidosonic sequences were still used, but most often to depict crowds of the dead, not a thriving people contiguous with ancestors and descendents with whom they concatenate to form an unbreakable chain.

NO STAGE AT ALL

Why this change in the "sound" of American programs? Any answer begins with the new place that radio occupied in the imaginary after Pearl Harbor. In the early years of the conflict, *Fortune* effervescently declared that international broadcasting created a global propinquity that might bring nations together, citing RCA's nine-hundred-acre forest of masts on Long Island as evidence of "man's conquest of space"; but at the very same time, the *Saturday Evening Post* worried about the reach of high-powered stations, explaining, "There is war in the ether on a much more extensive scale than on land or sea."[23] According to *Time*, this situation put consciousness at risk: "The tall short-wave antennae of Zeesen, Tokyo and Rome are beamed not only at certain areas of the earth," the magazine warned, "[but] at the delicately balanced human mind."[24] Others wrote of radio as a tool to awaken bellicose drives. In 1942, CBS executive William Lewis argued that radio could convey "action messages" to the public, while *University of Chicago Round Table* producer Sherman Dryer suggested that radio drama offered "the richest promise of eliciting concerted action" in recruitment, and radio critic Bob Landry proclaimed that radio ought to "instruct, order, explain, [and] beguile."[25] Many admired the efforts of the BBC, which at its height would monitor 1.25 million words in thirty-two languages and send more than thirty thousand words in "flash messages" to government departments each day; propaganda analyst John

Whitton even argued that Americans should emulate the persuasive techniques in Nazi radio dramas.[26] In 1942, radio writer and antifascist Arch Oboler gave a speech at Ohio State University in which he advocated a "radio of hate" to foment bloodlust toward the enemy, and within months Archibald MacLeish made his oft-cited remark that the principal theater of the war was not on a battlefield but in "American public opinion."[27] Waging war in the subconscious, emotions, opinion, and "will," 1940s radio aesthetics had to confront the mind perforce, a shift toward interiority that would require all of the rhetorical tools developed in 1930s drama, even as it tended to occlude that era's goals and assumptions.

On December 7, 1941, it was unclear how the commercially driven mechanics of American radio could facilitate the activities in which the medium would distinguish itself—news, morale shows, recruiting, and fundraising. When Pearl Harbor was attacked, most stations were airing news content only as commentary. That night, NBC did not even bother to preempt the unfortunately titled "Island of Death" on *Inner Sanctum Mysteries*. When WOR New York interrupted Sunday's football game to give emerging details on the attack, the station received complaints from disgruntled listeners for spoiling the game.[28] Federal offices initially showed little interest or savvy about how to use radio to promote the war effort. After the government immediately shut down fifty-five thousand amateur shortwave operators, many broadcasters assumed that it would simply take over the airwaves and were surprised to discover that the Office of Censorship's first set of directives to radio stations was no larger than a church bulletin.[29] Another problem was sheer disorganization. "Nobody knew what anybody else was doing," recalled advertising executive Raymond Rubicam. "Advertisers and radio networks were getting five and six government requests for the same radio time on the same day and program. There were no priorities. Nobody had authority to settle anything."[30] In response—and to secure his interests—Rubicam helped found the War Advertising Council (WAC), a clearinghouse for such requests. Within weeks the air filled with news formats and morale programs promoting recruitment as weather reports were withheld for fear of aiding the enemy, all with the tacit public relations aim of ensuring that ad firms would avoid heavy tax increases. By February, the United Fruit Company was advising Americans to add grapefruit to their "War diet," and in May the networks were turning over transcribed programs to the newly minted Armed Forces Radio Service.[31]

In June of 1942, the WAC's clearinghouse role was partly taken over by a new federal organ, the Office of War Information (OWI), which hired CBS luminaries, including commentator Elmer Davis and writer Robert Sherwood. The OWI consolidated a number of prewar agencies—the Office of Facts and Figures, the Office of Government Reports—and was charged with allocating spot requests, partnering with networks on specials, issuing program advice, and vetting material. Soon the OWI was shortwaving 2,688 shows a week in twenty languages over fifty-four transmitters around the world and was beginning to help work war themes, recruiting efforts, and bond appeals into shows—in one episode of *Easy Aces* in October of 1942, *Variety* found no less than six government-requested themes folded into the script, from rumor squelching to first aid.[32] Lowly genre shows such as *Lights Out!* and *Young Widder Brown* contributed to grand national purpose, encouraging the American public to give some thirty-eight million pounds of waste kitchen fats to help manufacture pharmaceuticals.[33] Prior to the founding of the OWI, antifascist programs often consisted of dramas about corrupt Nazi trade rules, but afterward the public began to hear about German policies for forced starvation and sterilization. By the end of 1942, the major networks had donated some $140 million in time to the effort; CBS boasted to have aired 6,471 war dramas, 3,723 announcements, and 4,158 news shows that year.[34]

Despite success, the OWI would be short-lived. Congressional Republicans slashed OWI funds from the 1944 budget, killing the office. But the OWI had already paved the way for agencies to propagandize for the remainder of the war, including the Treasury Department, the army, trade groups, artist unions, and especially the networks themselves.[35] On September 21, 1943, during a span of just eighteen hours, radio star Kate Smith raised more than $39 million for War Bonds, nearly half a billion in today's currency; according to one estimate, by the end of the war, CBS had given 38.7 percent of its entire airtime to shows about the conflict, more than an entire year of continuous time.[36] Too much of this material has been lost. Recordings exist from *The Man behind the Gun* and *You Can't Do Business with Hitler*, but little is left of *Womanpower*, *Spirit of '42*, and *They Live Forever*, just a few of the programs that aired in March of 1942. And shows "about" the conflict were just the beginning. Psychotic Nazis were on the loose in programs as diverse as *The Mayor of the Town* and *The Strange Dr. Weird*, while characters either served at the front or in war plants on *Stella Dallas* and *Joyce Jordan, Girl Interne*. Appeals for men to join the merchant marines ran on *Cavalcade of America*, while *Mr. District Attorney* and *Hop Harrigan* told teens to join the Ground Observer Corps ("to watch the coasts") or the Air Training Corps ("so you'll know what to expect"). In the first six months of 1942, Superman captured four enemy submarines, rescued an airplane in midair, and defended a Cairo hospital from bomber attacks, all while reminding parents about the nutritional value of Pep cereal and encouraging youths to "turn in scrap to slap the Jap." In 1943, *New York Times* radio critic John Hutchens cataloged many more overseas exploits:

> At this writing, Dick Tracy is engaged in underground activity in France; Terry, of "Terry and the Pirates" is knocking Japs out of the sky over Burma; Chick Carter is tracking down Nazi agents; . . . Captain Midnight and his Secret Squadron are performing feats of derring-do in the south Pacific; Hop Harrigan is taking care of the situation in Italy; Jack Armstrong and Captain Silver, master of an extraordinary vessel called the Sea Hound, are separately combating enemy influence in South America.[37]

In the evening, adults could enjoy a blackly comic sketch on the exigencies of rationing or an entreaty to accentuate the positive from Bob Hope, Jack Benny, George Burns and Gracie Allen, Kay Kyser, and Edgar Bergen.

At the acme of its reach across and into everyday life, American radio produced some of the most ambitious and complex narratives yet conceived. Although this new theater would have possibilities beyond those offered within the slim intimate-kaleidosonic continuum, many true believers abhorred these developments. In his introduction to a 1944 volume of radio plays, MacLeish lamented that the development of radio as "a stage for the word" was over.[38] New efforts aimed instead to develop radio "not as a stage at all but as an instrument." It is a dazzlingly ironic grievance. The poet helped edit FDR's speeches and engineered programs to raise money in as instrumental a use of the medium as was made by federal officials in American history.[39] He was even still nominally on the public relations staff of the State Department. More importantly, MacLeish's complaint misses the truth about the exploratory era because it rests on a false opposition. All stages are instruments, particularly those imagined and designed by artists in pursuit of an aesthetic agenda sophisticated enough to carry hortatory content, build dramas of space and time, and ultimately sell products. When audioposition became a tool to sketch images of "common grounds" in the air, the medium was immediately makingbelieve to make belief, favoring some points of view over others with the aim of coaxing listeners to react in a certain manner. The conventions developed in that process gave structure to the persuasive act. When intimate plays take us on journeys, it is so we can experience empathies that structure our ideas about the world. Were it otherwise, all of the care taken in selecting the listener's guide would be for naught. To draw pictures in the imagination is to invent the imagination as you want it to be, and that is inherently an instrumentalist enterprise.

This was especially true of the kaleidosonic programs of the 1930s, which could act as an implement for a range of movements seeking to create cultures of unity. Indeed, kaleidosonic rituals offered a screen on which collectivist fantasies could be projected by bringing together the voices courted by unifiers ranging from New Dealers and Standard Brands executives to radical political groups, each of whom had a different idea about what course the nation should take but also fancied themselves to be captain of the *USS Same Boat, Brother*. Kaleidosonic radio was not an "expression" of New Deal unity exactly, but New Dealers could listen to it and hear themselves, like any other groups seeking a mechanism to build cultural coalitions to replace the disgraced individualism of the 1920s. For some leaders, the whole purpose of "building a stage" was but a pretext for building a seating area, which seemed less like a national storehouse of vivid human impressions and more like a cage. As Theodor Adorno put it, radio demagoguery tended to substitute "the concept of the movement itself for the aim of the movement."[40] But this had always been the risk of the aesthetic of the 1930s. Nothing protected it from instrumentalism, because the kaleidosonic style responded to a nation that was less a plurality than a series of ways of thinking about plurality. The kaleidosonic scene of the nation-coming-together was an energizing rite that could be staged over and over again, always leaving the feeling of the public fungible to be shaped in another way later on. Radio sought no coalition of its own, but it leased the frequencies on which they could be dreamed by others and invented rules for that dreaming. Kaleidosonic America had always built solidarity to accommodate fantasies, and MacLeish's complaint is merely a gripe that radio offered this effect to clients other than those he preferred.

As American radio drama grew out of the Depression, it would still use space and time to trace vivid pictures in the minds of listeners. But there was also a modification in mood that marks a new act opening, as many plays no longer denied they were mounted upon machinery that manipulated perception. Indeed, with the passing of the dramas of space and time, many radio plays reveled in just the manipulative quality of "drawing pictures in the mind" that the field as a whole had spent the 1930s both pursuing and disavowing. In the 1940s, the plasticity of dramatic space lost significance when compared to the plasticity of those minds perceiving it. If the old theater had used plays like Corwin's "We Hold These Truths" to hear pluralities coalesce in grand exterior worlds, the new conventions would use them to shape interiority. In late wartime radio, correspondents no longer perched on terraces to witness gory conquerors fell great cities as they did in MacLeish's "The Fall of the City," nor would they mingle among a multitude to build profound solidarity with a listening public. In the world of 1940s radio, the mere presence of a microphone on the scene would indicate that the City had become but a node in a signal network, and it had therefore fallen already.

* 2 *

Communication and Interiority in 1940s Radio, 1941–1950

CHAPTER 5

Honeymoon Shocker

In October of 1939, aspiring writer Lucille Fletcher married radio composer Bernard Herrmann. Herrmann had been a protégé of several of Columbia's pioneering directors during the heyday of experimental radio, doing sound for Irving Reis's "The Fall of the City" on *The Columbia Workshop* and Orson Welles's "War of the Worlds" on *The Mercury Theater on the Air*. He would also soon score Norman Corwin's landmark "We Hold These Truths," before going on to do music for such films as *Citizen Kane* (1941), *Psycho* (1960), and *Taxi Driver* (1976). Herrmann met Fletcher at CBS New York. Hired to a clerical job right out of Vassar, Fletcher worked her way into publicity, later inaugurating a writing career by pitching Norman Corwin a scenario about a singing caterpillar and his wily agent. "My Client Curly" aired in March of 1940, and Fletcher was soon adapting stories for *The Workshop* while predicting changes to the aesthetic agenda associated with the show. Her forecast first appeared in a feature published in the *New Yorker* that April.[1] Relating studio lore about sound-effect gimmicks, Fletcher wrote that directors no longer used effects just to set scenes, but also for more challenging purposes, like using oscillations to signify that a character felt faint or seasick. "Your real sound effect artist," Fletcher explained, was delving into "the abstract."[2] By expressing affective states instead of cueing locations, radio seemed to be outgrowing Reis's overwrought sets and Welles's segue flurries, what I described in part 1 as "dramas of space and time." Fletcher surmised that the future lay in what she called "psychological drama," a theater that could use sound effects to explore interiority.

The prediction proved astute. During World War II, radio began to earn its sobriquet as a theater of the mind as American airwaves filled with psychological themes and new techniques to convey them. Pseudopsychology had been a fascination of many serials long before the war—the Shadow's power is not to *be* invisible to the eye, after all, but to "hypnotize men's minds so that they cannot see him."[3] But it was not until the 1940s that evening playhouses and serials really began to work on the theme of the mind systematically. *The Columbia Workshop* and *Mercury Theater* swapped the populist plays of Pare Lorentz and Carl Sandburg for gothic fables by the likes of John Galsworthy and Wilbur Daniel Steele as new suspense anthologies appeared everywhere on evening schedules. Sound tricks were no longer innovated for experiments in space such as "Broadway Evening" or to evoke the ornate scenes and restive mobs of "The Fall of the City." Instead, engineers finessed filters and actors used sotto voce to convey passages of stream of consciousness for "The Path and the Door" on *The Workshop*, "Johnnie Got His Gun" on *Arch Oboler's Plays*, and "Under the Volcano" on *Studio One*, while directors began to favor tales set in cloistered spaces in which action was close, atmosphere heavy, and boundaries rigid.[4] Corwinesque cantatas went out of style, and many broadcasters seem to have agreed with one-time radio writer Arthur Miller, who declared in 1947 that scenes of two voices were more artful than casts of hundreds, and scenes with just one voice were "best."[5] Dream sequences proliferated on shows as diverse as *The Weird Circle* and *The Lux Radio Theater* as anthologies from *Lights Out!* to *The Ford Theater* aired plays in which characters vocalize thought as they converse externally, aping Eugene O'Neill's *Strange Interlude*. On morale programs such as *Wings to Victory* and *The Man behind the Gun*, events often involve soldiers with inner self-doubt that is either spoken as asides, alienated from action, or enunciated by an external narrator. Brave triumphs coincide with the recuperation of this alienated voice, a restoration that resolves affect and action at once. Whatever we make of such sequences, it is hard to read them without a registry of psychological categories, understandings, or interpretations, one that would not have been so crucial just a few years earlier. To borrow the framework of critic Robert Heilman, the American radio play had gone from melodrama, in which "whole" characters act in externalities and lack inner conflict, to tragedy, which thrives on the "urgency of unreconciled impulses."[6] As a result, the problem of representation in radio necessarily became focused on how to draw pictures *of* the mind *in* the mind. At the same moment that radio as a social institution took on a role as the war's "nerve center," radio stories thematized the power of influence upon inner life.[7]

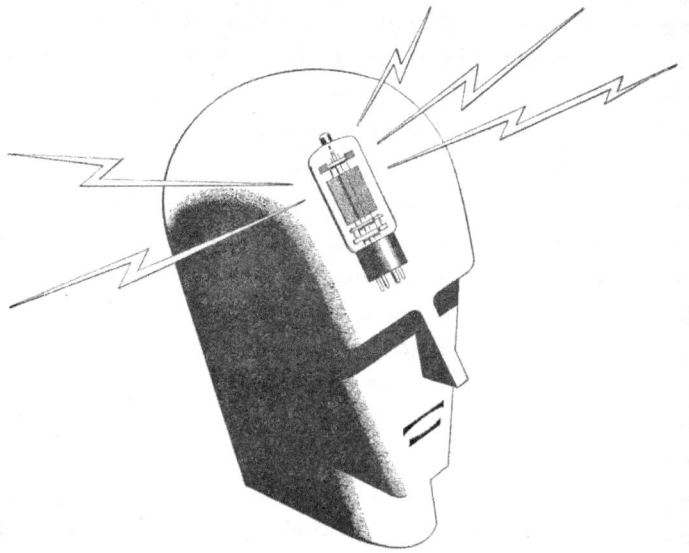

FIGURE 5.1. An advertisement from *Variety*, January 6, 1943, touting RCA's wartime service.

In part 2, I explore this urge to articulate mass communication and consciousness, or, more precisely, to view each of these ideas in the terms of the other. I will show that just as 1930s shows were structured around space and movement, wartime dramas often hinged on messages, and characters interacted with (and often succumbed to) systems of signals. In 1940s radio, the mind was imagined to be a receiver or transmitter of messages, the target of mass media campaigns, a repository of ciphers, or a node in an information network. Many plays contain acts of communication that are offered as if they are sufficiently illustrative of a psyche to make behavior intelligible, which reveals how the mind was understood in an era of vigorous mass propaganda.[8] Part 2 concentrates on three aspects of this theater of the mind: the semantic techniques that give it voice, the characters that embody it, and the narrative situations that stress it. In this chapter, I consider the radio plays of Lucille Fletcher to show how in the 1940s many radio dramatists abandoned the question of the nation and began to explore the human capacity to persuade, inspire, and instruct, a development that took place in thrillers in which interpersonal "talk" is replaced by abstract "signal." In chapter 6, I examine dramas in which identity is defined by whether a character tends to direct transmissions or to be directed by them, reflecting an era in which major studies were undertaken into media effects. In wartime, circuits of messaging seemed to create passive "receiver" and active "transmitter" positions that shape plot outcomes and test models of influence in a direct reply to contemporaneous communications research. Finally, in chapter 7, I discuss 1940s stories of eavesdroppers, ventriloquists, and signalmen, three archetypal figures that reflect the listener's own experience and also push the logic of the theater of the mind to the brink of failure by focusing on areas where the connection between media networks and the imagination led to error, breakdown, and metaphysical jeopardy. As a whole, these chapters account for a period of popular entertainment in which the human mind seemed to become knowable through—and bound by—signals, instructions, and warnings that circulate through the scene of the fiction. It is in such tales that we can hear popular culture reshaping commonsense accounts about how the mind operates through the emblematic form of entertainment of the 1940s.

RADIO WAR

The preeminent fact about wartime radio is its saturation of quotidian life both at home and abroad. By the time that manufacturers quit building radios for civilian use in April of 1942, sets were already in at least 82.8 percent of all

US homes, more households than could boast telephones or plumbing.[9] That year, network ledgers showed annual gross time sales reaching $180 million; tallying this sum with figures from the independent stations, and with spending on repair and electricity, *Variety* reckoned that the commercial radio industry was worth over a billion dollars a year.[10] Federal officials estimated that up to twenty million words were uttered each day over nine hundred domestic radio stations; and according to some surveys, most Americans considered their radio broadcasters more reliable than not only their newspapers, but also their clergy.[11] Meanwhile, manufacturers were shipping the US Army Signal Corps as much broadcasting equipment every two weeks as had been used in the entire Great War; by 1945, the US Army used more fixed radio equipment than had existed in the whole world prior to 1939.[12] The Corps was responsible for creating the Army Command and Administrative Network (ACAN), a radio-based command system run out of Washington that integrated large networks with the rest of what army historian George Thompson calls "the fabric of World War II victories"—a worldwide system of automated equipment, radioteletype, spiral-four field cable, microwave radar, FM relay, even mules and carrier pigeons.[13] With major stations in Reykjavik, London, Algiers, Accra, Cairo, Tehran, Karachi, and Hawaii, ACAN arguably created the first global communications network with uniform control and standard signal character and quality. With it, Washington could communicate with any field command almost in real time. In an emergency, signalmen could even flip a switch at the Signal Center on Governor's Island, New York, instantly commandeering the entire NBC radio network on the Eastern Seaboard.

ACAN emblematizes the ambitious scale of communications in these years, in the light of which writers routinely call World War II a "radio war" and describe Americans of the period as a "media generation." Yet even very prominent studies analyze no broadcasts in detail, and many discuss programming only as a coda to a passage on war films.[14] This approach is misrepresentative. Prior to the war, most Americans went to the movies two or three times a month, but they listened to radio for four or five hours per day, and in one survey more than 70 percent told pollsters that they would give up the former before the latter.[15] By the 1940s, the big networks were earning almost half of the combined gross of the eight big film studios; in the summer of 1942, Hollywood shot some 38 war films to aid morale and drive recruitment, but radio was already airing up to 603 war shows a week.[16] For some time, historians put these programs aside to focus on the institutions sponsoring them, emphasizing how war offices weaponized "truth" to moralize coercion, mobilized funds, and abandoned New Deal ideals in order to secure them.[17] Only in the last

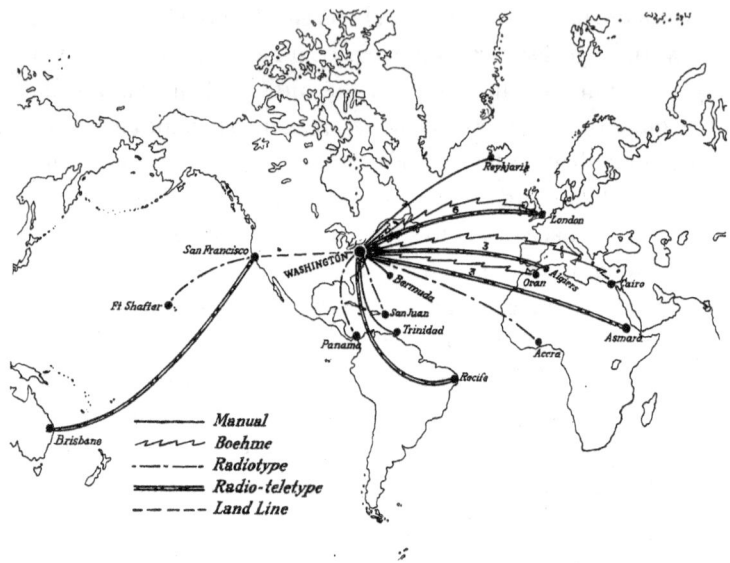

ARMY COMMAND AND ADMINISTRATIVE NETWORK OVERSEAS, DECEMBER 1943

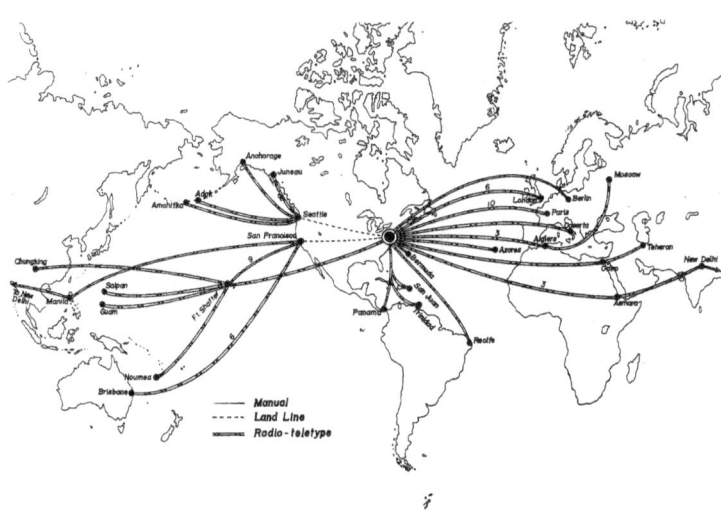

ARMY COMMAND AND ADMINISTRATIVE NETWORK OVERSEAS, JUNE 1945

FIGURE 5.2. The Army Command and Administrative Network (ACAN) in 1943 and 1945. Source: NARA. Image courtesy of the Center for Military History.

few decades have writers begun to grasp that the role of radio in war culture exceeded the policy-centered impression conveyed by memoranda. Historian Michele Hilmes has led the way, studying how patriotic programs brokered popular concepts of imagined community, and subsequent writers have used programs to learn how the war transformed notions about subjects ranging from dictatorship to the civil rights agenda.[18] These efforts still struggle to do justice to terms like "radio war" or "media generation," in part because communication itself is rarely the main question. I will argue here that if 1940s culture indeed featured an excess of messages, and radio was the premiere manner in which these messages were relayed, then one way to make the most of this situation is to figure out how words and sounds "work" in the wartime radio plays, because the genre is a site in which theory meets practice. Part 2 engages the eminence *of* radio communication in the period by engaging the eminence of acts of communication *within* radio of the period, an approach that leads directly to the conventions behind Lucille Fletcher's "psychological" sound effects.

A DRAMATIC SUGGESTION

The idea for Fletcher's first original dramatic script came just a couple of weeks after she published her *New Yorker* piece.[19] That summer, with radio programs on hiatus, Fletcher and Herrmann decided to go on a belated honeymoon. According to legend, the couple was just beginning a road trip across the country when Fletcher absently noticed a pedestrian on the footpath of the Brooklyn Bridge. Several miles later, she spied a second figure that appeared identical to the first, this time thumbing a ride near New Jersey's Pulaski Skyway. The incident would inspire "The Hitch-Hiker," a modern folk tale that made Fletcher a doyenne of suspense radio.[20] Her play consists entirely of the tale of a young writer named Ronald Adams. From an auto park on Route 66 near Gallup, New Mexico, Adams relates an account of a journey that had begun in Brooklyn six days earlier. We segue back in time, to a scene in which he reassures his fretful mother that the open road is perfectly safe nowadays, then packs up his Buick and heads for Hollywood, where he has been promised a job. Like Fletcher, Adams spots a hitcher on the bridge and experiences déjà vu at the turnpike. He dismisses the episode and continues west, until he sees the same figure again. Several hours have elapsed and a rainstorm has dissipated, but the man reappears, still shouldering the same overnight bag and showing fresh droplets on his overcoat. As the journey continues, the phantom calls out—"Going my way?"—from a gas station in the Alleghenies, a cow pasture on the prairie, and a reservation in the mesa country. Adams soon succumbs to his panic and tries to run the hitcher down. This has little effect. No sooner has the phantom fallen under the Buick than he reappears elsewhere. At last, Adams reaches the empty New Mexico auto park and phones home, hoping that his mother's voice can bring him back to his senses. But he can only reach a neighbor, who says that his mother is indisposed, mourning her son Ronald, who died several days before in an auto accident on the Brooklyn Bridge.

Like many broadcasts discussed in part 1, this drama conveys settings using keynote sound effects and sound color. The car's rumbling motor "says" that we are moving, a cowbell "tells us" that we have reached the prairie, while volume and reverberation "describe" the distance that separates Adams from the hitcher, a scheme on which the effect of suspense is greatly reliant. "The Hitch-Hiker" also fulfills one hallmark aim of 1930s radio—to allow our audio-position to move through deep space just for the sake of it—by arranging the tale as a cross-country journey. Fletcher's play draws upon exploratory 1930s radio yet also slyly alters it by falsifying representations of external events and movement. The play is full of impediments. For instance, although the hitcher ostensibly seeks a lift, Fletcher associates him with a series of impasses and breaks in the roadway. He appears at a railway track, under tunnel arches, on a bridge, at a state border, and leaning against a detour barrier. Adams even calls him "drab as a mud fence," and it is perhaps significant that the journey ends far from its destination but close to the continental divide. On these badlands, Adams sees barriers all about, explaining, "I'd see his figure, shadowless, flitting over dried-up rivers, over broken stones cast up by old glacial upheavals." At the great divide, the land itself seems to desiccate conduits and hoist barriers, as if refusing to be traveled upon, an effect complemented by changes in the manner in which the play offers exposition. Early scenes use sound effects to "set the stage," but later sequences rely on Adams's speech, a change in tactics that gives the setting an unreal quality. We lose our sensory grip; the land does not self-illustrate, denying us that resolute sense of place vital to the style of the 1930s. This denial foreshadows the revelation that the narrator is dead and that we have in fact traveled no further than the bounds of his posthumous imagination.

Fletcher builds a world in order to tell us that it is not there, using a theater that creates landscapes to actuate a theater that imagines the psyche as itself a landscape. This is why the sound profile of the play offers only trivial aural relationships between the listener and most of the depicted characters. In the 1930s, directors had contrived plays so that listeners are intimately positioned close to a character who has strategic spatial relation to representatives of benighted social groups, each of whom approaches us from deep in their natural locale. This structure encourages us to sympathize with a "near" character so that we may empathize with "far-off" characters that emblematize social problems in ongoing situations. In "intimate" radio plays of the 1930s, it is usually possible to imagine the drama related from another character's position, so the world of the play seems to be a pluralistic land of "the people" that offers copious vantage points. But things are quite different in "The Hitch-Hiker." The play has but one audioposition available. This gives it a solipsistic quality, since we cannot employ our hero as a proxy to access scenes that have much internal coherence. It is not even clear that the play occurs in chronology. All events from the point of crossing the Brooklyn Bridge occur outside of natural time, in a rupture in the line that connects the span of narration to that of the tale being related. The situation is stressed by the manner in which the listener perceives important story material. Although no other character in the fiction can hear the hitcher's voice, the listener can, which suggests that we are in Adams's sensorium rather than simply near it. This represents a decisive break with kaleidosonic drama and also challenges the whole notion of intimacy. In 1930s dramas, many directors position us nearby characters exploring worlds, but in works from the 1940s, that nearness is frequently replaced by engulfment as worlds become states of mind. Lost is the sort of stage that is able to comment on issues affecting public life. There are no social problems in "The Hitch-Hiker" because the play does not take place in society as such.

But how does this process make use of "psychological" sound effects, and what insight can it provide on the "radio war"? To complete the reading, it is necessary to situate Fletcher's play in a longer tradition of supernatural radio, a tradition that sets "The Hitch-Hiker" apart from 1930s fare as a matter of genre and also created opportunities in which 1940s aesthetic conventions were nurtured. The most enduring performance of the "The Hitch-Hiker" came on *Suspense*, a thriller anthology helmed by *March of Time* producer William Spier, who collaborated with composer Bernard Herrmann and gave Fletcher's script to John Dietz, who had once been a sound-effects man for Orson Welles. One of the signature shows of the period, *Suspense* began in 1942 and ran for two decades under the sponsorship of Roma Wines and Auto Lite Spark Plugs, boasting such stars as James Stewart, Joseph Cotton, Gregory Peck, and Lucille Ball. *Suspense* has recently interested scholars: film critic James Naremore cites it as a venue for disseminating noir taste in mainstream culture; drama historian Richard Hand notes that it fuses gothic film and the "woman's picture"; and cultural historian Allison McCracken argues that many of its plays destabilize prevailing cultural norms.[21] According to media historian Douglas Gomery, *Suspense* provided all that is worthy about broadcast drama, "the power of a complex script, skilled writers, great voices, sophisticated music, and top special effects."[22] *Suspense* also led a trend in which a number of thriller playhouses were inaugurated or revitalized in the 1940s, including shows such as Arch Oboler's *Lights Out!* and Himan Brown's *Inner Sanctum Mysteries*, as well as dozens with similar fare, such as *The Whistler*, *The Sealed Book*, and *The Strange Dr. Weird*.

These anthologies drew on radio's long association with the paranormal, a connection that predates the network period and is rooted in qualitative listening experience.[23] In the 1920s, many radio receivers suffered from weak reception, while stations disobeyed guidelines about transmitter wattage, leading to a chaotic situation in which it was very difficult for many listeners to know whether the broadcast came from the next town, state, or country. Several radio scholars have written of the excitement of listening in this period, explaining that enthusiasts took pride in picking up far-off transmissions, while artists and spiritualists attempted to use the medium to build new sorts of music or to speak to the dead.[24] Paranormal hokum aired on network programs of the late 1920s as well, as radio preachers tried to induce automatic writing in their listeners and evangelists used radio to heal the sick. Entertainment critics associated such gimmickry with novelty acts, but for many theorists today, radio's associations with the supernatural have also proven to be a powerful symbol of the existential ambiguities of experiencing modern media, which seems to transmute souls, disconnect the voice from self-aware subjectivity, and reveal that mediated interconnection is in many ways communion with a void.[25] After 1927, the aesthetic experience of the unidentifiable sound source began to diminish, as the Radio Act regulated station power and licensing in earnest. Yet the fantasy of communicating with the dead would never be far from how many Americans conceptualized the medium as a cultural form. Twenty years after the Radio Act, the *Saturday Evening Post* proposed that the most fitting symbol for broadcasting would be an image of a man with his mouth open, "the ghostly heart of the ghostly web of voices which weave across the wave lengths unceasingly to blanket the world in sound."[26]

The Depression years solidified such impressions, as fantasies of supernatural radio migrated from experience into dramatic storytelling.[27] In the early 1930s, CBS and NBC each aired versions of the Edgar Allan Poe story "The Tell-Tale Heart." The author was adapted at least thirty more times on anthology programs from 1935 to 1953, while many purloined his themes. Police show *Calling All Cars* aired an episode patterned after Poe's "The Black Cat" in 1939; *Dragnet* did the same thing twelve years later. Other gothic authors soon came to the airwaves. In 1937, NBC hired Stephen Vincent Benét to adapt Washington Irving's "The Headless Horseman of Sleepy Hollow" into an operetta; a year later, Orson Welles inaugurated his *Mercury Theater* on CBS with an adaptation of Bram Stoker's *Dracula*. The vogue also inspired radio writers. In 1936, both New York's WHN and WBNX featured *Ghost Walks*, broadcasting tours of the supernatural around the city, each station unaware that the other had the same title and premise. Meanwhile on WEAF Manhattan, Alonzo Deen Cole's *The Witch's Tale* became extremely popular.[28] Youths gathered around the campfire of "Old Nancy, Witch of Salem" to hear ghost stories in the same years that their parents came to the fireside of President Roosevelt to hear policy.[29] Such fare drew opposition. In 1935, a "National League for Decency in Radio" sent out mailings warning parents that blood-and-thunder radio promoted "sex delinquency."[30] Nonetheless, lucrative advertising contracts ensured that ghost stories remained bread and butter to Depression radio.

And in wartime, these shows became meat and potatoes. Allison McCracken notes that the number of "paranoid gothic" thrillers tripled during the war, many of them very successful—*Sanctum* was listed among the top twenty radio programs for fourteen years.[31] The craze was apparent as early as 1942, when *New York Times* critic John Hutchens noticed a sudden increase in evening ghost anthologies, detective mysteries, and gothic romances, what he called "the shockers."[32] Shockers plundered the work of such authors as Cornell Woolrich, whose romans noir were adapted for at least thirty-eight broadcasts between 1941 and 1953 on *Suspense* and *The Molle Mystery Theater*, *The Lux Radio Theater*, *The Globe Theater*, and *The Phillip Morris Playhouse*. "Has the War made Americans blood thirsty for crime and detection?" asked the University of Georgia when it awarded *Suspense* a Peabody Award in 1946; the insinuation is echoed today by critic Richard Hand, who suggests that the violence of the war increased public desire for escapist fiction.[33] That notion rings true but seems imprecise, since escapism is not exclusive to shockers. Such a claim can also be made about slapstick with commensurate success, and the voices of the dead were powerful dramatic elements well beyond the thrillers, structuring even patriotic 1940s dramas such as Robert Ardrey's "Thunder Rock" on *Studio One* and Norman Corwin's "On a Note of Triumph." Allison McCracken takes a more grounded view, showing that *Suspense* dramatized anxieties about women's wartime roles while at the same time subverting gender norms. This "subversion thesis" is connected to well-established struggles precipitated by changes in the wartime labor market and fits radio plays among films of a similar genre.[34] Critic Tzvetan Todorov has argued that tales of the fantastic are always in some deep way about identifying a law by transgressing it.³⁵ But neither escapist nor subversion theses situate the shockers in the larger context of radio entertainment. As a result, these interpretations tend to comment more on what gothic attributes mean in the abstract rather than what they mean in the program and period, giving shock credit for a degree of exceptionalism that it may not merit. After all, the people behind shockers also helmed programs in every kind of genre. Arch Oboler also wrote *To the President*; Wyllis Cooper wrote *Good Neighbors*; Himan Brown produced *Terry and the Pirates*; and Bill Robson directed *The Man behind the Gun*.

A simpler approach is to ask what may have motivated people in the industry to gravitate toward shock plays. Sponsors may have favored shock effects because many ads of the period marketed their products as anxiety tonics. *Sanctum* funded its tales of doom by hawking Carter's Little Liver Pills, "the best friend to your sunny disposition," and Bromo-Seltzer, which promised to cure "nerves"; *Lights Out!* sold Ironized Yeast tablets, which were sworn to relieve "that tired feeling" when "the war got you down." By attaching such slogans to tales of terror, these programs fomented unease before plugging a nostrum for it, a process that produced an interesting feeling of flow and ebb. For staff, the shockers were also political havens. Directors Brown and Robson each suffered a Red smear in the 1940s but continued to helm *Sanctum* and *Escape*. These programs also offered experimental scope to many writers. *Lights Out!* used stream of consciousness under Wyllis Cooper and Arch Oboler, two masters of the style, while *Suspense* often used retrospective narrators: of the 365 plays aired on *Suspense* in the 1940s, 230 begin with some variant of a fatalistic refrain: "I face disaster. You see, it all started when . . ." John Hutchens argued that there are pure stylistic advantages to the grim atmosphere of which shock radio programs liberally partook: "Doors squeak, footsteps pace off the measured step of doom, triggers click, doleful music wafts through the night. Those sound effects men know their business. The story may be dubious and the characters dull, and the violence so immoderate as finally to be only ridiculous but the atmosphere is always 'right,' proving once more that radio's power of suggestion is unequaled."³⁶ Because they were skimpy on substance but heavy on suggestion, the shockers became rich in feeling. "They are not art," Hutchens confessed, "but they have a certain effectiveness upon which you may choose to ponder before progressing to more advanced courses in public delinquency and the checking thereof."

Hutchens was on to something. Plays like "The Hitch-Hiker" rely on atmosphere in a way that sets them apart from the plays of Archibald MacLeish and Norman Corwin, in part because the "vocabulary" of the shockers is closer to that of war melodramas than it is to kaleidosonic fiction. Any description of the soundscape of the 1940s is impossible without the buzzes and clicks of war machinery, along with the insistent aural materials of battle—unrelenting gunfire, pounding on doors, dragging bodies, barking dogs, approaching sirens, speeding trains, and bombers overhead. These domineering and impatient noises resemble the squeaking doors and pacing footsteps of shock programs at least in the way they marshal Hutchens's "power of suggestion." Writers such as Fletcher were not just after a novel repertoire of sounds, after all, but also new semantic tasks that effects began to perform in tales in which they not only control the actions of characters but also shape the very "reality" of scenes. In the 1940s, bells and whistles were more than mere bells and whistles. They became sounds that foreshadow and warn, urge and compel, achieving levels of signification beyond the indications of actions and objects that they immediately describe. The shockers are at their core plays about this kind of perception. They grapple with a newly prominent layering in aural signification emerging in the craft of dramatic broadcasting, one in which sounds come across as haunted.

WARNING SYSTEMS

To illustrate how sound effects work in shockers, I would like to return once more to "The Hitch-Hiker." Above, I argued that Fletcher's play abandons the public space of Corwin's America in favor of a theater of interiority. In conjunction with this move, Fletcher also portrays Adams's inner world as a zone in which spoken communication undergoes a kind of failure and is replaced by an abstract psychological language decoupled from the human voice. The play explores a mind that becomes well articulated as it seeks to define itself through conversation but ends up being defined by what I will call "signals," a type of sound effect that seems to take over the play and, by extension, the mind.

Throughout Fletcher's play, Ronald Adams compulsively pursues interlocutory talk, frequently expressing an acute desire to "talk to somebody" as a means to soothe his frazzled nerves and achieve equilibrium. The play's direct-address structure is just such an attempt at palliative conversation. "If I tell it," Adams says with convincing certainty, "maybe it'll keep me from going crazy." Though desperate for talk, he is also wary that dialogue may fail, explaining that he must hurry or else "the link may break." The root of this anxiety soon becomes apparent, as we segue back in the story and hear Adams try to engage with a series of characters that he meets along his journey in troubled conversations that always fail to achieve satisfying exchange. Halfway through the play, for instance, Adams offers a lift to a young woman outside Oklahoma. While she coyly hints at an erotic interest, Adams learns that she cannot perceive the hitcher, which makes it hard for him to express his predicament to her. She soon misunderstands Adams's talk and questions his sanity, eventually fleeing.

As the narrative continues, Adams seems to fear that we, the addressees of his tale, might misunderstand him as well. Twice during the play, his narrating voice interrupts events in order to reiterate how the hitcher looked to him, as if he is afraid that he has been unclear in previous efforts to convey these details. Despite his trepidation, Adams feels compelled to keep trying to establish a connection, citing the great loneliness of his journey. This loneliness is surely a clue to the final revelation. Because the play takes place only within his mind, true rapport is precluded. Adams's loneliness is also ironic, since the hitcher's calls are a constant offer of remedy. As Adams reflects, "The thought of picking him up, of having him sit beside me was somehow unbearable, and yet at the same time I felt unspeakably lonely." This bind is sustained throughout the drama—the incessant desire to produce talk, and the anxiety that talk may not be produced successfully—as Adams loses every attempt to achieve an interlocutory exchange that is both bearable and "speakable."

With talk breaking down, the play introduces a new element into Adams's bind. Almost every time that he *does* try to address another character, the colloquy is both marked and pestered by a competing sonic motif, a "warning system" that sounds off when the conversation fails. The first expression of this sound occurs when Adams meets a gas-station attendant in Pennsylvania. As Adams tries in vain to express his predicament, a series of bells ring out in the background as the gas pump measures off the fuel that it dispenses into the car. The background chimes return in a new form further down the road, just after Adams fails to convince the owner of a diner to open up after hours. Driving off, Adams arrives at a railroad track, whose safety barrier descends across the road, accompanied by a repetitious dinging that strongly resembles the sound of the fuel pump. Toward the end of the journey, the play introduces a tolling bell deep in the background audio of Herrmann's score, as if the warning system has broken free of sonorous objects and now tauntingly registers in the atmosphere of the drama, in a strategic slippage between the underscore and the story that it supports. The repetitive quality of the chiming even influences Adams's exposition. Early in the play, towns evoke detailed commentary, but later they "tick off, one by one." The warning system finally thwarts talk when Adams calls home from Gallup, "to hear the even calmness" of his mother's voice (as opposed to the warning's hectoring staccato). In this scene, a long-distance operator instructs Adams to deposit a number of coins into the telephone, an activity that is extraordinarily overillustrated. For nearly a minute, Adams does nothing but follow the operator's instructions to fill the change box, which sounds as if it had been initially empty, creating fifteen impact sounds that resemble the pump, rail warning, and bell effects. Adams's attempt to return to the mother represents a catastrophically failed conversation, and the voice of the operator tells us so by devolving into yet another mechanical repetition: "Your three minutes are up sir . . . Your three minutes are up sir . . . Your three minutes are up sir . . ." The operator's dialogue intrudes like an annoyance, cauterizing the temporal span of the retrospective narrative, then fades like a siren or warning bell receding from earshot.

Fletcher's chime motif differs quite starkly from virtually all of the sounds discussed in preceding chapters. During the Depression, radio often set scenes with keynotes that suggested locales, an appropriate tool for directors trying to mark space by drawing on extant associations in the listener's mind to evoke a backdrop of prairies, churches, or foreign nations that we can believe to be "really there" in the context of the drama. Keynote sounds belong to a naturalistic theater, the sort championed in the 1930s, in which a sound exists to tell us about action and obdurate objects in space, indicating movement or illustrating locale. There is a certain "invisibility" to which such effects aspire, a total subordination to speech, which secures the world of the fiction and serves as the "final word" on what is happening. In 1940, CBS radio director Earle McGill explained the approach this way: "A sound effect should be so instantly recognizable or so rigidly and shrewdly prepared for or flow so naturally and realistically from the text that the listener accepts it without questioning the aesthetic justice of including it in the drama."[37] Critic Elissa Guralnick considers this to be a significant part of craftsmanship in the genre, explaining that in radio "playwrights make sights emerge from sound so straightforwardly that the connection between what we hear and what we see must appear unimpeachable."[38] Even if a drama is a work of pure fantasy, according to this school of thought, it is intelligible because sounds are tame things that draw upon sonorous life through a code of resemblance, pointing outward in a referential manner to lived aural experience and anchored with a verbal label if needs be. If a sound effect puts a picture in our mind, we can trust that the picture is not going to be falsified later on.

Perhaps sound effects can never escape that corroborative relationship entirely (that's how we understand them), but this naturalistic theater is also precisely what Fletcher was attempting to exceed with devices like the chimes in "The Hitch-Hiker." Because they are a motif and not an event, the chimes decouple from the individual sonorous events that express them, just as sound disengages from the objects that produce it. Thanks to this double process of abstraction, even though the sound may literally illustrate action through resemblance to sound that the listener recognizes, that cannot be the end of the reading, since the sound also has significance beyond the purported events that occasion it, offering a supplemental level of meaning that that tells us as much about the perceiver as it does about his perception. We know this from context too: naturalistic effects take place in an exterior space, but Fletcher's play has no such space, and so the chimes cannot be anywhere but Adams's mind, which makes it very unclear if they are truly "sounds" at all. Keynotes draw on cultural memory to create a cliché scene. After doing so, they melt into a background. By contrast, the chime motif rises toward attention, growing more prominent over the duration of the drama, providing atmosphere and even becoming the aesthetic justice of the play.

Adapting a term used by acoustic theorists such as Murray Schafer and Barry Truax, I call Fletcher's chime effects "signals"—sonorous events to which our attention is especially directed, sounds presented as ciphered emotional truths, repetitions that "control" the form and substance of the drama.[39] In psychological radio, signals do more than simply flow from actions and illustrate settings. They reveal inner truths and hidden facts. They take on a life of their own, pestering dramatic structures, intruding upon action and motivations, and affecting inner dynamics—hence the extraordinary duration and detail in the pay phone sequence in Fletcher's play. In their repeated transposition from one scene to the next, signals become like what experimental composer Pierre Schaeffer once called "sonorous objects," independent entities whose ontology stands apart from that of the objects from which they putatively emerge.[40] That is one reason why signals prompt a compulsive form of listening, both for the character, who cannot resist a signal, and also for the listeners at home, who cannot control the degree of attention they devote to it.[41] A signal is a sign into which the reaction it solicits is partially collapsed. It would be a stretch to say that the chime motif is the "message" of "The Hitch-Hiker," but it is certainly the most critical stylistic innovation devised for this broadcast by Bernard Herrmann, whose scores are known for psychological intensities; by Fletcher, who concentrated on sound effects when she described the future of radio; by William Spier, whose expertise in classical music made him acutely aware of leitmotif; and by director John Dietz, who specialized in sound effects. As the motif migrates from fuel pump to railway barrier, from speech acts to change box, it is as if it rises from a subconscious level to burst into the open at the instant of epiphany, circling inward from an aural periphery to erode its own claims of material referentiality, restructuring the protocols by which sound is used to make meaning, mood, and ultimately dramatic narrative.

EXPERIMENT #87

It is significant that signal-based dramas became prominent at a time of war. In the prose of many of the war's best-known correspondents, there is a sense of signal that recalls Fletcher's motif, suggesting a whole aesthetic habit of acute sensitivity to aural perception that marked the "media generation." Ernie Pyle's writing is typical. Just as signals displace keynotes in "The Hitch-Hiker," Pyle imagined war layering supernatural qualities overtop of referential sound. "War has its own peculiar sounds," Pyle writes in his anthology *Here Is Your War*, "They are not really very much different from sounds in the world of peace. But they clothe themselves in an unforgettable fierceness, just because they are born in anger and death. The clank of a starting tank, the scream of a shell through the air, the ever-rising whine of fiendishness as a bomber dives—these sounds have their counterparts in normal life, and a person would be hard put to distinguish them in a blindfold test. But once heard in war, they are never forgotten."[42] Pyle's ferocious sounds extract the "voice" of an object and commandeer memory in the same way that chimes take over Adams's mind and distort "normal" aural life by filling it with vestiges of a remembered reality that is ever-present. Signals traumatize time and the psyche at once. Pyle's list of sounds could be a radio script: "The memory of [war] comes back in a thousand ways—in the grind of a truck starting in low gear, in high wind around the eaves, in somebody merely whistling a tune. Even the sound of a shoe, dropping to the floor in a hotel room overhead becomes indistinguishable from the faint boom of a big gun from far away. A mere rustling curtain can paralyze a man with memories."[43]

Pyle's signals decouple from objects that instantiate them just as Fletcher's motif is always more than the pumps and telephone operators that summon it.[44] Signals take over mind and reorganize memory, inducing Pavlovian reactions in listeners both real and fictional. Films of the 1940s also show an affinity for signal-based dramaturgy. William Wyler's *The Best Years of Our Lives* (1947) climaxes with airman Fred Derry (Dana Andrews) recalling wartime fears and virile experience while sitting in the cockpit of a rusted-out bomber

FIGURE 5.3. "... forever, Amen. Hit the dirt." A 1944 cartoon by Bill Mauldin. The sound of incoming airplanes was very frequently referenced by war correspondents. Mauldin's scene emphasizes one characteristic of "signals" by calling attention to the truncation of the lapse between perception and reaction. Signal-based communication is interpretation-in-a-hurry. One does not ponder a signal, one *responds* to it, compulsively and without delay. Copyright by Bill Mauldin (1944). Courtesy of Bill Mauldin Estate.

abandoned in a field. Rather than illustrating the memory of war with visuals, Wyler chose to present a series of sound effects mimicking takeoff, "fierce" sounds that stand in for psychological trauma and seem to engross the consciousness of a lost, demoralized Derry, a sequence that spoke to a public for which the ability of sound to embody trauma was in part conventionalized by signal-based radio drama. In a larger way, signal-based narratives matter because they depicted the mind as an organ especially sensitive to aural information and beset by irresistible instruction on all sides, a mind ideally imagined for the instrumentalist imperatives on which effective war exhortation of any kind depended.

And in radio, the signal-based narrative of "The Hitch-Hiker" became a prolific archetype across programs, which told hundreds of tales in which some form of talk is overcome by an irresistibly compelling signal that fills up a drama and overwhelms a character's inner being. In "Tension in Room #643" on *Radio City Playhouse*, for instance, a woman associates the death of her son with a chime sound and grows uncommunicative and even catatonic until bells trigger her memory. In "Voices of Destruction" on *Front Page Drama*, an author tries to write out his confession of murder, but he fails to do so because the sound of Blitz bombers has traumatized him to the point that his ability to express himself verbally is compromised, and his confession turns out to be just a blank page, as if the bombastic concussions have erased it. A similar sequence of events occurs in "The Brighton Strangler" on *Suspense*: after an actor has given up speaking before crowds, the sound of an air raid reorganizes his consciousness so that he believes himself to be the strangler he once portrayed. In "The Beckoning Fair One" on *The Molle Mystery Theater*, an author suffers writer's block and argues with his lover; as talk and writing fail, he begins to hear what he thinks to be a song behind the sound of a dripping faucet, an idea that obsesses him and drives him to murder. In "Wailing Wall" on *Inner Sanctum*, a man murders his wife and stuffs the body in the wall, then begins to hear a whistling that he fears is her undead spirit; after shunning human contact and living with the signal for decades, the hero realizes that the sound comes from a small aperture in the masonry. It is telling that "Wailing Wall" draws heavily from Poe's "The Black Cat." More than any other single writer, Poe prefigures the logic of the sound signal in that many of his stories culminate in a scene in which a monomaniacal listener strains to hear an overdetermined sound. This is especially true of "The Tell-Tale Heart," which is not coincidentally the most frequently adapted tale in classic radio.

Because signals are not just a registry of spooky sounds, but a way of turning suggestion into a dramatic entity, signal-based dramas are not limited to the world of thrillers. Morale plays use the same semiotic operation to mark resolve. Consider *Cavalcade of America*'s "Continue Unloading," which follows a press photographer and a medic during the invasion of Sicily. Although our audioposition moves from ship to beachhead in the deep space of the play, the PA system of the transport vessel constantly repeats "Continue Unloading!" everywhere in the space of the drama, sounding equidistant from us even if we have moved according to the information in the dialogue. No matter the tribulations of which the servicemen grouse on the beach, the signal under which they maneuver continues unabated, programming their minds. When the medic falls injured, "Continue unloading" is all that he can verbalize from his delirium. *Words at War* aired a similar drama that used an address system to frame action in the North Atlantic, this time with the title "Condition Red." As in "Continue Unloading," each scene in "Condition Red" begins and ends with a PA signal, which syncs action and binds together events, commanding the fiction by commandeering the energies of its characters. These plays are but two of many in which men die but instructions remain, a process that fetishizes those instructions to the point that they accrue outstanding imperative qualities. Signals tend to intensify anything with which they coincide, sometimes monstrously. In "The Panther Story" on *Out of the Deep*, for instance, a diabolical Indian trader whips his wife for disobeying his will, a scene in which the crack of the whip is accentuated by the cries of a nearby wild cat, intensifying the cruelty of the abuse.

Of course, Lucille Fletcher herself continued to pursue signal-based themes in plays such as "The Search for Henri LeFevre" and "Sorry, Wrong Number," tales developed along with the same team as "The Hitch-Hiker." But even more significant are the scores of other stories that followed her model in dramas that aired on *Suspense* throughout the 1940s. In "The Lord of the Witch Doctors," action turns around preventing the sending of a sonorous attack signal to a German gunboat hidden offshore at a far-flung British outpost. In "Heart's Desire," a bank clerk becomes so obsessed by the sound of an ocean liner's horn that he cannot remember the code word that he needs to recover the millions that he stashed away with a pawnbroker before serving a sentence for theft. Frustrated by this failure for most of the drama, at last the clerk recalls and utters the code, only to be killed by an opportunistic femme fatale as the liner's horn mockingly resounds in the background. "August Heat" opens with the sound of a pencil scratching along paper, as an artist automatically draws an image of his own death. Perplexed, the hero meanders to a strange part of London where he meets a mason who has inexplicably carved a headstone with the artist's name on it. The two men fail to achieve understanding through talk, and although no motive appears in the play, the drama ends with two sounds—the pencil scratch and a chisel sharpener—which seem to have their own metaphysical connection that compels murderous action. "The 13th Sound" begins with a series of effects that describe a man being shot, and then we hear the wood creaking under hands as he grips a window frame in his death throes. The local sheriff extracts a confession from the dead man's wife only when he confronts her with a re-creation of the sound sequence in a "modern symphony." When it hears a signal, the guilty mind cannot help but confess. By definition, signals are always overdetermined and most convey irresistible suggestion, but not all take the form of effects or speech. In "The History of Edgar Lowndes" the sound of a train is a mnemonic of childhood trauma and incites a series of psychotic breaks by a killer, but in "The White Rose Murders" the same result is associated with, of all songs, "The Beer Barrel Polka."

No play follows in Fletcher's footsteps more closely than "Donovan's Brain," a 1944 *Suspense* broadcast based on Curt Siodmak's novel, directed by Spier and starring Orson Welles.[45] The episode begins with Dr. Patrick Corey (Welles) reading the history of "Experiment #87" aloud from his notebook. We segue back to some months earlier, to find Corey in his lab, developing a way to keep monkey brains alive after death. An airplane crashes nearby, and the injured tycoon William Donovan ends up in Corey's lab. Seizing the moment, Corey allows Donovan to die and harvests his brain. As the drama proceeds, Corey falls under the influence of the surviving brain in its jar and becomes Donovan's surrogate. When Corey's wife urges him to stop the experiment, he puts her in a psychiatric hospital. When Corey's colleague tries to turn off the fuses that keep the brain alive, he becomes paralyzed. In the end, Corey even murders his own son at the behest of the brain, which is trying to find a new body with which to pursue world domination. We segue back to Corey in the present and hear him exact revenge, destroying the brain and its life-sustaining equipment. In a denouement, we hear a news report about the gruesome scene, and learn that despite all that we have heard, the brain had already been dead for months.

For such a simple narrative, the arrays of cues that support it are quite complex. Welles plays Corey with a mid-Atlantic tint, but he also plays Donovan with a slow, deep register. The contrast between these voices highlights scenes in which one blends into the other, moments that coincide with a signature phrase, "Sure, sure, sure." But these signatures are much less prominent in the broadcast than the sound effect that Spier chose to signify the activity of the disembodied brain. Whenever the brain reacts to events in the drama, an oscillation sounds to represent a "sonar device" that Corey has supposedly attached to an encephalograph that measures Donovan's "delta waves." Resembling the sound of a slide-whistle, the signal has a lot of personality, warbling crazily when the brain is threatened, and lolling along placidly when at rest. In this case, not only is a psychological sound effect tasked with representing interior dimensions, but the play dramatizes veritable neurological auscultation. Like the chimes in "The Hitch-Hiker," this signal also seems to disobey spatial structures in the drama, drowning out the keynote bubbling beakers that illustrate Corey's laboratory and exceeding walls and doorways that split the laboratory from the home. Whenever music swells to indicate a segue of time or setting, the brain persists in warbling according to its reaction to events in the concluded scene, as if it were an actor continuing to speak from behind a curtain that has already closed for a set change.

As in "The Hitch-Hiker" and most of the other plays discussed above, the signal attains force only because something has gone wrong with conversation. After all, the objective of "Experiment #87" is not simply to keep the brain alive, but to converse with it. "I know it has some great plan," Corey reports frequently in his diary, as he attempts to decode the sonar signal, translate it, or step up its power "as one steps up the power of a radio transmission." Yet even when the doctor realizes that his mind has been subjugated by "Donovan," he still attempts to "converse" with the brain. In fact, Corey never successfully "talks" to the brain-signal at all, but spends his time serving as a vessel for signals. In this way, the play illustrates the notion that one cannot truly interact with a signal, only submit as it overtakes consciousness, a situation that is made all the more ominous by the revelation that Corey's submission has taken place entirely internally. Like Ronald Adams, Corey's mind is more crucially defined by its relation to signals than by its relation to sounds, a predicament that is a hallmark of 1940s radio fiction, and one that becomes most horrific when the source of irresistible instruction is a hidden part of oneself. "Donovan's Brain" leaves us with the same conundrum as "The Hitch-Hiker": at the end of each we do not know if we have heard a series of *sounds* or else *the idea of sounds* within a character's mind. Is there a difference? If signal-based drama maximizes radio's power of suggestion, it is a result of this extraordinary capacity to engineer an ambiguity between the categories of sound and idea.

Perhaps more than any other single innovation, signal-based conventions define the dramas that succeeded Depression radio. In the 1930s, radio had used keynotes to portray deep locales in order to bring unheard voices toward a microphone as if from the margins of society into the heart of its coalition. Depression radio was like a town meeting in which there lay a sense of community threaded thick with interpersonal talk and illustrative effects. But as war approached, although *voices* filled the air with an ever-growing cacophony, drama was also emptying of *talk* in the ordinary sense. Dramas like "The Hitch-Hiker" and "Donovan's Brain" literalized this process. With talk disappearing, airwaves supplied orders, instructions, and sirens that warned, urged, and compelled, a class of communications material whose creators aspired toward the same instrumental power over people as signals had over the characters in shockers. "Radio in the 30s was a calm and tranquil medium" comedian Fred Allen recalled. "Oleaginous-voiced announcers smoothly purred their commercial copy into the microphones enunciating each lubricated syllable."[46] But in the 1940s, the calm was broken up with all sorts of imperatives, and drama responded by enhancing its protocols of referentiality in order to express sound events with such power that they could reorganize the way characters imagined and what listeners expected. In this way, Fletcher's signal-based "psychological" radio play both responded to and took part in the excess of mass communications that leads writers to employ such terms as "media generation" and "radio war." Radio dramatists were using deep aspects of their craft to both question and reformulate how mass media organized signification in the world of the imagination, using vivid performances commanded by bells and whistles whose seeming simplicity masked their suggestive force. And signal-based dramaturgy was not the only component of 1940s radio fiction that asked such questions. In the next chapter, I consider the characters who actually heard these signals in the context of a story. As we shall see, for many radio dramatists, the process of listening was just as overdetermined as the psychological sound effects that it could not escape.

CHAPTER 6

Dramas of Susceptibility and Transmission

As my study of Lucille Fletcher's "The Hitch-Hiker" demonstrates, 1940s radio drama tended to neglect the spatial preoccupations of the 1930s and link concepts of the mind to the idea of communication through sound effects that both signify and signal at once. While this signal-based dramaturgy is indispensable to psychological radio, its component elements are also subtle, requiring repeated and intensive listening to unpack, which is one reason why it has been so easy to perceive the compelling dramatic atmosphere of wartime radio in general yet difficult to pinpoint the aesthetic devices that provided it. But if the "theater of the mind" hypothesis has explanatory force, it also ought to help explain more overt conventions, such as the behavior of characters in these dramas. In this chapter I turn to this question, arguing that many 1940s dramatists tended to use a character's status and orientation vis-à-vis key aural transmissions to inform us about his or her motives and identity. Characters seem to have adopted "transmitter" and "receiver" roles, in a system that resembled mass communication models that wartime social researchers were then using to conceptualize radio audiences. To introduce the issue concretely, another of Fletcher's plays can help, "Sorry, Wrong Number." Perhaps the most famous of all American radio dramas, "Wrong Number" aired six times on *Suspense*. Jack Benny satirized it on *Mail Call* before Paramount's 1948 film version, which was adapted back to radio for *The Lux Radio Theater* in 1950. The play has drawn new interest today; one major history of broadcasting has gone so far as to append the entire script.[1] Fletcher's tale has usually been read as a piece on interconnectedness in modern urban life or as a critique in which protagonist Mrs. Stevenson (Agnes Moorehead) represents deviation from gender norms.[2] Here I would like to consider something a little more elementary and representative of 1940s dramatic conventions in radio—the way that acts of communication underlie how we understand technology, Mrs. Stevenson, and her dire situation.

In the opening of the play, we find Mrs. Stevenson telephoning her husband Elbert at his office. She asks the operator to dial on her behalf, as the line has been inexplicably busy, even though it is after normal business hours. A professed invalid, Mrs. Stevenson is characterized as a chatterbox by her unending prattle and Moorehead's characteristic style of dropping an octave in mid-syllable, which makes it unclear if she is speaking to her interlocutor or to herself under her breath. Within a moment of meeting her, we feel certain that this is one of many telephone calls that the invalid typically places to Elbert each day. But this call will be different. A seemingly unrelated conversation intrudes on the telephone line, and Mrs. Stevenson overhears two men specify the time, place, and means for a murder. The protagonist tries to interject but finds that although she can hear the men clearly through her earpiece, she cannot issue sound into their conversation, evidenced by the fact that the men give no sign that they perceive her series of shocked vocalizations. In this scene, two technological malfunctions have transpired. Something seems to have gone awry in the telephone switchboard (Stevenson remarks that there must be a "crossed wire") just as something has also gone wrong with the transmitting diaphragm of her mouthpiece. At the end of the play, we learn that the first malfunction did not really occur. One of the plotters is indeed at Elbert's office (the intended victim is Mrs. Stevenson herself), so there has been no misdirection. But no alternative explanation is offered for the handset malfunction, which is a clue to the larger reality of Mrs. Stevenson's situation. Just as the transmitter of the phone seems to fail, events conspire to deny Mrs. Stevenson the compulsive talk that marks her nature. After hearing the plot, she makes a series of calls to the operator, the police, and a hospital, trying to convince others of her story, but even though her handset is now functioning properly, Mrs. Stevenson cannot articulate her problem in a way that will spur officials to act. Mrs. Stevenson's "say-so" does not compel the chief operator to trace a call. A policeman opines that the overheard call must have been "phony." Like Ronald Adams in "The Hitch-Hiker," Mrs. Stevenson calls her situation "unspeakable." This vocabulary reflects how profoundly acts of speech shape Mrs. Stephenson's conflict. Speech is her anodyne, a self-authorizing force that overcomes and justifies her status as an invalid. Denied this gratification, she is infantilized and seems stupefied by the problem. "I haven't had one bit of satisfaction out of any call I've made this evening," she complains, in the most revealing line in the play. As a result of this repeatedly thwarted attempt at satiation, the play takes on an odd resemblance to torture.

In the concluding sequence, all of Mrs. Stevenson's efforts to be heard have failed, and she begins to hear forced entry downstairs. The hired killer approaches, and like our protagonist, we struggle to hear him move up the stairs in a prolonged passage of acute reception that is seldom rewarded with detailed sound. The erstwhile pathological talker has become a rapt listener, hiding from her assailant and unable to speak to the police on the line above a choking whisper. "I'll be quiet, I'll be so quiet," she whimpers, as if cowering from a parent coming to punish her, before she is rendered aphonic by the murderous hands of the killer. Her dying scream is drowned out by a passing train, which seems to collude with the choking that it masks. Strangulation is a very common form of murder in old radio, a removal of life that is also a removal of voice. In "Wrong Number," that second removal is particularly stressed, because dramatic action does not conclude with Mrs. Stevenson's death, but instead with her killer seizing the telephone to tell the policeman on the other end that he has a wrong number.[3] The killer both covers his crime and commandeers the device that has given his victim's life meaning, an act that has force thanks to efforts to call attention to the dynamics of speech and hearing throughout the drama. By ritualistically denying a woman her voice and compelling her to listen, "Wrong Number" symbolically transforms Mrs. Stevenson into her malfunctioning handset. Lucille Fletcher famously claimed that she intended "Wrong Number" to be a play in which the telephone was the chief protagonist, but she may have achieved the opposite, a drama in which the protagonist is a kind of broken phone.[4]

The reading above is based on the observation that at every point in the narrative, it is Mrs. Stevenson's attempt to speak or to listen—to transmit or to receive—that directs how we imagine the perils of her situation. Here I take this observation a step further, positing that we should conceptualize wartime radio itself as a kind of laboratory for the theorization of communication by pointing out similar radio stories in which the soul of a character stabilizes around the fact that he or she issues or receives an aural transmission that shapes the world of the fiction, something that a dialogue-rich dramatic genre proved adept at portraying. I begin by seeking a motive for this process of characterization in 1940s audience research models dedicated to understanding the schematic relationship of a consumer to the messages that surrounded him or her. As we shall see, both writers of radio fiction and writers of social research tended to think about people as "active" transmitters or "passive" receivers. While researchers viewed these positions as fixed, dramatists often depicted characters who transform from a transmitter to receiver across a narrative arc (or vice versa), much as Mrs. Stevenson does.[5] In this way, the trials of Mrs. Stevenson reflect a whole mode of understanding psychological life during the war years that couples the idea of the mind with the forms of media that it encounters and employs, embodying a proposal at the core of radio storytelling and indeed the modern imaginary.

INVENTING VULNERABILITY

The theme of this chapter is rooted in a question as concrete as any in the broadcasting industry, one on which the monetization of radio depended: who *is* the listener? Throughout the radio age, writers attempted to understand this social entity as an atom of mass society, using it as a hypothetical figure to determine whether or not radio was beneficial for democracy. The answer was often in the affirmative during the 1920s and '30s, when several intellectuals portrayed the listener as a dynamic figure in space. The listener was much "closer" to leaders; barriers between them "collapsed." With such language shaping the conversation about radio as a cultural form, it is appropriate that directors emphasized the dimensional quality of sound. But the musings of critics and theorists were but one (relatively minor) resource used by broadcasters to grasp their public. Historians have recently proven that letters, phone calls to stations, boycotts, and other forms of listener response helped broadcasters to shape radio content directly.[6] As the 1940s progressed and audiences grew, however, few sponsors could be persuaded only by the superior prose of avid fans. By wartime, the issue of understanding the listener had been taken up by a group of scholars including Gordon Allport, Bernard Berelson, Edward Bernays, Hadley Cantril, Herta Herzog, Harold Lasswell, Paul Lazarsfeld, Robert K. Merton, and Wilbur Schramm.[7] These were not the philosophers who had idly mused about radio in the past, but professional social researchers employed by advertising firms, research institutions, and federal agencies. It was not until the war that the interests of these three entities became conjoined in a way that made it possible for communications to become a distinct area of research, in part because such work promised to provide an understanding of auditory appeals that would help agencies to craft them more effectively.[8] General Mills tried to understand its listeners for decades, but this project did not become an agenda until Uncle Sam needed it too—as the *Nation* put it in 1940, those trying to grasp the listener now had "the hot breath of the times" at their necks.[9]

During a decade John Durham Peters calls "the single grandest moment in the century's confrontation with communication," these researchers made communications studies into a field by channeling research funds and framing paradigmatic arguments; their biographers call them "founding fathers" of communications, the "disseminators" of public relations dogma or "the very model" of a wartime intellectual.[10] Bernays and Lasswell authored the most widely read books on media relations of the era, later advising war agencies and major advertisers. Berelson analyzed Nazi broadcasts for the OWI, directed behavioral science research at the Ford Foundation, and established the Center for Advanced Study in the Behavioral Sciences in Stanford. Allport served as president of the American Psychological Association and handpicked students for the clandestine services. Schramm wrote the first textbook on mass communications and founded the first PhD program in the field at the University of Illinois. With funding from the Rockefeller Foundation, Cantril conducted audience research at the Princeton Radio Project, while serving as pollster to President Roosevelt and editing the key journal in the field, *Public Opinion Quarterly*. Merton helped invent the "focus group" and spent a career shaping the entire field of sociology. Herzog worked for advertising goliath McCann Erickson, where she became chair of a market research unit. Lazarsfeld collaborated with CBS executive Frank Stanton and advised Archibald MacLeish at the Office of Facts and Figures. The Austrian émigré also directed the Office of Radio Research (ORR) at Columbia's Bureau of Applied Social Research, an office that almost singlehandedly invented survey-based research into media effects.

For these researchers, as historian Susan Douglas has put it, "the challenge was to get at what radio was doing to people's heads—as individuals and as members and shapers of a society."[11] Depression-era authors had eminence but wartime scholars had data, and with it they set out to make the American listener less of an abstraction. Prior to the ORR, just one question really vexed the industry: how many listeners tune into a given show? As early as 1928, Harvard's Daniel Starch used surveys to get an answer, and by the 1930s, ad firms were hiring prominent pollsters such as George Gallup.[12] As sponsorship expanded, so did the requests for reliable measurement, since money was at stake. In 1935, NBC charged sponsors eighty cents per thousand listeners estimated to be "reachable," and with audiences ranging between two and sixty million, the sums involved could become extraordinary.[13] Two services emerged to offer measurement data, the Crossley service and the Hooper ratings. Both used estimates of signal range, questionnaires, telephone interviews, and other means to gauge "program circulation." These calculations were a constant source of controversy.[14] It was unclear if surveys using "aided recall" or "unaided recall" could better reveal what listeners had tuned in several days before; neither service could satisfactorily explain how it corrected for households without telephones; and when client firms compared how each service rated the same program, discrepancies appeared that cast doubt on both measures. "As a matter of fact, nobody really believes in radio's statistical holy of holies," explained radio writer Robert Tallman in the *New York Times* in 1942; that same year, an industry analyst summed up frustration among executives, writing in *Printer's Ink* that the ratings were no more reliable than a "glass swami's bowl."[15]

To mitigate this skepticism, efforts were launched to evolve the paradigm of audience research. In 1937, *Advertising Age* reported that the French were testing a system in which announcers instructed listeners to turn on their electric lights when they tuned in a given show, afterward measuring the load change at the electric company; in 1943, the *Saturday Evening Post* profiled psychiatrist Louis Berg, who conducted a study in which he measured the pulse of women listening to *Young Dr. Malone* to see how their blood pressure reacted to "adultery, insanity, suicide [and] *psychopathia sexualis*."[16] Lazarsfeld's group tried several methods, in one case attaching recording devices to sets (an early version of the Nielsen devices destined to become standard later on), along with an analyzer that enabled test subjects to respond positively or negatively to show content.[17] These devices seemed rigorous but results proved mixed. Few negative responses were ever logged, and when a listener responded positively, it was not always clear what aspect of a show elicited this response. According to *Time*, one test subject reacted positively to a villain just because she appreciated that he always closed the door quietly.[18] Despite dubious results, the analyzers at least evolved the paradigm, bending show business decision-making around detailed audience data to an unprecedented degree. Never again would it suffice to know how many listeners were out there. The ORR now wanted listener *opinions*, fielding surveys that attempted to divine how men and women (of one region or another, of one educational level or another) listened in different situations (at what time, alone or with family, in homes or in cars), and what they thought about diverse topics. The results poured in: respondents with "higher cultural levels" liked reading, while those with "lower levels" preferred radio; more than twice as many women liked daytime quiz shows compared to men; 41 percent of all radio listeners liked mysteries; in 1940, voters who changed their mind in favor of Wendell Willkie tended to be newspaper readers, while FDR voters cited radio.[19] Tables on such subjects form the sheer stuff of 1940s media research. Cross-classifying the data, researchers made responses predictable and manageable. As philosopher Jacques Ellul would later argue, these techniques made propaganda "modern," because appeals were no longer subject to the predilections of an individual propagandist but were instead based on scientific claims, definitions, and controls.[20]

In recent years, scholars have pointed out that this research benefited some of the most cynical aims of commercial capitalism. Critic David Jenemann calls Lazarsfeld's paradigm "a shift from knowledge as an end in itself to the instrumentalization of thought at the service of business," and Jackson Lears explains that public opinion measurement let ad firms claim that they honored the sovereignty of individual choice, while setting boundaries to it—surveys were "an instrument for rendering public debate more manageable and predictable."[21] Historian Christopher Simpson argues that in order to flatter the assumptions of their patrons, Lazarsfeld and his peers degraded the topic that they purported to frame:

> The "dominant paradigm" of the period proved to be in substantial part a paradigm of dominance. . . . The key academic journals of the day demonstrate only a secondary interest in what communication "is." Instead they concentrated on how modern technology could be used by elites to manage social change, extract political concessions, or win purchasing decisions from targeted audiences. [Such work] reduced the extraordinarily complex, inherently communal process of communication to simple models based on the dynamics of transmission of persuasive—and in the final analysis, coercive—messages.[22]

These critiques are undeniable. Lazarsfeld's analyzers were indeed fielded by ad firms, and his findings were coauthored by broadcasting executives in what he called a "triple alliance of research, vigilant criticism, and creative leadership."[23] Critics in the 1930s wrote of the "social effects" of radio in order to understand the nation more thoroughly; 1940s scholars discussed the "psychology" of radio, although the only psychological detail deemed noteworthy was the set of circumstances that increased one's readiness to be convinced about elections and purchases. As theorist James Carey has pointed out, Lazarsfeld and his students did not seek representations of the communication process, but rather models for the enactment of the communications process, and that search surrendered the scholarly habit of critique too readily.[24]

Deserved or not, Lazarsfeld's drubbing obscures one effect of his work. By studying audiences, the ORR implicitly answered what it *meant* to listen, a philosophical matter of some importance to contemporaneous radio drama and 1940s popular thought. While survey research seemed much more concrete than the Crossley or Hooper ratings, it also had its own underlying assumptions. In Lazarsfeld's paradigm, a "listener" was a container of opinions and allegiances—more or less conscientiously held—and research consisted of surveying these opinions in order unveil how campaigns altered the content of mental repositories across populations. Rather than dispelling abstract models, the paradigm created one: the passive listener. Although Lazarsfeld based his work on the question of "who says what to whom and with what effect," he left speakers and their messages uninterrogated and instead poked and prodded the recipients of address, who are in this schema always in a responsive position, subordinate to a message and merely a measure of its effect.[25] Meanwhile, Edward Bernays described listeners using litanies of verbs whose passive construction matches the impotence that it denoted: "We are governed, our minds molded, our tastes formed, our ideas suggested, largely by men we have never heard of."[26] For his part, Lasswell defined propaganda broadly—"the direct manipulation of social suggestion"—but presumed little critical ability in listeners, describing housewives "bombarded by sounds" and citizens who become "innocent dupes."[27] In his study of the "War of the Worlds" panic, Hadley Cantril attempted to measure a quality in listeners that he could only describe as "susceptibility-to-suggestion-when-facing-a-dangerous-situation."[28] This unlovely criterion might sum up how the discourse as a whole defined the listener, as a creature whose most essential characteristic is its degree of vulnerability to a message, something that could hopefully be expressed as a quotient.

The passive listener trope even appears in the prose of the sharpest critic of this research, philosopher Theodor Adorno, who began to develop his critique of the "culture industry" while working at the ORR. Adorno excoriated Lazarsfeld for how his analytic approach constructed listeners rather than illuminated them, and he blamed radio for what he called "atomistic" listening, an individualized appreciation of music as mere quotation, theme, or tune.[29] As Martin Jay has pointed out, the critic often chose language that suggests the modern listener resembles a child, one particularly characterized by a kind of malnourishment—in papers written during his unhappy tenure at the ORR, Adorno calls listeners "childish but not childlike," "regressed," "technique-minded children," who "demand the one dish they have once been served"; he describes radio preachers as "carnivorous preying beasts" like the anthropophagous sirens of antiquity.[30] Elsewhere he writes of radio as a dish of food, "crispy gold-bond bread" or "lovely soup" that is always withheld, and in one essay, he coins the term "culinary listening" to describe how listeners isolate details of music.[31] These themes persist in later writing. In "The Culture Industry Reconsidered," Adorno argues that listening "impedes development"; elsewhere he writes that Third Reich listeners suffered the drudgery of Sisyphus, as American listeners resembled Tantalus, the never-gratified consumer of food.[32] Adorno's disdain for radio is rooted in psychological categories, yet in order to voice this critique, he drew from an argot of nourishment shared by many writers when they confronted the medium. In these years, General Mills executive Walter Barry described his advertising approach with language not far from that of the critic: "The pores of the radio audience are pretty well opened for a little sales talk after a few doses of romance or song. As a result, Wheaties has spreadeagled the breakfast food field in eight years."[33]

TRANSMISSION AS DRAMATIC SITUATION

Brainwashed, governed, bombarded, duped, susceptible, endangered, childlike, starved, drugged, spreadeagled: in their attempt to make the listener more concrete, these writers almost invariably employ a model in which listening is a state of excessive susceptibility defined by animalistic lust for products and lack of critical judgment. This abstraction inverted the utopian discourse of the 1920s and '30s, when writers argued that listeners were empowered by their newfound connection to world events. After Lazarsfeld's research, listener identity came to be synonymous with passivity, and scholarship was also soon couched in a binary of active/passive, a paradigm that is only reinforced by attempts to introduce nuance into it. In 1956, for instance, Lazarsfeld and Elihu Katz published *Personal Influence*, a landmark study that posited the famous "two-step" model of mass communication, in which media messages are "filtered" by local opinion leaders. "Ideas, often," the authors explain, "seem to flow *from* radio and print *to* opinion leaders and *from them* to the less active sections of the population."[34] Networks reigned, but word of mouth ruled. To make decisions, people looked to gatekeepers at strategic social locations in their peer groups. Although this model diminished the power of radio propaganda, when it comes to listening itself, the two-step model merely expanded the list of those before whom a listener is feeble. Addressees heed advice uncritically, whether it comes from a network spokesperson or an esteemed chair at the local PTA, because in this discourse speech is considered to be an act, while listening is only a state—to speak is to do, to listen is to be. Scholars may bicker about *what makes* the listener passive, *to whom* the listener is passive, or *the degree* of passivity involved among particular social categories, but these disputes reconfirm the appropriateness of a highly reductive binary of active speaker/passive listener in the first place.

The explanatory prowess of this binary is less important to the current argument than its place at the center of 1940s research and its indirect impact on radio narrative. Lasswell and Lazarsfeld were widely read in the radio trade, and they advised the very firms and war boards that hired Himan Brown, Arch Oboler, William Robson, and the many followers of the stylistic trends they set. When wartime radio staff thought of "psychology," it is partly the work of the ORR writers that they imagined. As a consequence, it is not surprising to find a close analogue of the active/passive binary in radio dramas such as "Sorry, Wrong Number," which in this context we can consider to be the story of an active speaker forced to become a passive listener, one who turns from a mouthpiece into an earpiece. This binary opens up a framework for thinking about subjectivity in radio plays, one focused in part on how tales spotlight characters who do an awful lot of listening, such as secretaries, reporters, stool pigeons, and detectives. The plays of the 1940s are often structured around such a character performing some great feat of aural perception. In "Footfalls" on *Suspense*, a blind man recognizes the gait of his son's murderer nine years after the fact. In "The King of the World" on *The Sealed Book*, a crook takes an immortality serum that gives him superhuman hearing—ticking clocks, dripping faucets, and the beating of his own heart drive him mad.

Just as the cinema cherishes the scopophile, who mimics the film viewer, the compulsive listener has power because his or her activities synchronize with the experience of the auditor at home. Indeed, many shockers are explicitly about characters uncritically obeying a supposedly supernatural sound that they later learn to be issuing from a hidden radio in the scene—"Voice on the Wire" on *Inner Sanctum*, "Dark Waters" on *The Lux Radio Theater*, "The Body Wouldn't Stay in the Bay" on *The Whistler*. Fictional worlds also featured characters who behave like garrulous radios. Fibber McGee was famous for extended alliterative sentences delivered at breakneck speed, and no late episode of *Amos 'n' Andy* was complete without a visit to chatterbox lawyer Gabby Gibson. Dialogue itself seemed to speed up during the war, as airtime was occupied by characters defined by excess of speech, from a fast-talking Hollywood agent in Arch Oboler's "Mr. Ten Percent" to a quasi-dictator who forms a youth league in "Poison Peddler" on *Counterspy*. In "The Voice of Death" on *The Strange Dr. Weird*, a woman disposes of her husband by screaming at the pitch needed to bring an avalanche down upon him. Just as some plays showcase aural apprehension, others dramatize prodigious loquacity and vocal might, both marvelous and horrible.

Because pathological talkers and compulsive listeners are so prevalent in 1940s radio dramas, these works yield to an active speaker / passive listener distinction with an interesting ease. Indeed, the radio drama profits from Lazarsfeld's grounding question—"who says what to whom and with what effect?"—in a far more precise way than his own data. Considered in this way, dramas rehearse the fantasy behind the ORR paradigm. If a character is dominant, then it is his or her capacity to send persuasive transmissions that indicates as much; if a character is subordinate, then it is a persuasive message that he or she is supine before. Characters simply lack consequential interiority prior to persuasion. Walter Lippmann once wrote, "Under the impact of propaganda, the old constants of our thinking have become variables."[35] Absent an essential preceding self, with characters operating as transmitters and receivers, radio plays promulgated a sense that acts of communication shape infinitely malleable minds incapable of offering any resistance. Radio plays are fictions in which people are very much like communications apparatuses through which persuasive transmissions flow perfectly. In this way, radio storytelling was a testing facility for the theory, practice, and even the very possibility of aural coercion.

What forms did these "transmission dramas" take? Many follow the binary of the ORR paradigm exactly, featuring receivers who are highly vulnerable to external suggestion. Consider John Dickson Carr's version of "The Pit and the Pendulum" on *Suspense*, in which two voices are added into Edgar Allan Poe's original text. As he suffers in the pit, the narrating hero Jean hears his wife Berenice, who encourages him to resist his fears, and he later hears inquisitor Fra Antonio, a torturer who entreats him to recant. Each voice resembles a radio message because neither character inhabits scenic space. While echoes that reverberate from the stone walls of the pit color Jean's voice, Berenice and Antonio are muted, so we know that they are in Jean's mind, a point that is confirmed by dialogue. At times these voices seem like rival ad campaigns that the hero "tunes in and tunes out," until he is saved by the sound of soldiers accompanied by brassy trumpets. It is a sound we have heard before as an underscore to Berenice's exhortation, signaling which advertising campaign has carried the day. No rudimentary sequence of events is more common in 1940s radio than this scenario: a man liable to hyperacousia is assailed by a voice from beyond in the dark and must choose whether or not to obey.

Several dramas feature listeners who are psychologically "infantilized" by such nocturnal voices. One such case is Arch Oboler's "Kill," which tells the story of Professor Darryl Hall. Sitting in a courtroom at the outset of the play, the thoughts in his mind "seethe and swirl." Hall imagines that he hears the thoughts of the jurymen deliberating, repeating "guilty" and "not guilty" endlessly. This is the first in a series of repetitions of overheard speech in the play, each of which is taken in by the hero as if the voices were assaulting him. Action segues back in time, and we learn that Hall has lived his whole life without dreaming. Then one night, not long before his wedding day, the voice of a demon comes to him in slumber, murmuring one word over and over, "kill . . . kill . . . kill . . ." The demon's sharp teeth and fearsome lips mark her as a "carnivorous" broadcaster, and the events of the play literalize Horkheimer and Adorno's observation that in the culture industry a recommendation tends to become an order.[36] Hall's dreams soon persist in waking hours, until they drive him to murder his fiancée. Having rejected the prospect of a mature relationship, Hall learns that the only way to escape the demon is to die by hanging, cutting off speech to end the vertiginous chain of repetition. A similar scenario takes place in an adaptation of Guy de Maupassant's "The Horla" on *Mystery on the Air*. In the play, an unnamed protagonist (Peter Lorre) is plagued by a voice that comes into his room. Lorre is afflicted with "the fear that he is no longer master of his own actions, even his own soul." Just as disembodied radio voices were thought to influence listeners, Lorre comes to believe in an invisible being that "dominates" him: the Horla, a creature whose presence in space is illustrated by a theremin. Night after night, the Horla comes to the bedroom as Lorre regresses to a savage, childish state. By the end, Lorre's calm voice becomes a guttural growl that breaks the frame of the narrative, continuing to speak in character even after the play is over.[37] The tale of the Horla reflects the core notion of communication held by propagandists, showing a gradual overcoming of the mind by external forces from beyond. In the 1940s, versions of de Maupassant's tale proliferated, airing on *Weird Circle*, *Hermit's Cave*, *Inner Sanctum Mysteries*, and more.

"The Pit and the Pendulum," "Kill," and "The Horla" all model passive listening using protagonists dominated by nocturnal voices. At the same time, however, the idea of passive listening is also part of the purpose of these programs, which after all exist to convince us to invest in Victory Bonds, drink Roma Wine, take Ironized Yeast, and smoke Camel Cigarettes. It is tempting to suggest that the play "subverts" the ad, inasmuch as the former warns of the idea of influence on which the latter depends. But this reading does not fully explain the situation. While ads are mass suggestion aimed at mass audiences, plays are about individual suggestion aimed at individual listeners. A better way to understand transmission in dramas is to consider how the concept of persuasion behind advertising is implicitly refined and expanded when single voices "suggest" as effectively as ad campaigns. We need look no further than narration to find instances of this effect. In shock anthologies, it was common for program hosts to interject in the middle of scenes to speak directly as if to the subconscious of a protagonist, goading him or her from outside of the story to make the very decisions that will later lead to downfall inside the story. The narrator becomes a "voice in the night." In "Final Decree" on *The Whistler*, for instance, a man who has recently murdered his wife's lover himself suffers a series of near-fatal "accidents" as the couple vacation at a hunting lodge. After several tense scenes, the score performs a cutaway, and the host of the show speaks:

> The pattern is clear, isn't it Gordon? The struggle between you and your wife, Joyce.... She wants to kill you doesn't she, Gordon? Somewhere, somehow, someday, she'll kill you—if you don't kill her first! It's been touch and go for so long now, hasn't it? Between two would-be killers who appear to everyone else as a happily, comfortably married couple. But now another way occurs to you, Gordon. A frame-up against Joyce. Something to get her out of the way. If only you can make it seem as if she tried to kill you . . .

It is precisely by following the Whistler's suggestion that Gordon is undone. Sequences of this sort are one of the conventions that distinguish shock radio from thrillers in other media, as it is rare that a horror film's third-person narrator speaks directly to a hero, goading him into crime. But in 1940s radio, interjections of this kind occurred in many broadcasts of *The Strange Dr. Weird*, *Hermit's Cave*, *Inner Sanctum Mysteries*, *The Diary of Fate*, and *The Sealed Book*.

The Whistler breaches Gordon's unconscious from outside the narrative frame, but the content of dramas also frequently features irresistible transmitters. Arch Oboler's heroes exert mind control in "The Battle of the Magicians," while Carleton Morse depicts telepathy and mind transference in "Bury Your Dead, Arizona" and "Land of the Living Dead." In 1946, *The Columbia Workshop* celebrated its ten-year anniversary with Artie Shaw controlling children with his clarinet in "The Pied Piper of Hamelin." *Suspense* featured many dramas in which characters achieve the same suggestive power as telepathy or hypnosis, but without benefit of supernatural powers. In "The Angel of Death," for instance, an imprisoned murderer uses hints to give his cellmate suicidal ideas "by the simple power of suggestion," and in "The Doctor Prescribed Death," a psychiatrist sets out to prove that he can talk a suicidal person into murder instead. *The Molle Mystery Theater* followed suit. In "Talk Them to Death," a carnival hand plants suggestions in the minds of the irascible owners of his sideshow, thereby fomenting jealousy and pitting owners against one another. I think that all of these dramas certify John Durham Peters's hypothesis that the modern dream of communication—a person-to-person meeting of minds that would leave no remainder—is not only a utopia of language, but also a fantasy of wholesale thought-transference precipitated by the advent of the mass media era.[38]

Because wartime radio is often about transmitters bearing irresistible messages and receivers submitting to them, story arcs frequently trace the career of a single act of influence.[39] In a version of Ibsen's *An Enemy of the People* on *Studio One*, a doctor's message about the danger of unchecked public opinion is the framing device for a riveting hour-long narrative. On *The Mayor of the Town*, Lionel Barrymore delivers many stirring speeches on local radio about pitching in for the war effort, but as a local opinion leader he also personally persuades fathers to let sons enlist from his front porch, illustrating the interconnectedness of mass culture and word of mouth in a way that suggests that these two categories of persuasive speech are not usefully distinguishable. That idea is also raised in *Counterspy*'s "Murmured Millions," in which a con artist runs a protection racket against corporations. When the head of the Double Circle Consumer Products Corporation refuses to pay up, mobsters fan out in rail stations, subways, and other public areas, where they stage loud conversations about the inferiority of the corporation's products and the possibility of "some kinda metal poisoning." Rumors fly and sales plummet, thanks to this "network ad campaign." Similar events take place in "Labor Pirates" on *Mr. District Attorney*, in which agents provocateurs sew false rumors among war workers that their plant is going to close. In these plays, not only do receivers possess no critical screen, but individuals also exhibit the same power as demagogues, as the principle that empowers mass persuasion obtains in all intercommunication. It is perhaps not surprising that in the golden age of propaganda (and of fantasies of propaganda) listeners are depicted as bombarded, duped, and childlike; but what is curious is that other human beings—not huge media systems—are often the ones that bombard, dupe, and infantilize, most often by dint of personal charisma or "will" alone.

AN EVERYDAY CULT

It seems clear that many dramas affirm the active speaker/passive listener binary, but they also challenge the paradigm. As noted above, "Sorry, Wrong

Number" is in many ways a chronicle of the transformation of a character from one of these identities into the other, suggesting that neither is fixed. This move was relatively common. On one episode of *The Mayor of the Town*, for instance, a reporter comes home from the front with an orphan boy named Ronnie, who has lost his parents in a bombing raid. Muted by his trauma, Ronnie arrives in Springdale sullen, panicking at the sound of airplanes overhead. Only when the mayor sings the boy a lullaby at night does Ronnie find his voice again. The day after nocturnal listening, the orphan is gabbing away in a sign of psychological recovery. A similar receiver-to-transmitter switch takes place in Norman Corwin's "Gumpert," in which Charles Laughton portrays a haberdasher in Passaic, New Jersey, who hears a mysterious voice in the night. Gumpert is more than persuaded or dominated by the voice—it reprograms his persona, compelling him to become a series of historical figures. Told that he is Niccolo Paganini, Gumpert begins to play the violin so masterfully that he is invited to Carnegie Hall; told that he is Julius Caesar, Gumpert begins to organize a coup d'état among workers in a city water department; told that he is Sigmund Freud, Gumpert begins to analyze the doctors called in to treat him. The tale caricatures the trope of the receiver who lacks defense from external suggestion, but it does so with a twist. At the outset, Mrs. Gumpert (Elsa Lanchester) characterizes her husband as a poor talker, citing his retiring nature, lackluster marriage proposal, and poor singing voice. "He was so timid," she testifies, "that he would practically faint if he had to get up before a group of salesmen." Laughton's halting delivery confirms this observation. Yet after each nocturnal session, Gumpert sheds his stutter and assumes an exaggerated, commanding tone. It is no accident that Corwin chose musicians, orators, and generals for Gumpert's assumed personalities, as all are masters of irresistible will. Only by becoming a receiver can Gumpert become a transmitter.

Other plays reverse that transformation. Consider *The Ford Theater*'s "It's a Gift." This comedy follows a war veteran named Grover, who can influence the disposition of others around him thanks to a magic piece of shrapnel lodged in his ear. Grover tries to capitalize on his talent, pairing with other transmitters to launch quasi-PR campaigns: with a street barker, Grover skews a betting racket; with an ad executive, he convinces consumers to buy "Grubbies" cereal; with a political boss, he helps to nominate a stooge candidate. Alas, each of these swindles fails because Grover cannot control his influence. After he falls out with his sweetheart, news reports describe an "epidemic of crying" up and down Broadway. In the end, Grover has the shrapnel removed and gets the girl, but in a comic twist, he learns that he has developed superhuman hearing. Other characters suffer as they switch from transmitter to receiver. A disastrous fate befalls the narrator of "Anonymous" on *Quiet, Please*. After giving a political speech, a transmitting politician takes calls from the audience. When one caller tells him to "drop dead," the words haunt him for days, and he begins to behave like a pathological receiver. At work, he obsesses over incoming messages and compulsively seeks advice. At home, he eavesdrops on his wife's conversations. Awake late at night, he fixates on the sound of a dripping faucet, which seems to repeat "drop . . . dead . . . drop . . . dead . . ." He also finds that he has lost his ability to transmit successfully. Searching for the woman who made the initial call, he dials numbers randomly, never connecting. The politician soon dies, victimized by the loss of meaning that accompanies the loss of microphone, just like Mrs. Stevenson in "Sorry, Wrong Number."

Some of the most complex dramas of transmission involve characters attempting to be transmitter and receiver all at once. This scenario plays out in Robert Newman's version of "The Tell-Tale Heart" on *Inner Sanctum*. In the opening, we meet Simon (Boris Karloff), who possesses supernatural hearing. He brags, "I can hear the grass growing, the sap moving in the trees, I can hear the stars moving in their courses." This receiver meets Oliver, a man with a highly developed power of sight. As the two travel the countryside together, Simon discovers Oliver's perversity when the latter kills an innocent swallow. Later, as the two turn in for the night at an old windmill, Simon takes it upon himself to kill Oliver in his sleep. The moment that Simon makes this decision, the mechanics of the drama suddenly change. Karloff begins speaking in direct address to the audience, a shift that is registered by a new tone in his delivery and the use of past tense. "He wasn't blind, I was! Blind to him!" Simon growls, as if reporting events retrospectively to some hitherto unaddressed third party. Simon the receiver has become Simon the transmitter, and he narrates the gory murder to us in a dissociated way, as if he is unattached to his own acts. When officials arrive to investigate later on, Simon's direct address ceases, as the pounding of the telltale heart beneath the floorboards transforms him back into a recipient of sound. In the end, we learn that Simon is in fact a deaf mental patient, and that his powers of hearing are a figment of his imagination.

In these fantasies about perceptive sensitivity gone berserk, characters may suffer for attempting to switch from transmitter to receiver, and they face danger as they attempt to be both at once. But because one's node in a transmission circuit defines one's psychic configuration, the worst situation of all is to refuse to transmit or receive anything at all. Such a refusal occurs in *Suspense*'s adaptation of Evelyn Waugh's story "The Man Who Liked Dickens." In the play, Richard Ney plays a jilted divorcée who is sick of both hearing and speaking to "the gossips." "I either bored them, or they bored me," he remarks, "one of the rare moments when the human equation is perfectly balanced." On impulse, Ney enlists in a journey on the Orinoco with a friend named (appropriately) Dr. Messenger. The expedition goes awry as its members flee or fall victim to misadventure and disease. Recovering from fever, our protagonist finds himself in the hands of a Charonesque recluse named (again, appropriately) Mr. Tod, who nurses him back to health. Afterward, Tod makes a seemingly innocent request. An illiterate with "a soft way of speaking," Tod asks his guest to read aloud from worm-eaten copies of Dickens left by his deceased father. Tod is among the most compulsive listeners in radio fiction; his tones become erotic as he describes Dickens's verbose prose: "so many characters and changes of scene and so many . . . so many *words*."

At first, Tod promises to help Ney return home, but it soon becomes clear that he has no intention of fulfilling the promise. Whenever Ney brings up the subject, a dog begins to bark in the aural background as if to underscore the point. Like Adorno's child demanding the same dish again and again, Tod insists that the reading must continue, a process that Ney describes as "the subtlest of tyrannies, the imprisonment of another man's mind." It takes Ney years to finish Dickens's oeuvre, after which Tod ominously says, "It's delightful to start again, each time I find more to enjoy and admire." At last, Ney manages to get a message to a search party, and then he begins to throw Tod's books into the fire. Though free from Tod's mind, Ney is transformed by his "Dickensian" encounter and begins to compulsively receive and transmit at once. In the finale, Ney reports to us the sound of the tree frogs, falling wood, and birds as if he has become hypersensitive to them, and as the burning pages of Tod's books rise into the air, words stream from his lips as if automatically. In the beginning of the drama, the human equation was balanced by mutual boredom, but by the end of the play an excess of hearing and an excess of talk are expressed by the most balanced and excessive of Dickens's many words, which Ney declaims into the jungle air: "It was the best of times, it was the worst of times, it was the age of wisdom, it was the age of foolishness, it was the epoch of belief, it was the epoch of incredulity . . ." Ney's character has been reintroduced to the world and to the circuitry of send-and-receive that defines it.

In the dramas above, the states of transmission and reception drive action, echoing Lazarsfeld's binary by showcasing characters whose internal dynamics are expressed as a drive to listen or to speak. Plays detail these drives as if they are all that is needed to frame story arcs: a man murders his fiancée because he is a listener and is repeatedly told to kill; a woman who talks incessantly is undone by overhearing voices; narrators coax characters toward misfortune; a superhuman listener alters persona as he begins to speak in direct address. These outcomes are difficult to comprehend unless they are considered in light of the fact that in this context "passive receiver" and "active transmitter" fit into a conventional fable subscribed to by dramatists and audiences, in which the key to any mind is its schematic relationship to a message, whether that message be a voice in the night, a conversation with the coercive force of propaganda, or a palliative sound event through which one transforms from receiver to transmitter.

The last might be explained as "false consciousness" in the Marxian sense.[40] Such a reading is easy to rehearse. As capitalist radio networks monopolized airwaves and disenfranchised the people, the argument goes, dramatists amplified compensatory fantasies in which ordinary people become "broadcasters." By airing such plays, writers invert actual relations and reinforce the status quo.[41] Audience members are like Gumpert, who *seems* to have a transmitter's stentorian voice, but only because it is given to him by godlike powers, before which his very being is in fact—and at regular intervals—enslaved. According to this view, transmission tales exist to distract listeners from the fact that they are targets of media campaigns, atoms commodified into publics for sale. Indeed, there seems no true difference between influencing one or many, between the fungibility of the individual and that of the mass.

Yet the false consciousness hypothesis neglects so much about these plays, including the interesting danger with which transitions of identity are often invested. It also pays scant attention to the more elementary move at hand. Before dramas *can* invert the passive listener into an active speaker, these subjectivities must first be made available, and the dramatic world is both defined by irresistible streams of information and divided into only two psychic situations. In the plays described above, characters are either like sirens of antiquity with irresistible voices or like mariners tormented by high thrilling song; no third position seems available.[42] Communication is so effective that it is indistinguishable from force. Only because the play is constructed in this way does danger erupt when a figure ceases to receive and begins to transmit (or vice versa), as these switches complicate or enfeeble a transmission circuit, the basic convention of the conceptual universe imagined on radio for wartime. Transmission dramas therefore do not merely work to reproduce a power relationship but reframe that relationship as in the first instance communicative in nature by creating an account of the psyche that emphasizes its ceaseless saturation in mediated experience. Drawing on the work of Jacques Derrida, theorist Briankle Chang has written of a "postal principle" when it comes to thinking about intercommunication, a law in which all messages presuppose clarity and deliverability from a unified sender to an equally unified receiver through an eminently reliable telesystem linking person to person.[43] It seems that such a system of utopian equivalence thrived in transmission dramas built around the models of 1940s communication theory; using it, media researchers and radio dramatists collaborated to turn the everyday act of verbal exchange into a kind of cult. But dogmas of power, unity, and deliverability did not stand unimpeached in wartime radio either. In the next chapter, I will look at ways in which the law of communication frayed at the furthest edge of its circuit, when it reached outward toward the listener at home and passed from plays of commentary to those of metacommentary.

CHAPTER 7

Eavesdropper, Ventriloquist, Signalman

Throughout part 2, I have explored two proclivities in wartime radio plays that set them apart from a great deal of preceding broadcasting. First, programs aired psychological "signal-based" dramas in which stylized sounds not only spell out events but also harass the perceptive apparatus of a character and ultimately collapse response into stimulus. Second, these plays frequently feature characters that take on "active transmitter" or "passive receiver" roles, and so by listening with Paul Lazarsfeld's question in mind—who says what to whom?—it becomes clear that momentous acts of aural transmission provide characters quiddity and explain their behavior across a dramatic arc. Both of these conventions employ communications to put interiority on exhibit in scenarios in which it is difficult to fathom the decisions that lead to a pattern of action without coordinating them with some kind of sonorous exchange that is also occurring in the story. That schema mirrors the situation of the listener at home, so it is clear that these dramas intuit implicit beliefs about broadcasting that were shared by dramatists and audiences alike during the years in which radio listening was most prominent in American life. For instance, several plays discussed above presume or even substantiate models of coercive persuasion, in which individual skepticism is implausible and any code of belief seems to be little more than the sum of external appeals that have accumulated over time. It is easy to see why such a trope might appear within the framework of the commercial mass media model, which is always persuading us to do one thing or another. Yet the visions of the absolutely impressionable human are seldom dramatized without disquiet. In these plays, it is dangerously theurgic to take on a role defined by a coercive interchange, and there is always something dismaying about "talk" that becomes "signal." In wartime radio, before an act of mediation is virtue or vice, before it is weak or strong, it is in the first instance profound—too profound. Throughout the 1940s, many plays seem to "feel their way through" the expanding plenum of mass communication as if it were a dangerous but compelling enigma.

This chapter completes my study of that unusual habit with readings of plays concerning three archetypal characters that each pull the aesthetic framework of wartime radio until it begins to fray: the ventriloquist, the eavesdropper, and the signalman. The first two archetypes challenge theories discussed in chapter 6, in part because eavesdropping requires an "active" receiver while ventriloquism requires a "passive" transmitter. These characteristics create situations in which the unified identities of sender and receiver become implausible. Acts of communication are represented as charades, and the capacity of mediation to define interiority seems dubious. Eavesdropping and ventriloquism are confessions of the inadequacy of the whole notion of using acts of communication as components for structuring a theater of the mind, as transmissions reveal their incapacity to guarantee distinction between one consciousness and another. Signalman tales pose a similar challenge but from another direction. By amalgamating with an information system, signalmen do not merely interact with media, they act *as* media, and their experiences reveal the joist of mind and medium to be an interstice lacking reliable boundary and thus always managing existential jeopardy. Of course, I group these three archetypes together in the belief that they are three faces of the same phenomenon. In the plays, each of these figures exists slightly outside itself, providing a way for the dramatic conventions discussed in chapters 5 and 6 to become self-reflexive. Ventriloquists, eavesdroppers, and signalmen also stand in for (and fantasize about) the unobtainable object of radio aesthetics: the listener at home. The black magic of these plays is to unravel our own listening, and in doing so the radio play reached limit-cases that frustrate 1940s style, putting its assumptions under new strains that would fascinate dramatists into the 1950s, at which point their work would have less of a role in substantiating models of inner life and more of a role in pushing any such model to the point of deterioration.

THE MISSING HIMMLER

Eavesdropping sequences are not unique to radio narratives, but they have several special attractions in the genre. Characters typically eavesdrop on dialogue, an element on which radio relies heavily, and it is often an illicit act coinciding with a moment of exchange.[1] Eavesdropping has the trappings of sense-specificity. As filmmaker Dziga Vertov once remarked, "The ear does not spy, the eye does not eavesdrop."[2] Writer Carl Van Doren thought that overhearing showed radio's intrinsic strengths. "No matter how clearly we see what is happening out of earshot," Van Doren wrote, "we can seldom help wishing we could hear what is being said; whereas when we are eavesdropping we can be interested without particularly caring whether or not we can see."[3] The remark highlights the mischief of the practice. One listens *to* music, but one listens *in on* conversation, and that sense of trespass both mimics our own synchronous overhearing and adds a sense of transgression to it, as if the radio is winking at our own inordinate focus upon it. Like eavesdroppers, radio listeners tune into speakers who evince no awareness that they are being overheard, which is precisely what makes listening to a radio play different from listening to a music program or speech. In the 1940s, overhearing may have had particular appeal because it challenged the conventions of transmission-based plays. I have argued that in wartime "receiver characters" often become "transmitter characters" and vice versa, but the messages themselves seldom come apart or deviate from their appointed courses. One can think of these narratives as a simple circuit in which signals proceed with dispatch from a legitimate sender to a designated receiver and back again without diversion or distraction, which is why these stories have a fatalistic quality. The sound of the Horla overcomes all physical and psychic barriers as certainly as fluid in a syringe enters a vein, to borrow a simile frequently used to describe propaganda.[4] John Durham Peters has argued that such a model of communication is a veiled modern longing for an unreachable utopia where intervening media assure that nothing is ever inhibited, concealed, or misunderstood. When modern people use media to achieve "good communication," Peters explains, they seek a version of interpersonal dialogue that produces a degree of understanding that is tantamount to "mutual ensoulment."[5] But that promise never really matches experience, as mediated discourse is more often marked by dissemination, scatter, and breakdown. These problems take center stage in eavesdropping stories owing to a malevolent presence precipitating that breakdown, a third being in the listening situation whose hidden perception suggests that messages are clumsily handled and liable to go astray. By dint of the chance of diversion, eavesdropping destabilizes the circuitry of the drama, transmissions veer off course, new parties appear, and informational asymmetries develop, often becoming the new basis of action. Thanks to an act of aggressive listening, these plays emphasize the tendency of communication

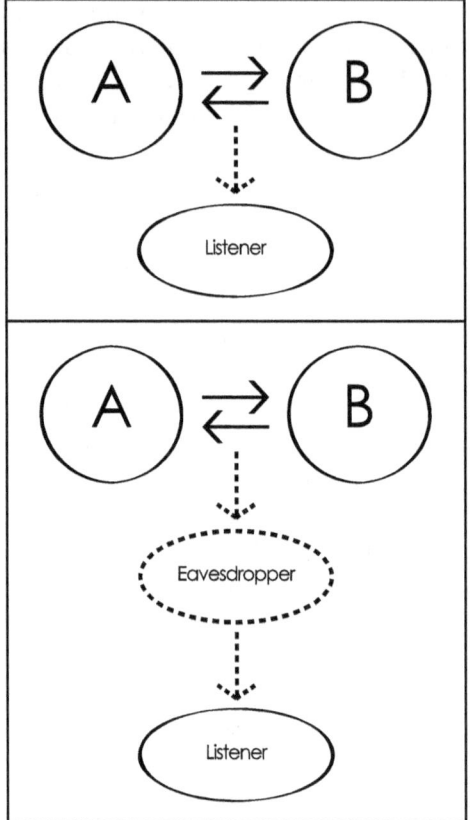

FIGURE 7.1. Radio listening as eavesdropping and as eavesdropping *on* eavesdropping.

toward an uncontrollable dispersion that even challenges our own locus as tacitly approved auditors of speech acts provided by the diegesis.

In the 1940s, these tales of "diverted messages" commonly took the form of espionage tales, most of which involved messages that came perilously close to antagonists while passing through hidden circuits that seemed to be everywhere, all of a sudden. In the months following Pearl Harbor, broadcasters aired scores of tales of underground Nazi networks and Japanese "Black Dragons" out to steal war production information on such programs as *Superman, Dick Tracy, The Lux Radio Theater, The Strange Dr. Weird,* and *David Harding, Counterspy*. As 1942 continued, *Words at War* adapted atrocity reports "smuggled out" of occupied Norway, France, Holland, China, and Yugoslavia; Hop Harrigan stole the blueprints of a cannon to be used on the new "Messerschmitt 110" fighter; Nazis hid directions to a bomber plant in a wax museum exhibit in "Menace in Wax" on *Suspense*; an antifascist gave the formula for a new synthetic to an American spy by humming notes that corresponded to chemical elements in "The Man Called X" on *The Globe Theater*; and on *The Mayor of the Town*, idyllic Springdale was infiltrated by a spy ring attempting to steal a cane containing evidence about a German concentration camp. The career of a piece of information motivates these dramas, each of which portrays an unbounded circuit of exchange. These plays are not about a single forceful phrase full of replete meaning directed at a person alone in the dark (the primal scene of a transmission drama), but about complex reports, blueprints, and communiqués circulating along shifting polygonal courses. We rarely meet the originator or final recipient of the transmission. Instead, the drama usually takes place at a bottleneck in the career of a message, when control over it is temporarily fumbled. And seldom can the existence of that message help us to ascertain identity or read motivation in the way they can for characters in other situations. Quite the opposite. Parables of lost information *always* contain characters of uncertain identity. Spy stories are full of characters whose core allegiance is as malleable as the trajectory of the messages that they convey. That shared indeterminacy of both agent and object of action is an underpinning of the genre.

To use eavesdropping to depict such diversions, radio dramatists faced a challenge. While eavesdropping directs our attention at speech, that speech is the secondary business occurring in the scene. Eavesdropping is not *dialogue*, but *action*—the activity of listening. Like all action in radio, overhearing requires something sonorous to indicate it. Low oscillations are used in "The Voice Machine," a 1941 *Superman* adventure in which the hero has a device that can listen in on any conversation in time or space. Such contrivances are hardly the norm. Usually, dramatists indicate eavesdropping with narration prefatory to the act. In *The Diary of Fate*, for instance, an announcer often explains that a character is listening in on the bookie next door or a businessman in the next airplane seat, a situation that inevitably leads to turpitude. In *Chandu, the Magician*, cliffhangers often follow secret conversations between the heroes, after which the announcer reveals that operatives from the Brothers of Jeopardy have been lurking around the corner all along. Where appropriate, dialogue indicates overhearing. In "Unknown Source" on *Mr. District Attorney*, a colloquy at the outset indicates that a moll will work her way into a district attorney's office and pass on secrets to criminals later on. Dialogue is also used to highlight eavesdropping after the fact. Consider the following passage from an episode of *Captain Midnight*, in which villain Ivan Shark and his henchman Gotto run into a room from which they have just overheard

Midnight and his protégé Chuck. Unaware that Midnight has escaped, Gotto draws his gun:

SHARK: Fire, Gotto, fire!
GOTTO: There ain't nothing... There ain't nothing to fire at, chief.
SHARK: What do you mean? I heard Captain Midnight's voice, Gotto, I heard him come into this room. Don't tell me there's no one to fire at, Gotto.
GOTTO: Yeah, yeah, so did I, chief. Why, I heard that door open, and I heard them walk into the room.
SHARK: And I heard him talking to that kid, Chuck Ramsay, in the dark just a few feet away from me.
GOTTO: That's for sure. So did I. Why, they were over there by the main door, looking at the light switch, but there's nobody to shoot at now.
SHARK: I'm positive they were there.
GOTTO: This ain't right, chief. There's something spooky about it. Maybe we didn't hear them at all, maybe it was parrots we heard.

Befuddled dialogue continues, reiterating the facts once again. The delay is all the better for Midnight, who has by now made good his escape.

Captain Midnight's approach comes across as heavy-handed. Speech informs us of the presence of the eavesdroppers, but it does not let us sense eavesdropping from an audioposition shared with one engaged in the act. In other words, in *Diary of Fate* and *Captain Midnight*, we know that eavesdropping is under way, but we are not listening to the act of listening, undertaking the "metalistening" that makes eavesdropping such a savvy commentary on the medium. To create a sense of our participation in the eavesdropping, dramas relate overhearing by coloring overheard sound differently from other vocals. For instance, in "The Walkie-Talkie Stickup" on *This Is Your FBI*, we are aligned with a schoolboy who overhears peers committing robbery through a walkie-talkie, a mediating device signified by removing bass from sounds that are conveyed through it using a high-pass filter.[6] Sound coloration of this kind was used to signify the use of radios, telephones, air-to-ground systems, and wires stretching from bathyspheres to the surface. By the same principle, low band-pass filters were used to attenuate high frequencies, and the effect suggests that tangible materials in the scene are hampering clarity—walls, gags, doors, floorboards. Radio writer and historian Erik Barnouw called the process "shunting," pointing out that it could signify any kind of intervening medium, thereby evoking a scene in which bodies are divided by barriers or mediating devices while sound is not.[7] Shunted voices could both construct scenes in the mind of the auditor and dismantle them in a single gesture. In "Death Bound" on *Sanctum*, a thug overhears his lover's muffled voice as she telephones his rival from the next room; in "Jealousy" on *The Whistler*, a woman overhears her dastardly sister downstairs, plotting to switch her medications; in "The Man Who Talked" on *Calling All Cars*, the FBI eavesdrops on a suspect through a wall shared with another apartment. In each case, the scene of eavesdropping is always outgrowing its boundaries because characters make erroneous estimates about how far their voices carry. That error encourages us to imagine a room, house, or building, but it also indicates clearly that as a matter of audioposition, a room is actually a floor, a floor is actually a house, an apartment is actually a building—and in the porous world of the pulp radio play, nosey landladies, cuckolded husbands, and unscrupulous reporters are always around the corner, awaiting a slip of the lip. In the 1930s, dramatists had joked that radio used footsteps to signify space so often that no room seemed to be carpeted anymore. By the 1940s, the carpeting had returned to the décor, but walls seemed stripped of any insulation whatever.

While speech and shunting were available to dramatists, the tendency was often to favor the former, particularly in plays tasked with encouraging the public to keep a lid on war secrets as part of the "Don't Talk" campaign promoted by the Office of War Information, alongside the well-known poster series. The approach reflected the fact that in war plays, eavesdroppers tend to be located very far away. When loose lips sink ships, it is because a voice carries not from one room to another, but from one continent to another, a point also brought home in many of the posters. An example of a typical "Don't Talk" radio play is "Death Talks Out of Turn" on *The Molle Mystery Theater*. At the outset of the show, the program host didactically reminds us of the disproportionate way our own voices can amplify in the context of global war, explaining, "One American war secret means a thousand dead Americans." In the drama, we meet narrator Lt. Andy Blake, stationed in Washington, DC, as a staff officer. Blake discovers that his plucky spouse Arabella has come to the capital on her own, to seek clerical war work out of patriotic zeal. As soon as she arrives, Arabella overhears a conversation on the bus about a rooming house at which sixteen women live, each of whom works at one war office or another all around DC. Suspicious, Arabella takes a room at the house, and Blake eavesdrops on her from the street to be sure that she is safe. A scream sounds out in the night, and Blake tangles on the lawn with the proprietors of the rooming house, Mrs. Fielding and Mr. Jones. Later that night, the lieutenant breaks into the house and listens in the dark for any sign of mischief; then he hears footsteps descend the stairs, indicating that Jones has been eavesdropping on him. Just as Blake

FIGURE 7.2. Herbert Morton Stoops, "Careless Talk Got There First" (1944). A poster in the "Don't Talk" series launched by the Office of War Information. In these posters, GIs usually fall victim to "talk" in moments of passive vulnerability. Image courtesy of Northwestern University Library.

is about to be caught, he discovers that Arabella herself has been listening in all along, and she intervenes to knock Jones unconscious. The couple flee the scene, and Arabella explains that she had intentionally let Jones eavesdrop upon her earlier that day, in order to mislead him into believing that she had recorded important war information in her diary, just to see his reaction. By the time that the couple reach Blake's apartment across town, there are shadowy figures outside eavesdropping on them. Arabella takes up her pistol and runs

FIGURE 7.3. Eric Ericson, "The Sound That Kills: Don't Murder Men with Idle Words" (1942) OWI posters often contain reminders that sound is dispersive in nature. There are no such things as "idle" words in this poster, as "the end of a rumor" proves that speech is not indolent at all. Also note that the recipient of the intelligence in Germany is absent, as if there is a cell missing in the cartoon that ought to show how information got from the subversive agent to the U-boat captain. Image courtesy of Northwestern University Library.

into the street, while we remain with Blake inside as he overhears shots from beyond the door.

Of the eight acts of eavesdropping that have occurred in this first segment of the play, only the last allows us to experience it through shunting. In every other case, we rely on narration or dialogue. After a cutaway, we learn the substance of the plot: Jones and Fielding have stitched together the idle chat of secretaries (one office is teaching staff Arabic, another preparing narrow-gauge rail, etc.) to discover that the Allies are preparing to land in North Africa. The agents plot to communicate this plan to SS Chief Heinrich Himmler by way of a spy submarine offshore. In this scenario, total war has radically transformed how far voices may disperse in space. Shocker characters discover that a room is a whole floor, something that we can perceive alongside them, but the secretaries in "Death Talks" make an error of another order of magnitude, as the context of their home falls into the earshot of a global enemy. And unlike eavesdroppers represented aurally, this global enemy is not present in the drama. The "receiver" in this scenario is Himmler, an imperceptible absentee whose status in the drama itself is a projection. Eavesdropping scenes are ghost stories of another kind, turning on a being situated at the other end of a geodesic, as far from the depicted scene as we are. This practice renders the eavesdropper unknowable, a boogieman whose existence can be neither proven nor disproven. The eavesdropping fable is naturally political because in it our own senses cannot be trusted to identify the reach of our vocalizations, and only authorities in the form of narrators know the true audience of our speech acts. This disempowering idea would be exploited in coming years, as tales of diverted messages became the armature of Cold War programming in the 1950s. In plays that succeeded "Death Talks Out of Turn," the missing Himmler would be replaced by "Uncle Joe" Stalin, J. Edgar Hoover, and other faraway panauditory presences in the echo chamber of Red Scare paranoia. Over time, eavesdropping went from being a commentary on the folly of "mutual ensoulment" to become what theorist Jacques Attali once called it—a technology of power.[8]

During the war, however, this neurotic account of overhearing opened up vistas for conceptual play. Consider "Secret Agent 23" on *Ceiling Unlimited*. As the play begins, we hear a Nazi secret agent speaking in direct address from inside a US bomber plant:

> I hope you are hearing this. This is Secret Agent 23. Am inside American war plant, one of America's greatest aircraft manufacturers, the Vega plant. Am as a worker disguised. This secret broadcast is made through concealed

microphone worn on lapel, cleverly concealed behind Vega workers identification badge.

The premise is that we share a node in a stream of information alongside Agent 23's contact on the other end of a closed two-way radio. Promising "facts and data of American production of aircraft," the agent passes by coworkers speaking in "untranslatable" catchphrases like "That cat's groovy as a two-cent movie." Soon, Orson Welles leads a group of G-men to seize the intruder. Welles takes the microphone:

> Hello, hello Hauptstation, hello Berlin, hello Tokyo, hello Axis. Your Secret Operative 23 didn't know it but his secret broadcast had a coast-to-coast hookup. CBS to be exact, on the Lockheed and Vega program. Don't worry; there are no spies at Vega, we see to that. For your information it was all arranged. We thought our listeners might be amused. Hello Americans . . .

Agent 23's "facts and data" had evidently been astray from the get-go, and we discover that we are not truly in the Hauptstation, only addressed at the same time as the Nazis. Gloating, Welles shows off the work of industrious free people for all to hear: an Italian-American sculptor working in the plastics department; a woman at the drill press, who says she has been "paying back in rivets, double, for every bullet they put into my Joe." The rest of the drama continues to take the form of a broadcast, with one end of the information circuit missing from the drama—there is no "other side of the wall" depicted—so information is diverted to a receiver who is entirely hypothetical. It is another case of the missing Himmler, a structuring absence whose position in a stream of reception is attested to only by authority figures. If the eavesdropper in a story is our surrogate, inasmuch as his activity is identical to our own, "Secret Agent 23" not only aligns us with our adversary but also leaves him strangely imperceptible. We face the extraordinary prospect of overhearing alongside ears that might not be listening, or even exist. Of course, one presumes that the real Himmler was not listening, yet potential listening remains the center of the exercise, the fear it is intended to underline. In an age in which the act of listening had the power to fix characteristics and concretize relationships, eavesdropping was an extreme dramatic move because acts of overhearing and fantasies of metalistening lead to a third process that undermines both. The real goal of "Secret Agent 23" is to produce a pure impression of a second inaudible presence, reception without receiver.

THE WORDS OF OTHERS

The unusual feel of eavesdropping plays was complemented by contemporaneous dramas that had a metalevel of another sort, ventriloquist plays featuring characters who challenge communicative speech with its own simulation. Voice-throwers came to network broadcasting during an amateur variety show craze in the early 1930s.[9] By wartime, ventriloquist Edgar Bergen and his dummy Charlie McCarthy hosted a popular Sunday NBC program, while Dick Tracy, Charlie Chan, Captain Midnight, and the Shadow routinely threw their voices to stymie pursuers. Ventriloquists are often suspects in mystery plays such as "Jane Arnold" on *Broadway Is My Beat*, "Dead of Night" on *Out of This World*, "The Marvelous Barastro" on *Mystery in the Air*, and others. Ventriloquist plays typically use two actors whose voices have differing weight and timbre to play ventriloquist and "dummy" until the climax, at which point these two voices unify, a maneuver that one would expect to unify a dispersed identity. Yet this unification rarely gives us clarifying access to motives. Instead, ventriloquist dramas usually end in tableaux of catatonia. In Ray Bradbury's "Ria Bouchinska" on *Suspense*, for instance, a ventriloquist named Fabian explains his murder to a group of suspects through the voice of his beautiful dummy, who is presented to us as another character entirely. Fabian explains that the doll is not in his throat, but in his heart:

> Sometimes I'm powerless. Sometimes she is only herself, nothing of me at all. Sometimes she tells me what to do and I must do it. She watches over me, reprimands me, is honest where I am dishonest, ethical where I am wicked as old sin. She lives her life, I live mine. She's raised a wall in my head between herself and me and she lives there.

The usual position of a mediating apparatus—its in-betweenness—has become perverted in this scenario. The doll fails to mediate the ventriloquist to his audience and has colonized him instead.

Our own experience of the dialogue is similarly distorted. We listen to a set of secondary characters train their attention on a simulated oral fissure that we ourselves cannot perceive, since we have no choice but to listen to the ventriloquist behind it and use that voice to imagine what the doll might look like. Characters stare at a dummy to localize a voice, while we "stare" at a voice to imagine a dummy. The confession concludes with the doll explaining that she "overheard" Fabian committing the murder and cannot live with that knowledge, thereafter undergoing a kind of death. At the end of the confession, the ventriloquist seems to run out of words and finds himself working the aphonic jaw of his doll, trying "make it sound again," until he realizes that her persona has been sequestered "behind the dark wall" of his mind forever as a segment of his own identity withers away. As a message drifts from ventriloquist to dummy, neither being is discrete. The ventriloquist whose dummy goes mute is one whose full interior recoils beyond our reach, becoming as unknowable as the invisible ears of faraway eavesdroppers. In transmission plays, a message originates in one mind and terminates in another. Ventriloquism throws the reliability of that exercise into question as the unity of transmitter disaggregates and communication cements more barriers than it removes.

Is the radio itself a kind of ventriloquist's dummy? Has our mind been colonized by the medium rather than extended by it? Do we stare at this device with the same idolatrous fixation as the characters in "Ria Bouchinska"? There is a tradition of theorists arriving at such a conclusion.[10] The radio is like a mask, at once concealing a body but liberating a voice. Pierre Schaeffer associates radio listening with the pedagogy of the philosopher Pythagoras, who was known to speak to his pupils from behind a curtain, as a pure "acousmatic presence" whose words became more alive because they seemed to come from nowhere.[11] It is fitting that the philosopher who founded music theory is the center of this primordial scene of radio; like the Shadow, Pythagoras was rumored to have traveled to the mysterious East to learn secret powers that included superhuman hearing and the ability to be in two places at once.[12] Theorist Michel Chion has written of films that feature Pythagorean "acousmêtres," hidden beings who seem to possess powers of ubiquity and omnipotence until the moment of "de-acousmatization."[13] In Chion's view, broadcast radio is an acousmatic medium par excellence because it does not allow for such an unveiling. Radio voices can never remove their masks, so the magic of the medium is also its trap. Perhaps the unavoidable ventriloquial quality of radio is rooted less in the properties of the medium and more in how it has been socially shaped to allow commercial exploitation. Acoustic theorist Barry Truax has suggested that consumerism alters the very nature of sound.[14] With spokespeople for shaving cream, bootblack, or dentifrice waiting at every intermission, radio plays *exist* to convey a set of "secondary" messages, so any broadcast is not quite what it announces itself to be. A sense of ulterior motivation thus pervades the work, becoming obvious at regular intervals whenever Ozzie and Harriet brag about the sturdiness of their dinnerware or Jack Benny contorts a joke so as to include a superfluous mention of Jell-o.[15] While we traditionally associate the source of a sound with the self of a communicator, modern audio products are in fact merely the "outer voice" of hidden interests. The acousmêtre is not the Shadow at all, but the Blue Coal Company sponsoring *The Shadow* in order to promote the sale of Pennsylvania anthracite.[16]

On the other hand, there is also an explanation for the ventriloquial quality of radio plays that has nothing to do with either medium-specificity or how the medium was socially shaped: ventriloquism underlies all forms of speech and is only especially manifest in this speech-based genre. Whenever we speak, our words are riddled with phrases that originated elsewhere, suggesting that any message implies a plurality of messengers. Critic Mikhail Bakhtin made the point some time ago: "The transmission and assessment of the speech of others, the discourse of another, is one of the most widespread and fundamental topics of human speech. In all areas of life and ideological activity, our speech is filled to overflowing with other people's words, which are transmitted with highly varied degrees of accuracy and impartiality."[17] Bakhtin is interested in how externally produced discourses—what "people say"—turn into "one's own word." The degree of this personalization varies enormously, and ventriloquism is less a variety of performance and more a consequence of the innate heteroglossia of modern languages and a measure of the extent to which the words of any given speaker are a repetition not yet claimed with a signature.[18]

This multiplicity of origination is a useful insight for unpacking radio dramas, which often seem obsessed with exploring how it works. Adaptations like "The Count of Monte Cristo" on *Mercury* and "The Black Curtain" on *Suspense* contain scenes in which proxies vocalize on the behalf of mute characters, who blink their eyes or tap messages in code. Proxies are also featured in plays about ventriloquial expressions of affection appropriated for radio, such as "Cyrano de Bergerac" on *The Theater Guild on the Air*. *The Lux Radio Theater* adapted several stories in which composers "speak through" singers, including "Naughty Marietta" and "Break of Hearts." In the winter of 1948, *The First Nighter* broadcast a cycle of plays in which stooges become "mouthpieces" of gangsters and women writers "speak through" men to reach an audience—"Help Wanted, Female," "Love is Stranger than Fiction," "A Writer in the Family." In a version of John Hersey's *A Bell for Adano* (1944) on *Words at War*, action is not punctuated by the officers charged with pacifying a conquered Italian town, but by the town barker, a go-between who relays commands from officials to the locals. A year later, "Rainbow" focused on another intermediary, a Russian who works as the spokesperson for the Nazis in his occupied town. Each of these plays depicts a relationship that corresponds to the schematic of ventriloquist and dummy, yet we are never sure of the degree to which a "dummy" personalizes a message, and this uncertainty can be tasked to many dramatic purposes. Just as explicit ventriloquist stories show a mind that is internally plural, ventriloquial relationships between characters rupture the integrity of singular identity by multiplying the authorship of any statement, slogan, edict, or command. That puts into disarray any simplistic model of the connection between uninhibited mass communication and any unified mind.

It also raises quandaries of where identity ultimately inheres. Consider another of Ray Bradbury's radio scripts, "Killer, Come Back to Me," which aired on *Molle* in 1944. In the play, a crook named Johnnie robs a bank in Colorado, in the process meeting Julie, the lover of a deceased gangster named Ricky Wolf from Central City. Julie takes Johnnie under her wing and sets out to enable him to assume Wolf's identity and take over criminal rackets. A ventriloquial relationship is established as Julie rehearses Johnnie in how to deliver the command, "Okay, you guys, reach! This is a stick up." Julie tells Johnnie who to meet and what to say, constantly reminding him that both in crime and in love she is in charge: "I'm giving all the orders. I'm the one who says when." The play contains a series of set pieces that highlight the symbiosis, scenes in which a public address voice or an amplified singer fills the aural background, while Julie whispers to Johnnie in the foreground. Once Julie believes that Johnnie is ready to become Wolf, the plan goes awry almost immediately, as Johnnie decides to take over the racket as himself. Bakhtin might call this twist a moment of "ideological becoming," in which Johnnie "selectively assimilates the words of others," much to the consternation of his erstwhile Svengali.[19] Johnnie soon takes on more than he can handle in racketeering and women, which leads to an argument with Julie, who is accidentally shot in the ensuing scrum. Moments after the death of his ventriloquist, Johnnie the dummy faces a godlike voice on a loudspeaker similar to those that have been hounding the pair throughout the aural landscape of the play. Police, who have mistaken Johnnie for Wolf, surround the safe house. But the dummy does not have Julie to whisper to him now, and he can only cling to her body begging, "Come on, Julie, tell them, tell them who I am! . . . Julie, please, give me the *words*!"

Like the doll in "Ria Bouchinska," Johnnie is a medium requiring a speaker behind him to be animate and sustain identity. In plays like "Killer," the mind is a sieve whose retention of messages is unreliable once those words cease to flow. This situation obviously leads to an infinite regress. So long as our words do not reliably originate in our own subjectivity, then they may also not really come from our hidden source either, which may be yet another medium conveying the words of another, and so forth ad infinitum. This suggest that if acts of speech truly penetrate beings, as the logic of 1940s transmission dramas suggests, then it turns out that interiority is hardly interior at all. Not only is doubt cast over the possibility that a communicator is localizable and singular, but the contractual relation between utterance and transmitter is revoked. To put it another way, in the aesthetic framework of the 1940s, ventriloquist dramas are conceptually challenging because they suggest that whenever we attempt to grasp hold of the voice of a hidden speaker, we end up with yet another intermediary construct, another dummy, another vacancy.

NERVOUS CENTERS

Eavesdropping toggles us between states of listening and metalistening, and ventriloquism handicaps distinctions between speaker and mediator. Both situations erode the predictability that acts of mediation concretize in other conventionalized dramatic situations. But perhaps the most perplexing anxieties of the 1940s are expressed in dramas about characters who are responsible for managing signals in some way: reporters, broadcasters, press agents, messengers, Signal Corps officers, telephone exchange operators, and other information workers. The new prominence of these "signalmen" reflected their proliferation at home and abroad as a direct result of World War II. More than 35 percent of all members of the US forces were clerical staff dealing mostly with messages, reports, and communiqués; no other national armed force had nearly so many people managing information.[20] The official history of the US Army Signal Corps puts it this way:

> The commodity dispatched overseas in greatest quantity during World War II, and at greatest speed, was neither munitions, nor rations, neither clothing nor supply items. It was words, billions of words, messages of strategy and command, plans of campaigns and reports of action, requests for troops and schedules of their movements, lists of supplies requisitioned and of supplies in shipment, administrative messages, casualty lists. All of this and much more, such as services for the press (news dispatches, telephotos) and services for the soldiers (expeditionary force messages), poured over far-flung wire, cable, and radio circuits and channels—routine and urgent, plain text and enciphered, on a scale unimagined before the war by any communications agency, military or commercial.[21]

This emphasis on information had a parallel in a series of professions, particularly in journalism. According to one analysis, by the time Germany invaded Poland in 1939, some 10,000 reporters were covering World War II, many of them surrounded by agents intent on swaying coverage.[22] After D-Day, one

British officer joked that there were approximately 1.85 war correspondents per mile of front.[23] Encounters between signalmen and journalists are curiously common in wartime narrative. Correspondent Vincent Tubbs of the *Baltimore Afro-American* considered the communications post "as interesting a spot as can be found in a battle area"; Walter Bernstein of *Yank* described a battle by following spiral-four field cable around it; Scripps-Howard's Ernie Pyle added the names and hometowns of signalmen into his copy, to expedite transmission of his dispatches.[24] For those who participated in it, the news process had a strong resemblance to the signal-based work of the armed forces. Here is how *Time* described it:

> The copy gets beaten out on the portable typewriter, gets trimmed by the censor with his little loose leaf notebook of directives, gets whisked to the cable office, flicks undersea in dits and dots. And the cable editor fights it out with the city editor in the city room, where the phones keep ringing and the rewrite men step into the booths to take the stuff from the stringers in the corner drugstores, and the presses are booming downstairs on the early edition, and cigarette smoke hazes above the grey men with the eyeshades in the slot.... And the teletypes rap out their spasms of typing in rhythm; in the glass-enclosed room the announcer faces the mike; the newsman is timing his four minutes flat, with a minute commercial; the expert is typing his views of elastic defense; and there sit the bored technicians behind the dials, keeping the pitch of sound in hand, as it goes on the wires to a hundred cities and off the antennae to a hundred million ears. It blats in the taxi, it roars over the public-address system, it speaks in the mess hall, in the midnight coffee stand.[25]

This is how the radio age imagined its "information society," as a pulsing organism made up of separate and innumerable signalmen in a line of codes that turned the world into broadcasts. In its beats, flicks, booms, raps, and blats, information seemed alive in the folklore of the 1940s. It also seemed deathly. In *The Human Comedy* (1943), novelist William Saroyan centers his classic depiction of small-town California around young Homer Macauley, a modern Thanatos charged with bringing news of casualties throughout the town from its telegraph station, the only contact that citizens have with the faraway war.[26] Ithaca's boys go off to war, but only telegrams return, a transubstantiation that turns men into messages of regret.

Whether they were reporters, clerks, linesmen, or messengers, increasing numbers of Americans were in the business of informing, and listeners found this aspect of their experience reflected in programming, from signalman training dogs in *The Man behind the Gun* to telegraph operators on *Suspense*, and stool pigeons on *The FBI in Peace and War*. Actually, you can think of the 1940s radio play as a kind of human information system running through the backdrop of the pulp narrative as we know it: bellhops, lighthouse keepers, beat cops, bookies, bank tellers, circus barkers, ad men, and emcees, all with a secret to tell or withhold. In wartime radio dramas, many characters were tied to a media system or rumor network that had a nerve center of one kind or another. Such centers even became a preferred location in which to position the audience both within the drama and as a "mediating space" between other scenes. Back in the 1930s, the most popular radio superhero was the Shadow, a highly mobile figure who prowled bright nightclubs and dark cloaca in search of action. The introduction of the program referred to Lamont Cranston as an untethered "man about town."[27] But in the 1940s, the heroic model was more like Superman, who worked at the offices of the *Daily Planet* newspaper precisely because it was a centralized place to hear about ongoing plots prior to sallying forth after villains at work in the world beyond.[28] Other programming formats developed in the same way. In the 1930s, *Calling All Cars* often featured several peace officers from various districts cooperating to pursue criminals, but this effort lacked a locus from which this work was directed. A decade later, similar programs segued to action at a "headquarters" almost immediately, often illustrating these locations lavishly with background activities conducive to detailed imaginary reconstruction, such as PA systems, printing presses, telephones, telegraph keys, and news tickers.[29] On *The FBI in Peace and War*, the postwar structure of the program usually followed criminals up to no good, punctuated by segues to the voice of an FBI special agent narrating between scenes as a news ticker clatters in the background of headquarters. On *Counterspy*, listeners heard a whole central law-enforcement campus over the years, with David Harding addressing new recruits in an auditorium, rummaging through the records room, and practicing at a firing range. On NBC's *Battle Stations*, the battle of the North Atlantic is described through reports and testimony collected in Washington, not by depicting much local action. Other naval programs focus on audiopositions that work as a signal hub, such as the bridge in "To All Hands" on *Words at War* and the sonar room in "Submarine Astern" on *Cavalcade of America*. Most episodes of *Soldiers of the Press* start out in a nerve center, where we meet a United Press agent against a backdrop of typewriters. As the reporter goes into the field, we follow, yet remain at a distance from ensuing events because the reporter narrates to us in past tense, as if still seated next to his radioteletype. Nerve centers became prominent symbolic repositories in the 1940s in part as a reflection of the actual process of centralization of information and industry that was justified by the exigencies of war. Wartime America had *power centers*. So did wartime radio plays.

That mimicry could become critique. Many signalman plays associate nerve centers with social dysfunction. An early broadcast along these lines is Irving Reis's "Meridian 7-1212," which aired on *The Columbia Workshop* in 1939. The play begins with a journalist on deadline seeking a topic for a human-interest story. The writer hits on the idea of profiling operators who answer Meridian 7-1212, the telephone number one dialed to hear the correct time at the New York City switchboard. Along with the reporter, we listen in on operators as they repeat, four times per minute, "When you hear the signal, the time will be . . ." As the operators producing a vocal performance of real time, the conceit of the drama is that the ensuing scenes segue as if following each call back to its source outward from the signal center into its shared present. The play is about the switchboard, which links a suicidal man who calls to find the time (because his life insurance runs out at midnight), a group of expatriates drinking in London (making bets on the time back home), and a lawyer who learns that the key witness is prepared to recant on the eve of an execution. In a twist, the operator who gives this lawyer the time (too late) is the sister of the man set to die. The play is one of many that falsify the anonymity of a network and reveal it to be obscuring a preexisting social relation.[30] A similar commentary occurs in "Weather Ahead" on *Radio City Playhouse*, in which we meet a radio operator in the control tower of a South American airfield, who confesses that he is on the run from charges of murder. As dangerous weather descends, the signalman learns that the airplane that he is guiding into the runway contains a detective who has discovered his true identity and is coming to arrest him, a situation that leads to a murderous temptation that must be overcome. Both of these narratives suggest that signalmen exist in a state in which duty is pitted against self-interest. These stories use the junction of the individual and network as an encounter amenable to parables of social obligation that emphasize a greater good, which is the essential convention in many of the affirmative narratives of the war years.

The junction between human and network also became a scene for moralizing publicity. In radio plays, signalmen were often turning overheard whispers into "facts" for public consumption, a process that involved unforeseen effects. In "Correction" on *Radio City Playhouse*, for instance, a physician named Lundgren enters the newsroom of a newspaper with a gun, demanding a retraction of a story about the death of a girl on Long Island, a screed calling Lundgren "a fast and loose society doctor" who supplies local celebrities with narcotics. The allegation is false, but as a result of the article, Lundgren has been dismissed from his position. As the standoff continues, the head editor is called in, and we learn that years ago Lundgren's father had been smeared by the same newspaper with allegations of indiscretions with his students, rumors that turned out to be specious but ruined him nonetheless. The newsmen express no regrets, insisting that their duty is to "safeguard the public morals." Besides, argues the editor, in the whirlwind of publicity, no one will remember the Long Island story in a day or two, and Lundgren will no doubt have a chance to clear his name. Lundgren proposes to call strangers at random and ask if they remember the smear, in order to test the editor's theory. The editor bets his life. When phone respondents *do* remember, Lundgren executes the editor forthwith. "A paragraph of type can ruin a man's reputation," Lundgren explains before killing himself. At the dawn of the 1940s, programs like *Big Town* depicted journalists crusading against social injustice; in the middle of the decade, journalists became hardy heroes of worldwide adventure on *Words at War*; but by the end of the decade, playhouses used journalists to issue gruesome reminders that by virtue of the fact that they command a gateway of point-to-mass relationships, tending to official "public knowledge," reporters amplify perceptions that can have ruinous and far-reaching consequences that can neither be predicted in advance nor amended after the fact. Promising reliable information, the mass media age also became the age of deadly rumor, the force that Virgil famously described as a colossal winged beast covered in listening ears and speaking mouths, spreading truth and falsehood equally in the air.[31] Nurturing this hideous beast but unable to control it, signalmen had great power and yet none.

While questions about the responsibilities associated with information may be symptomatic of public concerns over transformations in 1940s society, the most self-aware signalman plays pass through these issues to address the metaphysical qualities of "network society." In the wartime American imaginary, signalmen are coextensive with a tissue of communications whose scope is beyond reckoning. They integrate with this sublime web, and their quotidian habits form around it. In that respect, they exist permanently in a state that we occupy temporarily whenever we switch on the radio set and invite its transmissions in. One dramatist whose work gets at this situation is Wyllis Cooper, who really invented stream of consciousness in radio on *Lights Out!* and told many signalman narratives in his Mutual Network series *Quiet, Please* at the end of the 1940s.[32] In "12 to 5," for instance, we meet Connie Duffin, the overnight disc jockey at a local radio station. As he sips his coffee on a night the same as any other, we hear Duffin's end of a set of conversations with callers requesting songs (including the theme song of *Quiet, Please* itself). But then he receives a series of visits from a stranger named Herbie, who brings in news reports to read over the air. The scripts turn out to be from the future. A reported murder has not yet happened, and the weather and time are incorrect. Angry phone calls soon come in, one from the man supposedly killed. Duffin attempts to play songs, but everything he spins airs as a death march, no matter how it sounds to him. One by one, the routines of the signalman at his nerve center go haywire. Soon the visitor from the future exits, and we get a report of his death in a taxicab accident. Herbie then reappears with a Teletype informing us that Duffin himself has lapsed into a coma a moment after the end of the broadcast, which concludes the instant the DJ finishes reading the report. A similar metaphysical trap confronts a signalman in Cooper's "Green Light," the narrative of an elderly railroad "brass pounder" named Phil, who tells us the tale of how he lost his leg working the rail line many years ago. We segue back to late one night in the signal shack, as Phil's lover Addie is encouraging him to take a real job and marry her. Phil agrees and seduction ensues, at which point an unexpected signal comes down the telegraph line before a train thunders through and wrecks. When Phil sets out to investigate, the wreck has vanished. Similar appearances persist over the course of a few nights until the brass pounder manages to stop the phantom train and sees that it is a ghost locomotive driven by Casey Jones. Phil loses a leg on the train tracks beneath the ghost train, ending his plans of winning Addie. The scene is surely a self-sabotage, if not outright castration, but the deeper point is that Phil's sublimated sexual anxiety is displaced into the railroad signal system itself, something achievable as a result of Phil's coterminous relation to a far-flung communication circuitry. The signal network works as Phil's prosthetic unconscious.

PERDITION AND MIRACLE

Where does the media system end and the mind begin? Are messages that we send true in their courses? Are speakers and listeners entities with boundaries, inside and out? These are some riddles that structure American feeling in 1940s culture according to radio drama. Just as eavesdropping extends listening until it has no listener, and ventriloquism undermines speech to the point where it has no speaker, Wyllis Cooper's plays take the propensity of mind and medium to define one another to a conceptual extreme, as the disc jockey and brass pounder disappear into signal networks, a vortex at the logical end of the theater of their minds. This process also describes what happens to us when we listen to a radio play, as we take up those sounds with which the medium lures us and then proceed to enter into the fiction, layering sound cues with assumptions and fantasies in the highly disciplined process that we call "imagination." In that sense, signalman stories reproduce radio listening at its most rapt, the moment when we forget all the work we do to understand and individuate sound and get carried away, intoxicated. Does this aesthetic opportunity redeem radio's theater of the mind from its seemingly compromised role as the hortatory component of wartime culture? That is an open question. It is tempting to judge wartime radio as a useful but degenerate framework whose only historical role was to underwrite a power-based model of communication invented to mobilize affect alongside industry. But using eavesdroppers, ventriloquists, and signalmen, wartime radio also achieved extraordinary self-awareness, staging questions unaddressed at any other time in the history of the medium.

For instance, these plays could be a means to explore the *appel du vide* effect within what we might call the "media consciousness" of the era. Consider an adaptation of Charles Dickens 1866 tale "The Signalman," which was directed for *Lights Out!* by longtime radio director Albert Crews in 1946, during the last summer revival of that storied shocker program.[33] Crews's play begins with a voice calling out into an ambient silence, "Hello, below there!" The utterance is an illustration of what sound designers call "perdition," the sense that a voice is leaving us never to return, a sound without destination, like a lost soul.[34] The first invention of the play is an abyss. As the voice drops into a chasm, a first-person internal narrator explains that he is standing at the precipice of a ravine in the middle of London, calling down to a man at a signal house near a railway tunnel far below. The signalman reacts with a curious horror and then waves the caller to come down to him. We learn that our narrator is a reporter after "people who work in little-known occupations" for the human interest beat, just like the magazine men in "Meridian 7-1212." As a telegraph ticks away in the signal shack, the signalman obliges with the tale of his life, but then he hesitates. He asks the writer to come back tomorrow, when he will tell him another story. On the second encounter, the signalman explains to the journalist that he has been having a recurring vision of a man by his signal light near the tunnel, a figure who covers his eyes and waves with one arm, calling out, "Hello, below there, clear the way!" The first time that this occurred, the signalman thought little of it, until a week later, when a woman died on the train right near his signal light. A second premonition foretold a train wreck. Now, the signalman explains, he has had a third premonition, and does not know what to do. "It's the responsibility that crushes me," he says, "because of this specter, because of this knowledge of what will happen along this line, I am responsible for every child, every mother, every person who rides upon it. I ought to warn them, but I don't know how to warn them." As it did for Connie Duffin, here a premonition deranges the signalman's defining literacy in the language of real time, in the "immediate" presence that characterizes his "mediating" place in the signal system. "You've got to live in the present," the reporter advises his friend. "Don't think about the future . . . just keep your balance in case something bad does happen." As the reporter leaves, the signalman feels relieved at having explained his predicament.

That night, the journalist dreams of the signalman's vision, absorbing its uncanny mood and projecting himself into it. Returning to visit the next day, the reporter learns that his friend is dead on the side of the track by his signal. The previous night, we are told, a train had barreled through the tunnel, but the signalman ignored it and stood in the way, as the engineer covered his eyes from the horror and yelled, exactly as had been foretold in the vision. Later that night the reporter writes up his story, reflecting that the signalman did not commit suicide, but rather "bore the responsibility for his fellow man so strongly on his conscience that he died from it." Unable to guarantee a network of exchange, the signalman absents himself from it, surrendering to the call of the void aesthetically set out at the beginning of the play. His drama epitomizes the precarious phenomenal state of men and women who receive signals from beyond and disseminate them to another beyond, in an infinite and authorless system. The final scene consists of the reporter finishing up the copy for his story as a messenger boy listens in, waiting to run it to the press room, where the story of one information system will be told on yet another. Vanishing from the rail network, the signalman reappears in a media platform as a ghost. By signing his name to the end of the tale—"Charles Dickens"—the reporter both absorbs and also transmits the discourse of another as he comes into full subjectivity. The signalman himself scatters into four directions: the rail signal system, the press story, the words of the reporter, and the imagination of the listener at home. A medium has become a message. The mood of the play ends as it began, swirling downward into a well of lost sound.

Yet in this vertiginous well there is also a sense of relief. Of the many tales of fraternity between journalist and signalman in 1940s radio culture, "The Signalman" is among the most poignant because in the course of telling a story of message breakdown, the drama also explores surprisingly deep connection between isolated human beings through communicative exchange, even "mutual ensoulment." Philosopher Paul Ricoeur has written that from an existential point of view, communication is "an enigma, even a wonder," since it trespasses over the "fundamental solitude of each human being."[35] He explains:

> My experience cannot directly become your experience. An event belonging to one stream of consciousness cannot be transferred as such into another stream of consciousness. Yet, nevertheless, something passes from me to you. Something is transferred from one sphere of life to another. This something is not the experience as experienced, but its meaning. Here is the miracle. The experience as experienced, as lived, remains private, but its sense, its meaning becomes public. Communication in this way is the overcoming of the radical non-communicability of the lived experience as lived.[36]

If signalmen resemble listeners at home, then the overcoming of noncommunicability between Dickens and the signalman suggests a fantasy latent in the dramaturgical world of wartime radio, a wish that the twin solitudes of the speaker and listener be not thus. Aristotle believed that because sound exists in the mass of air, its necessary condition was a void between bodies.[37] In striking a surface, we produce a brief union between resonant object and ourselves, an elapsing moment of linkage. Crews's play could be a sentimental study in that fleeting jointure. The trick of "The Signalman" is to parlay perdition into reply, overcoming the isolation of signalmen, who, despite their direct relations to modern communications apparatuses, always seem to begin their ordeals trapped by unrequited feeling. In this way, the "metabroadcasts" in this chapter gesture toward how media enable ecstatic transcendence of solitude. If the solitary vulnerability of the individual listener is a key construct of wartime, so is its overcoming.

Throughout the 1940s, many dramatists both promoted and challenged the link between the mass media and consciousness, producing tales about characters capable of self-deconstructing in a way that raises unanswerable questions that are ultimately about the listener, that unnamed presence most fully external to the drama, the being to whom the dramatist is ultimately trying to transfer experience, and the entity of whom any eavesdropper, ventriloquist, or signalman is but an avatar. That fact was suddenly clarified by wartime dramaturgy, but that clarity would not endure. In the 1950s, communication was no longer entertained as a basis for conceptualizing human interiority. As time went on, broadcasting would be transformed by technical changes ranging from television to FM, the transistor, and magnetic tape, as the industry was rocked by changes in network structure along with a complicated political upheaval. Under these circumstances, not only would the conceits that drove the theater of the mind vanish, but so would the possibility that radio *could* define the mind, an organ that was about to be reimagined to suit Cold War narratives of identity and psychiatric models of the self. Yet in its role as a laboratory for the collective psyche, radio drama also had surprising new energies. Nineteen-thirties radio had built a theater in the mind and 1940s radio became a theater about the mind, but in order to make sense of the transforming media culture of the 1950s, radio dramaturgy would, naturally, go out of its mind.

3

Radio and the Postwar Mood, 1945–1955

CHAPTER 8

Later Than You Think?

As the Second World War was ending, few writers were more deeply invested in radio than Arch Oboler.[1] Oboler began writing radio plays while he was still in school, submitting scripts to NBC and finally selling "Futuristics" in 1934. By the end of the decade, he was authoring up to sixty-five plays per year for NBC's *Grand Hotel*, *Irene Rich Dramas*, and *Your Hollywood*. In 1937, he made headlines for writing a racy skit about the Garden of Eden for Mae West and Don Ameche on *The Chase and Sanborn Hour*. After the broadcast, West was not invited to appear on network air for a decade.[2] His work was eclectic. Several plays use the kaleidosonic and intimacy structures that I have associated with Norman Corwin, while others exemplify stream of consciousness, a method that he adopted from Wyllis Cooper. Fan magazine *Radio Life* applauded how Oboler showed the "spiritual and psychological truth" behind the bang of a shutter in the wind, an observation that recalls Lucille Fletcher's penchant for fashioning psychological signals out of naturalistic sound effects.[3] As a radio director, Oboler was famous for cuing his actors from a tabletop and coaching a performance from mike-shy Joan Crawford by directing her to perform barefoot. He was also industrious. During the 1940s, Oboler helmed shocker *Lights Out!*, authored *To the President* and *Plays for Americans*, and collaborated on a series for the Treasury Department entitled *Five for the Fourth*. Although critics scorned Oboler's work, historian Erik Barnouw writes that many performers considered his plays highly "actable," and the *New York Times* ranked Oboler with Norman Corwin as a "Eugene O'Neill" of radio.[4] His commitments to aural narrative are legendary: Oboler reportedly employed a Dictaphone while writing in order to select words that "sound like radio"; he was known to vanish from a party, returning with a complete script in a hour or two; in 1940, he commissioned Frank Lloyd Wright to design him a gatehouse in Malibu with a brook running under it, just for the sound.[5] Tallying up Oboler's talents and boasts, critic Richard Hand writes of a brilliance that exhibited "no shortage of ego," while radio aficionado John Dunning assesses him a little more bluntly: "Genius . . . or show-off?"[6]

Whatever his artistic merits or social graces, by wartime, Oboler had become a leading voice among writers. In 1944, Oboler took out a full-page ad in *Variety* that challenged network heads to set aside two half-hour slots per week for "a radio of ideas" that clarify "the conflicting issues involved in our wartime national economy, in our wartime world and in the puzzling future."[7] His timing was poor. That year, experimental programs were being slashed everywhere, as a postwar dearth of newsprint sent ad firms hurrying clients to invest in radio, where there were sixty-five thousand units of airtime for sale per day.[8] With profit came tumult. In 1940, NBC was broken in half by the FCC under commissioner James Fly, creating the American Broadcasting Company in 1943, while the Mutual Network changed owners often, remaining mired in financial scandal throughout the 1950s.[9] And in what has been called the "single most decisive competitive development" in the history of radio, CBS boss Bill Paley launched the "Paley raids" of 1948, wooing NBC mainstays like *The Jack Benny Show* and *The Edgar Bergen Show*, collecting twelve of the top fifteen radio shows, and earning CBS leadership of the industry for the first time.[10] Technological developments also transformed the medium. With the advent of magnetic tape, broadcasters began to rely on prerecording, and transistor radios revolutionized sound fidelity.[11] Soon flagship programs such as *Escape*, *Dragnet*, and *Gunsmoke* all began to use a greater quantity of illustrative sound effects than their predecessors. Meanwhile, the advent of audiotape inaugurated the experience of sound as we have it today, with our multiplatform ways of storing, sampling, editing, fast-forwarding, and erasing—a registry of practices that theorist Friedrich Kittler calls "the musical-acoustic present."[12] Yet even this stylistic change pales beside another development, one that led Oboler to reverse his prognosis for the industry less than a year after his "Radio of Ideas" item. Writing again in *Variety*, he now issued a "Requiem for Radio," explaining that although radio was still "jingling for Super-bread, and rhyming for Super-cigarettes," the future flowed through the "moving electron rays" of television.[13] Few could have articulated the impending doom more ominously. The article seems to reverberate with Oboler's tagline from *Lights Out!*: "It [gong] is [gong] later [gong] than [gong] you [gong] think." For those who used such phrases to make theater of the mind, the hour was growing late.

Or was it? In part 3, I argue that not only did radio drama continue to flourish for as many as ten years after the war, but its conventions also evolved to express some of the cultural and political challenges of the postwar era. Broadcasters reinvented their métier, airing plays that exhibited innovative models of time, space, communication, and consciousness in order to tell tales about postwar life. To borrow some phrasing, part 3 describes "the theater of the mind in a conservative age," even a "paranoid style" that dramatists made out of the aesthetic components at their disposal.[14] By analyzing these components, I aim to prove that radio still played a role in how Americans grappled with the notion of mass mediation. In this chapter, I show that despite perceptions of its moribundity, radio drama actually expanded during the period that Oboler identified. I also show that to express Cold War anxieties, dramatists began to create plays with immobile protagonists and races against time—a new kind of "drama of space and time" utterly unlike the one that prospered during the late Depression. In chapter 9, I return to the theme of communication, arguing that crime shows of the late 1940s and early 1950s encouraged subordination to authority by using direct address and confessed testimony. In chapter 10, I look at stories about undercover men, tests of conscience, and self-repression, arguing that in each of these narratives we see new models of the mind emerging out of wartime patterns. Part 3 ends with a study of *Suspense*'s "Zero Hour," perhaps the last great play of the network era, and one that embodies many of the trends accounted for in the remainder of this book.

At its heart, what follows is a story of how earlier formal and thematic preoccupations gave new innovations aesthetic and ideological shape. Immobility is a curious spatial quality because it distorts the exploratory utopias for which radio had been organized in the 1930s. As wartime shockers became postwar cop shows, the defining act of the Red Scare—testifying—drew its authoritarian overtone from the fact that a greater variety of transmission had previously filled wartime plays. And since 1940s radio had tried to apprehend psychic states of being, when 1950s radio put minds in states of duress, these tests of fortitude signaled the birth of a new definition of interiority. Indeed, the late radio period is marked by nothing so much as a struggle on the part of writers to find a compelling new account of the mind, a struggle that had many results but few resolutions. In sum, part 3 considers postwar culture to be a third act in the story of classic American radio, an unexpected last unfolding of some of the aesthetic and conceptual issues "in the air." During this act, it may be that radio's capacity to put pictures in the mind was declining as TV sets became more popular, and it is certain that wartime models about the mind were transforming, but for a decade after the end of the Second World War, radio drama continued to work as a theater *for* the mind, a site for feeling through and working out important issues in ways that helped decide the mark that radio culture left on the twentieth century.

PROGRAMMING THE COLD WAR

If it seems odd to explore innovation in a section on postwar radio, it is because writers usually foreground radio's "decline" in the 1950s. Like Arch Oboler, many describe a "rivalry" between radio and TV in the postwar era, one that led to the "death" of the former as the latter increasingly defined American life. This rivalry is often expressed in rhetoric that makes it seem as if "radio" and "television" were grouchy siblings rather than complex technological systems and the set of mediated experiences they involve. Some writers say that radio was "scared silly" or underwent a "period of desperation," while others write of how radio sets were "relegated to other parts of the house" after TV sets arrived in the parlor.[15] This moody language disguises the fact that it is difficult to establish criteria by which comparisons can be made between these media. By 1947, RCA had invested $50 million in manufacturing TV sets, yet as late as 1950 there were only some five million in use, and CBS had just begun to broadcast content. In the immediate postwar years, getting into the new FM band was a much more pressing concern for many broadcasters than television. Within six months of V-J Day, the FCC had agreed to hear 279 requests for FM licenses.[16] Indeed, FCC licenses for TV stations were embargoed for four years during radio's so-called period of desperation. And although TV sets were in two-thirds of American homes by 1955, historian Susan Douglas has found an uptick in radio ownership over the course of the 1950s, and Christopher Sterling and John Kittross note that radio stations earned more money at the end of that decade than they did at the beginning.[17] Between 1949 and 1952 domestic radio listening increased 11 percent in the daytime and 5 percent at night according to a Nielsen ratings study; meanwhile radio played an increasing role in public diplomacy overseas, with international broadcasting organs like the Voice of America taking a place among the most important tools in waging the Cold War.[18]

Even a cursory analysis of content tells a similar story. Several radio shows persisted well into the 1960s, often as tie-ins to TV. *Suspense* and *Dragnet* aired on both radio and TV for years; Jack Webb boasted that he shot the first TV episodes of the latter directly from its radio scripts. In many ways, radio and TV were appendages of the same organism. As Michele Hilmes has pointed out, radio is after all what made TV possible.[19] Network heads procured branded talent from radio divisions to staff TV programs, which were produced using the capital raised during decades of *The Guiding Light*, *Jack Benny*, and *The Lux Radio Theater*. Even after *The Jack Benny Show* moved to TV permanently in 1955, marking the end of the classic era, the medium was not quite "abandoned like the bones at a barbeque," as comedian Fred Allen famously remarked.[20] Rather, the story of 1950s radio is in many ways about reinvention, as the medium became a music box for the Top 40 sound and rock 'n' roll, listening became more common in automobiles than in homes, and community-based content returned to prominence—in 1946 just 34 percent of radio revenue came from local advertising, but by the 1960s that percentage had doubled.[21]

The growth of television was not the only force transforming postwar radio. As Oboler predicted, radio also faced criticism during these years for its emphasis on commercialism, a critique that had been simmering ever since the 1920s and that has today become a touchstone for scholars interested in how the radio age prefigured postwar American consumer culture and its discontents. Historian Kathy Newman considers 1946 the beginning of the end of the radio age because in this year the FCC released the so-called Blue Book, a critique of the system of ad-based revenue, with recommendations to decline licenses to stations that relied too greatly upon it.[22] The Blue Book voiced the displeasure of regulators, providing data to prove that networks could afford considerably broader public service than the pittance that they claimed was commensurate with their mandate under the law. Newman explains that the Blue Book was part of a "backlash against radio's new cultural power," one also associated with Frederick Wakeman's 1947 novel *The Hucksters*, which portrayed ad men in an unflattering light.[23] The strength of this backlash is not clear. Although the Blue Book challenged the network system, there was little direct follow-up when it came to the commercial basis of the medium. Still, the Blue Book became the basis of FCC actions promoting local content, and it is also true that the number of independent stations began to increase for the first time in a decade.[24] In 1945, all four networks offered five or six solid hours of prime-time material every night of the week; but by October of 1956, CBS's Sunday night networked only three shows in a window between seven and eight, and the following May, Mutual's Sunday night offered just a half hour of *Bandstand USA* at nine. Soon the norm became local stations carrying syndicated music and quiz shows, a return to the idea of radio as a music system, "narrowcasting" to target local markets as part of what social historian Lizabeth Cohen calls "segmentation" of the postwar public into ever-smaller consumer subcategories.[25]

Of course, both television production and consumer backlashes are often framed within a third narrative of postwar radio, one that Oboler did not foresee but that made unworkable any "radio of ideas" worthy of the name. Like the film industry, radio was embroiled in a Red Scare.[26] In a very quick transformation, anticommunism replaced antifascism as the dominant outlook of American radio, as writers, actors, and directors were slandered by a very small group of opportunists—according to writer David Everitt, radio's Red Scare was driven by as few as five people.[27] Not only was the great Norman Corwin blacklisted, but proxy also damned any actor starring in his work. After appearing on Corwin's *An American in England* series, actor Joseph Julian was so sought-after that he worked on forty-nine programs in the 1948–49 season. But after he was blacklisted, he eked by on a mere $1,630 a year.[28] Historians of this period tend to center their stories on *Red Channels*, a publication that appeared in the summer of 1950, soon after Senator Joseph McCarthy's Wheeling speech brought him to national attention. A screed concocted by former FBI agents and a disgruntled radio writer, *Red Channels* named such luminaries as William N. Robson as sympathizers, alongside such heroic news pioneers as William L. Shirer.[29] As Thomas Doherty has explained, in time the radio blacklist became little more than a racket. After paying a blacklister to do a "background check" on staff, "the producer got a cleared cast, the sponsor got a huge audience, and the racketeer got a nice payoff."[30] Under these circumstances, the time was inopportune to use radio to sort out the "conflicting issues" of the "puzzling future," as Oboler had hoped. By the 1950s, dramatists had their own conflicts and puzzles, as many learned just how late the hour had become.

Whether on account of a red book, a blue book, or "electron rays," the consensus is that the golden age of radio drama fell into decline at some point between 1946 and 1950. Writers draw a bright line in cultural history that helps to throw into relief developments of the postwar era: the birth of the TV nation, the rise in segmented cultural markets, McCarthyism. By marking the end of the radio age, these stories characterize the postwar years as a special cultural-historical period disconnected from the past. But bright lines do not always clarify. In the first place, they risk mistaking coincidence with causation, making it seem that the Red Scare inevitably led to a decline in radio, although such a teleology ignores a great deal of pro-loyalty material that aired at the same time. Radio was not just a target of McCarthyism, but a part of it. Of course, there is evidence that some programs were canceled as a result of the scare, but *Red Channels* may not have affected programming in a concussive way. To show the problem, we can turn to radio schedules.[31] Despite the impression of rapid change in the twelve months after *Red Channels* appeared, CBS made only three changes to its Monday evening lineup, Mutual made no change at all to Saturdays, while ABC made just two changes to its Friday night. On the other hand, NBC shuffled its Friday schedule on eleven separate occasions, which may have been an effect of losing staff to rebuff allegations, but it is just as easily explained by network custom to roll out programs in prime-time weekends to build an audience, or a number of other factors. The trouble here is that writers neither convey nor analyze the complexities of what was being aired, which calls into question the notion that phenomena such as TV, a commercialism backlash, and the Red Scare resulted in a curtailing of radio production.

Comprehensively charting all these qualitative changes in postwar radio programming would be an immense undertaking. Here I can only make a start. Figure 8.1 shows the number of unique programs that aired on all networks in prime-time hours, by category, across five sets of twelve-month periods. Roughly following five-year increments, the graph focuses on time periods that are usually cited as landmarks for the industry: the year after the passage of the FCC Act (1934–35), the first full year of the war (1942), the first postwar year (1945–46), the year following the publication of *Red Channels* (1950–51), and the year after *The Jack Benny Show* moved to TV (1955–56). Not only does the graph show that Oboler's 1945 requiem was premature, it also suggests that any declination narrative is at best a partial description. Only a few radio formats declined in the late 1940s—music, sketch comedy, and variety—and each of these had been decreasing in number since the 1930s, probably as a result of the fact that comic performers tend to consolidate their audiences over time. The biggest names in comedy were almost the same in 1950 as they were in the late 1930s. Moreover, the graph shows that the dynamics of programming obey internal forces at least as powerfully as they respond to external events in the fits and starts that some of the scholarship suggests; all categories follow long-standing trend lines. Most importantly, radio dramas seem to follow a unique trajectory vis-à-vis other formats, exploding in production well into the 1950s. As late as 1956, American networks aired about the same number of prime-time dramas as they had twenty years earlier. This detail escapes portrayals of late radio in a state of unmitigated decline, which is not to say that radio was impervious to development, only that the hypothesis that dramatic content suddenly vanished is not supported. In an effort to bring balance, I propose moving radio away from the "receiving end" of postwar phenomena,

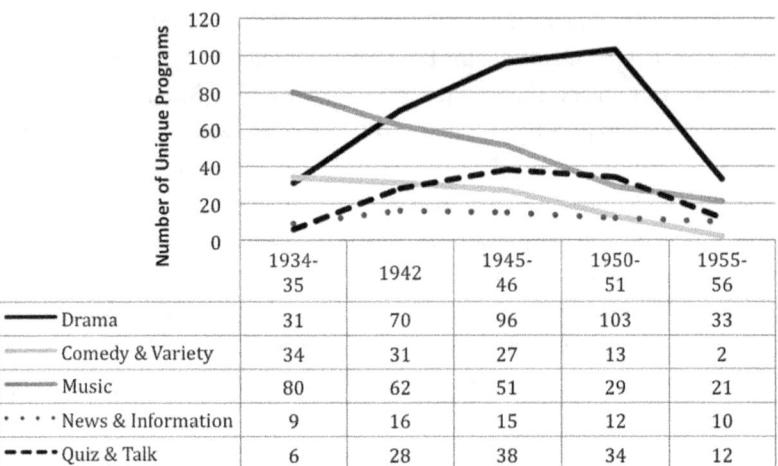

FIGURE 8.1. Number of network programs by category on evening airtime, 1934–56.

to ask how it gave meaning to those phenomena. This perspective is intended to suspend—but not discard—the rise of TV, the consumer backlash, and the Red Scare. Rather, I begin from a different place, asking not what postwar social changes took from radio, but what radio produced *for* postwar society using the spatial conventions and psychological materials with which dramatists crafted entertainment that Americans would continue to enjoy well into the 1950s.

Even with limited parameters, dramas of the late 1940s and early '50s are among the most challenging to research. There is little secondary literature to consult, as historians of broadcasting tend to pivot swiftly into television, along the way producing almost no critical scholarship on programs that aired exclusively subsequent to the war.[32] Another issue is the wealth of available material. Thanks to the advent of magnetic tape, the number of extant recordings bunches up from 1947 through 1952. A deeper problem is that most of the materials of this phase are episodic shows or serialized fiction, which make them formally disconnected from the work of writers like Norman Corwin and Arch Oboler, who aired work in anthology formats. To show the decline in anthologies and the rise in crime and mystery serial programs, figure 8.2 uses the same time frames and data set used above to consider what genres were driving prime-time drama. While the decline of anthology programs is clear, its ramifications for style require some speculation. Since anthologies asked for brief staff commitment, it had been easy to commission occasional scripts that tested innovations and made thematic excursions; you could conceivably ask someone like William Saroyan or Archibald MacLeish to contribute a script or two to a workshop program, but it would be difficult to bring them into a weekly contract. With serials on the rise, there was therefore an overall narrowing of the range of narrative techniques, a simplification that stymied some of the creative energy of the old experimental programs and shockers. Serials were also much cheaper to produce and more amenable to tie-ins with films and pulp novels, all of which earned them derision from critics seeking high culture from radio. It is understandable that dramatists such as Arch Oboler looked at postwar radio with disappointment. Their model of radio writing would not prosper in it.

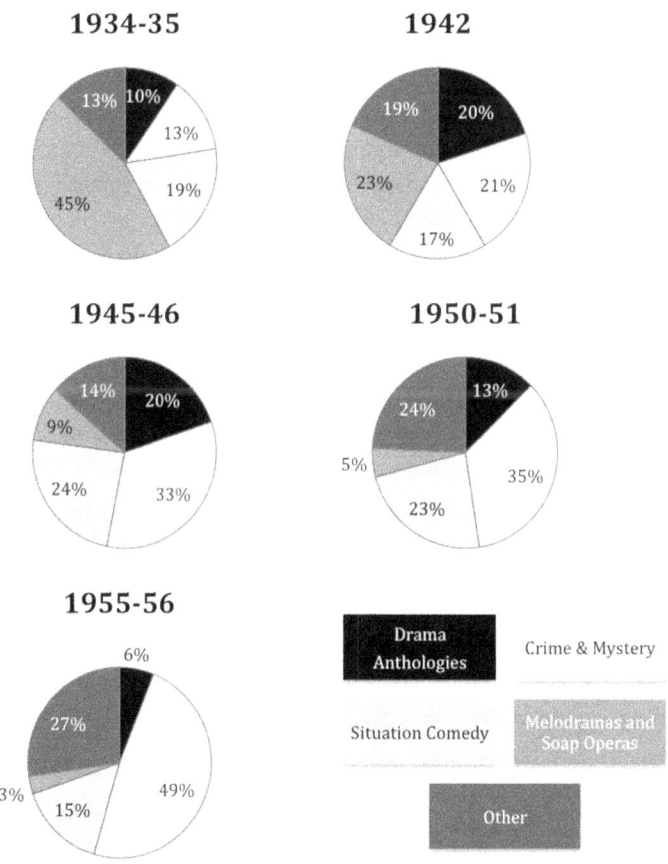

FIGURE 8.2. Percentage of radio drama shows by genre in prime time, 1934–56.

CLOSING THE WORLD

The decline of anthologies is just one postwar tendency that seems decoupled from the traditional preoccupations of American radio drama. For one thing, late radio is obsessed with technology. During the war, *Superman* prefigured the atomic age before it arrived, slating a series including an "Atom Man" villain prior to Hiroshima. On *Counterspy*, dramas featured special recorders to catch far-off conversations, glasses to decode images, and a sound-wave device that shattered steel. On *Dragnet* and *Mr. District Attorney*, the police used high-tech forensic techniques such as tire-tread analysis, invisible powders, and computers. *Hop Harrigan* showcased new developments in aviation that ranged from glass landing strips to crop-dusting. On *Mysterious Traveler*, a young professor used a fifty-ton computer in the "World Wide Business Machines" laboratory to calculate winners at the horse races. And a new slate of science fiction shows went far beyond *Buck Rogers* space westerns. *Dimension X* quizzed listeners on robotics and cybernetics; its broadcast "Destination Moon" told of a rocket breaking the atmosphere, an episode that was coincidentally interrupted by the start of the Korean War. In "When the Worlds Met" on *2000 Plus*, amiable aliens refused to meet with earthlings, citing as evidence of our backwardness not only torture in Russian prisons but also lynching in the United States. A similar scenario took place in "The Outer Limit," with aliens placing an invisible shell around the earth to destroy us all should any atomic bombs be detonated—the cautionary fable aired on *Escape*, *Dimension X*, and *Suspense*. Similar fears of nuclear disaster haunted "Adam and the Darkest Day" on *Quiet, Please* and "A Pale of Air" on *X Minus One*. Science fiction had come of age, as had other sorts of genre fiction. During these years, *Suspense* abandoned its commitment to thrillers, airing historical dramas, westerns, comedies-of-error, true-life "dramatic reports," plays based on songs, and public service stories about traffic fatalities. Cold War themes also filled the airwaves. On *Escape*'s "Two if by Sea," a diplomat uses a coded radio broadcast to stage a desperate gambit to rescue his wife from the clutches of lusty Russians, while "Rim of Terror" features the story of a young woman who picks up a defector on the highway. On *Suspense*, freedom-loving youths escape from behind the Iron Curtain in "Listen, Young Lovers," and hydrogen bombs are described in "The Case for Dr. Singer" and tested in "How Long is the Night." On *Dangerous Assignment*, special agent Steve Mitchell goes after an African manganese cartel sympathetic to the Russians in "Nigerian Safari" and fights leftist Sicilian bandits in "Relief Supplies." On *Counterspy* agents of "a foreign power" go after the recipe for high-grade mica in "Fabulous

Formula." In 1949, Vice President Alben Barkley appeared on *Counterspy* to entreat employers to hire the disabled. One year later the show fulfilled its duty to the public by asking listeners to turn in neighborhood subversives to the FBI.

Postwar radio was a new world. But to find our way through that world, it is necessary to look deeper into how some of its stylistic choices built on the conventions of the past. Consider director Elliot Lewis's drama "Pigeon in a Cage" which aired on *Suspense* in 1953, a play that at first appears to have little to do with Cold War–era structures of feeling. At the outset of the play, we hear an alarm buzzer being pressed, along with a present-tense first-person introduction from our protagonist: "I'm Gerald Brewer, I'm thirty-five years old, and I've got a wife. I've got a kid on the way and I'd like to be home now. Wish I were out of here." Brewer is a wallpaper-hanger trapped in a malfunctioning residential elevator after working late on the job at a mansion. There is no one home, so all that Brewer can do is ring the emergency signal and hope for a response. He complains of the small elevator, works the lever to prove that it is broken, and describes the cherry wood, iron gate, and square hole overhead. He calls out, rattles his cage, and lies down on the ground to peer into the living room through a sliver of exposed light. Transfixed, Brewer's thoughts drift, and he has a lapse of consciousness, awakening to voices. His narration changes to past tense, enhancing the dreamlike quality of ensuing events. Before Brewer can request aid from the newcomers, he overhears his employer Harry Rogers plotting murder with a girl named Janice in the parlor. Soon, footsteps approach the elevator, which now seems like a protective shell rather than a cage. By now Brewer is no longer pressing the alarm, instead preventing the elevator from moving so as to keep his presence undetected. Janice decides to go upstairs and look for a gun, approaching our audioposition as she calls the elevator car, and letting Harry know that the car is stuck. Brewer hears Rogers taking the stairs instead and breathes a sigh of relief. When Harry returns, the two conspirators almost convince one another to give up on their plot, but soon we hear a car pulling up in the driveway, as the victim has arrived. Moments later, we hear the report of a pistol that seems to foreshorten all of the complicated spaces that the play has traced for us. In the space between our audioposition and escape, the telephone rings. From deep in the shadows at the back of the elevator, Brewer hears Janice pick up the telephone to learn that Mrs. Brewer is looking for her husband. In this way, Brewer hears Harry and Janice realize that their crime has been overheard. The play has arrived at a reversal. Up until now, Brewer has projected perception outward, much as we do, producing the effect of a radio play within a radio play. But with the

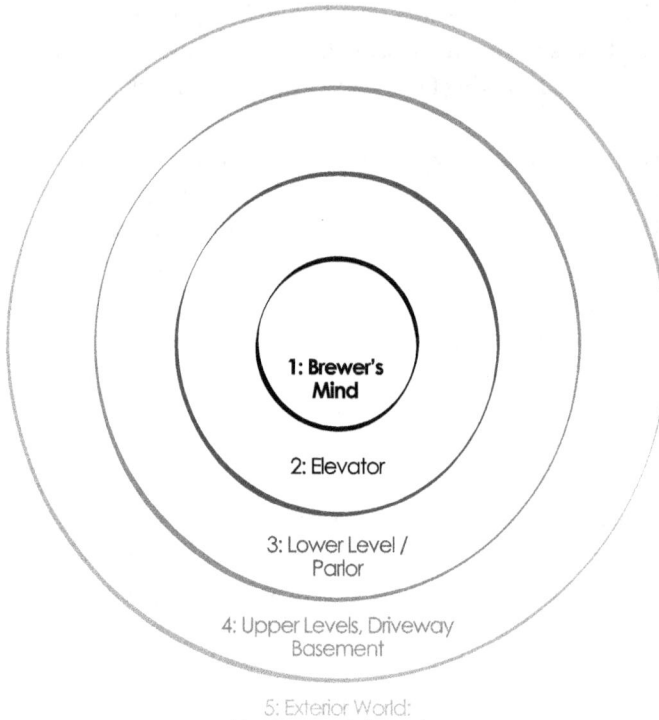

FIGURE 8.3. Setting as suggested by volume in the broadcast of "Pigeon in a Cage" on *Suspense*, May 25, 1953.

telephone call, everything reverses. Brewer is no longer just tuning in the "radio play" beyond the proscenium of his elevator. Now he is also being listened in on *by* the play within a play, producing the uncanny sense that we ourselves are overheard, too.

At this point, the space of the drama has been fully designed, and the play is revealed to be less about a paperhanger's desire to get back to his wife and more about the puzzle presented by a series of spatial tiers around him, which correspond to a series of nested spheres of earshot extending outward from a static audioposition, a radial model of relations that is cued by cooperating dramatic elements. Volume illustrates the proximities of these dramatic spheres to one another, while dialogue conveys the blocking of one body relative to the others in the system, also specifying height, width, and depth. Meanwhile, almost all of the sound effects are used to exceed spatial barriers, marking penetrations from one level to another by one or more agents in the drama—the alarm buzzer, the recorded music Harry plays for Janice, the elevator call button, the car in the driveway, the gunshot, and the telephone call. Around this highly structured armature, the action of the play is so overorganized that the stratification of its stage business seems to be the source of all tension. Consider the differing ways in which these characters hear and move around the tiers. Throughout the play, the listener can hear everything from tiers 1 to 4, while Janice and Rogers can hear everything from tiers 3 to 5. Brewer begins the drama speaking to us in tier 2, trying to issue sound to tier 5, although he fails to do so. In his lapse of consciousness, he retreats into tier 1, and then he awakens to finds that he is trapped in tier 2 by events in external layers. The second half of the play is little more than a series of attempts by Brewer to get from tier 2 to tier 5 without going through 3 and 4, as Janice and Rogers try to break into tier 2 from both below and above it, or to "get into" tier 1 by offering to bring Brewer into their conspiracy. In the end, the conspirators fall out, and Brewer is free to telephone his wife beyond in tier 5.

The set of events that takes place between these spheres is not as interesting as the fact that we can study them at some length. I would associate this play strongly with what historian Paul Edwards calls "closed world" narratives, a hallmark of the Cold War era.[33] Drawing on a metaphor first used by critic Sherman Hawkins, Edwards argues that many postwar tales feature "siege-like" stories that take place in bounded and self-referential spaces where every action spirals into a central struggle that draws in all aspects of the surrounding field. Edwards associates these implosive narratives with "containment culture," a concept based on the policy that diplomat George Kennan framed to "contain" the Soviet threat.[34] Historians have shown that models of closed worlds and contained space are reflected in postwar film and fiction, but by neglecting radio, their analysis tends to miss that these worlds seem all the more "bounded" as a direct result of a contrast with the capacious spatial dramas of an Irving Reis or Norman Corwin. In the experimental 1930s, it had been common to air dramas that use an audioposition at the center of scenic depth, and it is often profitable to map out implied dramatic space. But in most of these cases, audioposition exists to give coherence to a drama as we move across its big, unbounded landscape, whether intimately or kaleidosonically. Unlike the old experimental spatial dramas, "Pigeon" is defined by a static audioposition, an immobility that discovers some of the effects of contained stillness, which had not been exploited extensively in the past. The drama's world is not a geography but a prison. It is an expression of postwar culture of some ingenuity because it is a reinvention of the old dramaturgy of space and time, inasmuch as the play builds all its effects around the one positional formula that radio never really offered in the 1930s—entrapment. "Pigeon" is a play with the audacity to stand still in a dramatic form once known for its ability to bilocate, segue, and sprint.

The "radial immobility" of "Pigeon" was amazingly common in the postwar era. In *Dimension X*'s "Hello Tomorrow," we enter a future world ravaged by nuclear war, from whose radiation the survivors have fled into underground tunnels. Generations later, society has fallen under totalitarian rule, and star-crossed lovers must flee to the surface to escape genetic control police. After breaking free from their entrapment and negotiating a perilous journey, the lovers find an Eden-like scene on the surface, as the world has reconstituted itself. Years later, *X Minus One* aired Phillip K. Dick's "The Defenders," in which people live underground while a war between the West and the "Asiatic league" continues on the surface among robots. When a young professor falls for a POW from the other side and escapes to the surface with him, she finds that there is no war or radiation on the surface after all, and that the robots have struck a peace. Another sort of immobility occurs in "Three Skeleton Key," a horror-adventure that aired on *Escape* in 1950. Vincent Price stars in this tale of three lighthouse keepers on a rock in the southern seas that is overrun by a horde of rats from a ghostly sailing vessel. Over the course of the drama, the three men—the garrulous Auguste, the taciturn Louis, and our unnamed narrator—gradually lose floor after floor to the incoming horde, whose constant squeaking represents the threat of invasion from without. Once again the puzzle lies in an attempt to somehow escape from a fortified close position to an open distant space without passing through a perilous middle distance. In the final sequence, with both Auguste and Louis incapacitated, our narrator is trapped in the light itself, watching the eye sweep across the bodies of writhing rats until another vessel comes to take them out to sea once more. That entrapment stories proliferated in the postwar years is no surprise to students of the period, but I would submit that the feeling of a closed world in these dramas is aesthetically available only because of its contrast to a well-defined prewar dramaturgy. The mood of 1950s drama felt so stifling only because the 1930s drama from which it directly descended felt so kinetic.

THE ATOMIC PRECIPICE

Just as a new sense of immobilized position was invented for Cold War drama, so too was a new sense of time, one that grew out of a set of prewar preoccupations. To illustrate, let me come back to Arch Oboler's *Lights Out!* tagline cited at the outset of this chapter—"It is later than you think"—a phrase that is among the great cliché's of classic radio. Drawing on the real-time quality of broadcasting, the line was used to signal to the listener that the ensuing events take place in a time-outside-of-time in which our incredulity is suspended. Oboler's play on temporality is expected of a dramatist who came to broadcasting in the 1930s, a period that I have argued to be marked by a series of experiments in reordering "natural" chronologies for dramatic effect. Indeed, the notion of time-outside-of-time resonates through many plays discussed in this book, including two that I have analyzed most extensively. In Archibald MacLeish's "The Fall of the City," the Studio Announcer calls forth to "listeners over the curving air . . . /From furthest-off frontiers of foreign hours, mountain time: ocean time/Of the islands, of waters after the islands." While space is defined, the Announcer cites portents that suggest that time is not so determinately structured—in the City, "the wall of the time cracks." A similar sense of time-outside-of-time characterizes the "The Hitch-Hiker," whose narrator says at the outset that his ordeal began six days ago, a duration that we learn to have never elapsed in any mortal chronology. According to what the fiction tells us, only the time of narration is real, and not the time of the events that the narration describes. The trick is typical. Radio shockers work by rewiring linearity, looping chronology back on itself just prior to some culminating event, such as a trial sentence or the death of our protagonist. The first-person narrators of *Suspense* come into existence as they conjure a tableau and then abscond from the flow of time to tell their story as they remember it unfolding at the eleventh hour of the action, as if poised on a last precipice over which the telling of the tale will finally push them.

In these narratives, we can already perceive the time-out-of-time convention changing. In the 1930s, the sense of extratemporality limited the range of interpretations that could be made of depicted events, emphasizing the loose grip of natural chronology on the story. But in the later 1940s and '50s, the time-out-of-time idea had another signification, one of mood, as dramas were depicting time running fatalistically out in order to heighten suspense. In 1950, the announcer of *Dimension X* led us into the program asking, "Do you know what will come in a hundred years, or fifty, or the next minute?" But when the program returned in 1955 as *X Minus One*, the intro was replaced with a launch countdown, a nakedly mechanical sign of time slipping away toward a momentous event. Characters in radio plays were increasingly caught up in what Tom Gunning has called a "Destiny-machine," a trope in which the idea of time is less a flowing metaphysical concept and more a mechanism of modernity, something for which the clock is the central emblem.[35]

The lapse of time is strongly associated with a whole series of mechanistic forces in postwar thrillers. When a supposed murder victim is found alive in *Molle Mystery Theater*'s "Corpus Delecti," telephone exchange operators in Dallas, Chicago, Buffalo, Albany, and Rochester must swiftly transfer calls to get a stay of execution moments before the switch is thrown on an innocent man. "Long Distance" on *Radio City Playhouse* is virtually identical. Arch Oboler's last radio play before leaving the industry in 1946 was "Night Flight," the story of an airman running out of fuel over the Pacific. In "Special Delivery" on *Radio City Playhouse*, a woman frames her husband for her own murder, sending a judge a special delivery mail confession, and then racing to retrieve the letter when she reconciles unexpectedly with her husband. In "Statement of Fact" on *On Stage*, an ambitious prosecuting attorney tries to convince a woman to lie to avoid jail time for killing her husband, whom she murdered once she realized how their marriage had ruined her true spirit. As a clock ticks softly in the background of the interrogation room, it seems that time is running out on the murderer, but it turns out to be the attorney who discovers himself to fail the moral test, as he reveals himself to be just as imperious as the woman's husband had been. On programs of the late 1940s such as *The Clock* and *The Diary of Fate*, meanwhile, the "Voice of Time" from 1930s news shows was replaced by a "Voice of Fate" who told the tale of a character caught by moral failings that will inevitably lead to murder, alienation, or death. These narrators often introduce the drama by describing seemingly insignificant props that will lead from normalcy to ruin as the play elapses—as Fate explains, "A lost wallet, a rainstorm, these are the tools with which I work"—so that the listener can count each item off as time runs out for the protagonist. In *The Diary of Fate*, however, most of the victims of "Fate" are working-class men who err as they attempt to gain higher social standing through crime, emphasizing that one's fate is driven not only by the godlike narrator laying wallets and rainstorms in the path of life, but also by the many overstructured mechanics of modern society, from a fuel gauge to the postal system and rigid norms of class and gender.

From the end of the war until the middle of the 1950s, dramas portrayed Destiny-machines of all sizes. Wyllis Cooper's *Quiet, Please* tells the story of a man with a watch that can work as a personal time machine who makes the mistake of setting it forward past his own death—Cooper even titled the play "Later Than You Think" as a gag on the *Lights Out!* tagline. A few years later, on *Suspense*, races against time involve issues that affect a wider community. In "To None a Deadly," a child has been issued the wrong prescription, and a pharmacist must track him down; in "Barking Death," a rabid dog is loose in a town; in "Vial of Death," a case of bubonic plague has been waylaid by a researcher, leading to a breakneck chase. Races against time also have ramifications for whole cities and nations. In "Big Bomb" on *Dragnet*, Joe Friday must talk down a man with a bomb at city hall, telling the story of "fifty-eight minutes that stood between a city and total destruction." In "Sabotage in Paris" on *Dangerous Assignment*, agent Steve Mitchell must prevent a briefcase bomb from reaching a political meeting of European leaders. In its final season, *Quiet, Please* aired several plays about atomic disasters that play on the idea of lateness, including "If I Should Wake before I Die," in which scientists blow up the moon with an atomic rocket. The cautionary fable ends with the sound of a ringing alarm, as actor Ernest Chappell steps halfway out of character to remind us, "It's not too late to wake before you die." When it thematizes the awesome power of atomic weapons, the location at stake in the "loss of time" in a radio drama often spills out from the edge of the fiction being depicted and into our own world.

These plays show a gradual process in which the conventional usage of time-out-of-time that characterized 1930s radio became a time-running-out in the 1940s, and then was finally folded into the "nuclear apocalyptic" trope of the 1950s.[36] Actually, the catchphrase "later than you think" was by then the counterpart of the icon of the Cold War: the "doomsday clock" in the *Bulletin of the Atomic Scientists*, which Paul Boyer has called "one of the best known symbols of the atomic age."[37] A similar image appeared on the cover of *Time* in January of 1949, which showed purple clouds and a clock hanging above the head of President Harry Truman. One of the most popular songs that year was "Enjoy Yourself, It's Later than You Think," which was released in three versions within months of each other by Guy Lombardo, Doris Day, and Tommy Dorsey. And this is to say nothing of the many clocks that became central motifs in 1940s noir films like *The Big Clock* (1948) and *The Stranger* (1946). With so many countdowns, clocks, and races working as dramatic motifs across the media of the period, it seems clear that a mood of lateness emerged from suspense radio to become useful to the imagery of the era. Is there a politics to that mood? Richard Hofstadter suggests as much in his well-known essay on the politics of paranoia, explaining that to be an effective paranoid spokesman, figures tend to portray themselves as living forever at a turning point. "It is now or never in organizing resistance to conspiracy," Hofstadter writes. "Time is forever running out."[38] Oboler's cliché articulated the desperation behind such posturing. Clearly, even if the radio age was in decline in late 1940s and early 1950s, this decline is surely less interesting than the way that the idea of lateness itself could structure human thought and feeling as that deterioration took place.

Trapped in radial space, immobilized before incoming hordes, and with time sliding away mechanically, postwar radio had drawn an ingenious constellation of rhetorical equipment that would give shape to dramas, exacerbating an interrelated set of beliefs and moods held in common by many Americans in postwar society. But what made this set of conventions so vivid is how it contrasted with the tradition in which broadcasters and audiences were steeped. If not for the capacious and kinetic dramas of the 1930s, immobility would have little meaning. Had fatalism not been so metaphysical in the past, its mechanistic quality would not disorient. In the chapters that follow, I expand on how radio remade itself during this period in police procedural programs and other serials to adapt concepts of communication and interiority that reinforce the underpinnings of postwar society while also critiquing them. As we shall see, the aesthetic of panic outlined above was just the beginning of the paranoid universe of the postwar years, a period in which radio would also make a number of problematic assertions about the relationship of the perceiving subject to authority. Richard Pells has argued that as a result of the conservatism of the McCarthy period, many Americans suffered "inhibited thought" and "chose silence."[39] What follows attempts to articulate what this inhibited silence sounded like in the theater of the mind.

CHAPTER 9

Just the Facts

In part 2 I argued that during World War II, radio programs often seemed geared to ask what communication is and how it deposits beliefs into the mind. Because the spirit of this query informed both aesthetic structures and story themes, there was a conceptual substratum linking such otherwise disparate programs as *Words at War* and *Inner Sanctum Mysteries*, both of which can be readily analyzed by unpacking semiotic qualities or asking how key characters transmit signals, receive messages, or speak through proxies. So what became of this substratum as the 1940s waned? This chapter will show that just as radio reinvented spatial and temporal forms to fit Cold War anxiety, notions about communication also transformed, especially in crime programs from the end of World War II into the 1950s. Perhaps because of its idiosyncratic thematization of communication, radio excelled among its sister media when it came to the types of crime story that centered on direct address and acts of testimony. These two conventions reflect a persistent cultural fascination with coercion. They also helped many Americans process the modalities of authority, the pursuit of evidence, and the dynamics of self-incrimination during years when these ideas were of no small importance. Radio drama was at once asking how mass communication deposits belief in the mind and lionizing those who disclose it enthusiastically to officials. Many plays endorse such activities outright, as others warn that the compulsive production of testimony turns citizens into inchoate agglomerations of information—characters become "just the facts," as the tagline from *Dragnet* had it. Nineteen-fifties radio was retasking a familiar set of commitments and conventions to express a Red Scare culture in the midst of a political realignment marked by the performance and production of testimony for authority structures. To put it another way, by attending to the prevailing patterns of communication in the 1950s crime story, this chapter argues that shows about communication became "loyalty programs," in every sense of the term.

CRIME RADIO IN THE AGE OF NOIR

For decades, radio crime shows were eschewed as diversions, but nowadays scholars such as Kathleen Battles, Jason Loviglio, and Elena Razlogova have rediscovered programs of the 1930s and 1940s such as *The Shadow* and *Gangbusters* to understand evolving attitudes toward the mass media and crime.[1] Although my readings build on the insights of these authors, the account below differs in its focus on crime narratives in both anthologies as well as serials, and in my concentration on the postwar years, which might be the most prolific period of the genre—a given postwar year saw up to thirty-six unique crime and mystery shows air in prime time each week. These programs were tonally different than earlier fare. During the 1930s, Battles has argued, criminals were depicted as violent enough that they lost their romantic "social bandit mythology" and the listener could identify with righteous, manly police heroes, but seldom did these hardened criminals seem so maniacal as to be incurable through proper education.[2] Postwar programs took a different approach, producing qualitatively darker villainy to be first sensationalized and then rebuffed. In tales such as "Too Hot to Live" and "Can't We Be Friends," aired on *Suspense*, thinly veiled rape scenes feature breathy men who murder women during musical underscores that gruesomely climax with the act. On *Broadway Is My Beat*, plays such as "Helen Carrol" have incest subtexts, while we hear stories of pedophiles hanging around schools in *Suspense*'s "A Little Piece of Rope" and tracking down girls in *Dragnet*'s "Child Killer." *Dragnet* also aired stories of pornographers exploiting heartland girls fallen prey to the lure of Hollywood, drag queens robbing motorists, youth gangs headed by "young Hitler"-types who lead otherwise healthy boys down the road to ruin. And the popular cop show was not alone. Drag-racing teens suffer fiery deaths in *Mr. District Attorney*'s "Hot Rod Killer," while women sell babies to buy a new TV set in "The Baby Peddlers" on *This Is Your FBI*. Such themes correlate with drugs, another postwar obsession. In its first two hundred episodes, almost one in ten *Dragnet* broadcasts features narcotics. Postwar radio let loose what had been repressed, as if to prove that deviance is "containable" by muscular forces policing the edges of conformist society. Most episodes of *Dragnet* start with a call coming into our audioposition in a shallow room at headquarters

"And now a word to our friends in the underworld: If we, tonight, have been able to convince you that crime does not pay, and you are persuaded to square yourselves with society, please mention this program when you turn yourselves in."

FIGURE 9.1. Whitney Darrow Jr., "And now a word to our friends in the underworld..." A 1948 *New Yorker* cartoon lampooning the disingenuousness of crime radio's claim to operate as a public service as it profited from tales of violence. © Condé Nast and the Cartoon Bank. Used with permission.

as a crime is reported, and Detective Joe Friday drives out on a quest into a scene somewhere in the city that features a richly illustrated deep space—a film stage with a scene being shot in the background, a salvage yard with crunching gravel underfoot, a public swimming pool with splashing kids. From this uproarious public space, Friday retrieves a dangerous criminal, drawing him back into the tight sonic world of headquarters, as if eliminating a discordant note from the happy din of the bustling postwar society outside, sequestering the voices of madness and destruction in the silence of police power.

Such complexities suggest that scholars have barely scratched the surface of the role that radio played in the development of the modern crime story. Throughout the 1940s, radio programs promoted the theatrical releases of films noir and adapted them for the airwaves, while radio directors favored both the voices and authors associated with this cycle of films. *Suspense* cast Joseph Cotton, Rita Hayworth, Ray Milland, Robert Mitchum, and Barbara Stanwyck during years in which they starred in such films as *Journey into Fear* (1943), *Detective Story* (1951), *Lady from Shanghai* (1947), *The Big Clock* (1948), and *Clash By Night* (1952). Dana Andrews starred weekly on *I Was a Communist for the FBI* while headlining such films as *Where the Sidewalk Ends* (1950), and Van Helfin played private eye Philip Marlowe on air while appearing in *The Strange Love of Martha Ivers* (1946). Fred MacMurray, Dick Powell, and Edward G. Robinson had distinguished radio careers, also appearing in *Double Indemnity* (1944), *Murder, My Sweet* (1944), and *The Woman in the Window* (1944). And while character actors such as William Conrad, Vincent Price, and Richard Widmark were playing rakes and heavies in *The Killers* (1946), *Laura* (1945), and *Kiss of Death* (1947), one or more could also be heard virtually every week on *Gunsmoke*, *Escape*, *The Shadow*, and *Inner Sanctum Mysteries*. Just as film scholars have neglected this material, literary scholars have also ignored *The Adventures of Sam Spade* and *The Adventures of Philip Marlowe*, even in accounts of the hard-boiled novels and magazines of which these radio programs were franchises.[3] Raymond Chandler's "Murder in City Hall," "Spanish Blood," and "Pearls Are a Nuisance" aired on *Molle Mystery Theater* and *Suspense*, while James M. Cain's "Love's Lovely Counterfeit" aired on *Suspense* and Dashiell Hammett's "The Glass Key" aired on *The Mercury Theater on the Air*. Thriller author Cornell Woolrich was also featured on *Molle* ("Nightmare," "Two Men in a Furnished Room") and *Suspense* ("The Black Curtain," "Singing Walls"), while the 1944 film version of his story "The Phantom Lady" was adapted for the *Screen Guild Theater* and *The Lux Radio Theater*. *Lux* also adapted several other well-known crime films, such as *This Gun for Hire* (1942) and *Key Largo* (1948). *The Screen Director's Playhouse* broadcast "The Killers" and *Academy Award Theater* aired "The Maltese Falcon."

These programs represent an exciting challenge to the debate over noir, which is among the hoariest cases in genre studies, one so bursting with questions that critic James Naremore has proven that discourse on the genre is in many ways much more interesting than the actual films.[4] "*Noir* poses a fascinating thicket for film critics," as Tom Gunning has put it, "and the temptation, once one enters it, is never to come out."[5] It is curious that writers enter these woods without the compass provided by thousands of hours of recordings. For instance, ever since Chandler's "Simple Art of Murder" essay in 1941, scholars have pointed to the "wisecrack" in hard-boiled fiction as a "laboratory" for American language, a hypothesis that remains much more speculative than it needs to be by failing to use the aural media as a test for how such experiments reached American ears.[6] As Gerald Nachman has observed, radio private-eye dramas are "populated with cardboard heavies, wise-guy heroes, and slinky sloe-eyed babes with insinuating voices" that are familiar to the student of noir, and on radio one could quite frequently hear "hard-boiled Mickey Spillanish or ersatz Raymond Chandlerese."[7] Radio programming also makes extensive use of the trademark first-person retrospective voice-over found in many classic films noir, perhaps most famously in Otto Preminger's *Laura* (1944).[8] But is the origin of this structure really in genre film? In the early 1940s, "it all started when" was the opening of two out of three *Suspense* shows, and a history of the use of this type of temporal lapse cannot be told without taking into account radio programs such as *The Mercury Theater on the Air*.[9] Indeed, in the case of *Laura*, the narrator in question is not only a broadcaster by profession, but also a thinly veiled caricature of radio commentator Alexander Woollcott.

With such rich terrain available, it is small wonder that the relationship between film and radio is currently among the most promising emerging areas in the field of sound studies.[10] My own sense is that noir can be best revisited as a genre that was only made imaginable by a historically unique conjunction of several forms of media practice: it was a conceptual space in which several mass media "conversed" with one another about aesthetics in the most expansive manner. Here I would like to explore just one passage in that colloquy, to motivate a study of significant aspects of postwar radio drama, including its use of direct address and testimony, activities that were perhaps more common in broadcasting than in otherwise similar mediated narratives. A clue to the discrepancy can be perceived in the different evolutions of noir themes in these media. Paul Schrader has argued that there are three principal phases in the history of classic film noir.[11] The first phase began in 1941 and included films such as *The Maltese Falcon* (1941) and *This Gun for Hire* (1942), stories featuring lone-wolf figures. This was followed by a phase of films such as *The House on 92nd Street* (1945), "realist" movies that deal in police routine. Finally, the last part of the cycle brought narratives of psychosis and suicide, such as *Gun Crazy* (1950). This periodization is strikingly out of sync with radio. Broadcasters seemed to first embrace tales of police procedure in the "realist" programs of the 1930s, such as *Gangbusters*. Then, sometime around 1941, the wartime shockers introduced "psychotic" noir tales, which continued to air well into the 1950s; meanwhile, lone-wolf detective serials had a golden age between 1945 and 1951. The latter ranged from the comic *Nero Wolfe* (1943–51) and *The Thin Man* (1941–50) to the classic *Philo Vance* (1945–50), as well as hard-boiled programming including *Phillip Marlowe* (1947–51), *Sam Spade* (1946–51), *Boston Blackie* (1944–50), and *Pat Novak, for Hire* (1946–49). *I Love a Mystery* featured characters who had been "soldiers of fortune" during the war but professed to be "private detectives" after it. Even Hop Harrigan, who had spent the war in the skies over Europe, retired to a rural airfield where he played gumshoe, solving mysteries loosely related to aviation.

The lone-wolf phase was abbreviated as the Red Scare hit the radio industry in 1950. Part of the change is directly connected to Dashiell Hammett, whose affiliation with the Civil Rights Congress put him in the hands of Red-baiters in 1949. CBS had already cancelled *The Adventures of Sam Spade*, and *The Thin Man* was soon dropped. By 1951, almost all other lone-wolf detectives vanished from prime time. In September of 1948, CBS cut *Casey, Crime Photographer* from Sunday nights. Mutual dropped *The Falcon* from the same night in early 1950, along with *The Saint* five months later, and ABC axed *The Fat Man* and *I Deal in Crime* from Monday nights the next year. Many cancellations made room for police procedurals. When Mutual cut *Michael Shayne* from Tuesdays in January of 1947, it was replaced with *Scotland Yard's Inspector Burke* and *The Crime Cases of Warden Lewis*; ABC's Friday schedule replaced *Richard Diamond, Private Detective* with *Defense Attorney*, yet made no change to *The Lone Ranger* or *The Sheriff*. In 1949 producer Frederick Ziv syndicated *Philo Vance*, based on the tales of S. S. Van Dine's bon vivant detective, but by 1951 he was selling *I Was a Communist for the FBI*, the tale of an undercover man infiltrating subversives. By the height of the Red Scare, most of what remained in the crime genre were shows developed in the 1940s that centered on special agents—*The FBI in Peace and War, Dangerous Assignment, David Harding, Counterspy*—and a new crop of police shows and federal agent adventures, such as *The Man Called X, Sergeant Preston of the Yukon*, and *Tales of the Texas Rangers*, indicating that radio had come full circle back to the procedural framework that had prevailed in the early 1930s. If a given time slot was designated for a crime or mystery show, that time was probably filled by something like *Gangbusters* in 1936, something like *Inner Sanctum Mysteries* in 1943, something like *Marlowe* in 1946, and something like *Dragnet* in 1952.

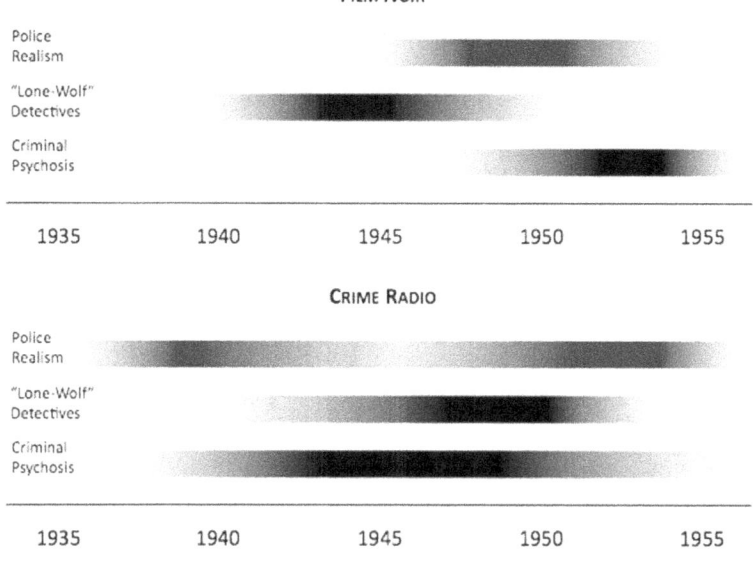

FIGURE 9.2. Evolving themes in the crime genre during the golden age of radio.

So by the end of the war, while crime films were entering their "psychological" phase, radio had been in one for years, and at the same time it transformed a parallel lone-wolf cycle into a police-centered formula. Here I concentrate on the latter development, which can be illustrated by close listening. Consider a typical play of the lone-wolf cycle: "Night Tide," an episode of *Marlowe* starring Gerald Mohr and adapted from Chandler by writer Milton Geiger, a veteran of *The Columbia Workshop*.[12] The play opens with the voice of the hero about to recount a retrospective tale. This move differs from a *Suspense* thriller in that the time of narration does not overlap with the duration of action. "Night Tide" is not told by a character who is poised at the last moment of the fruition of events that are still happening to him or her, but rather by a detective who had participated in the main action some time ago but is now at some unspecified moment in the future beyond the conclusion of the tale about to be told.[13] Our narrator is not in midst of things and cannot be affected by them, but speaks in a mode of recollection that reminisces how Chandler framed his stories on the page:

> When it started the tide was high on the San Pedro waterfront, and a hot-tempered kid had murder on his mind. But there was a knife at my throat, a

beating under the piers, and a corpse on the beach before the tide went out again . . . and the kid was finally stopped.

After a crash of music, announcer Roy Rowan identifies the show: "From the pen of Raymond Chandler, outstanding writer of crime fiction, comes his most famous character in [drum roll] 'The Adventures of Phillip Marlowe'!" Another crash of music fades into the droning foghorn and slapping water that will underscore the drama as keynote effects. We also hear a cruising car, suggesting that we move into this soundscape from elsewhere. Marlowe continues:

> It all started in San Pedro, the harbor of Los Angeles. The lights on the piers were fuzzy through the wet mist that creeps up out of the ocean every night. I drove slowly looking for the establishment of Mike Basso, my new client. One side of the crooked street was nothing but the smell and the sound of oily salt water sloshing through the pilings beneath the piers. And the other side was a tangle of warped dingy buildings equipped to satisfy the thirst of reckless men who never get beyond the waterfront of any port. The foghorn out on the breakwater began to bellow as I parked near a U-shaped pier labeled "Basso Docks, Private," and walked past the line of moored fishing boats to a squat two-story office. The bottom floor was dark, but the second floor had lights on. So I started up the steep wooden stares and was half way to the top when I caught the voices . . .

The preface employs words and sounds associated with film noir to evoke a cliché environment in which the denizens of crime fiction exist and their acts can take place, the kind of Janus-faced place that Chandler once theorized—a place where "the nice man down the hall" is really a gangster and "a judge with a cellar full of bootleg liquor can send a man to jail for having a pint in his pocket."[14] Chandler's proscriptions for crime fiction flow from this move of theorizing the existence of a "realistic" landscape full of characters that each possesses a second, underworld self. In a city with an underbelly, people have dark sides, ulterior motives, inner polarities, and second selves; Chandler scorns policing structures because this underworld can only be accessed by figures marginal to it. As Warren Susman has noted, it is by virtue of detachment from the social fabric of rules and regulations that hard-boiled detectives prove able to "impose order" on a disordered world.[15] That disorder is surely the inspiration for Geiger's imagery of scummy, stagnant water in a city of crooked streets that lead from the upper world of Los Angeles to the

San Pedro underworld, a Hades-like scene conveyed by first-person opening narration, the hallmark mode of address in lone-wolf shows. As critic Edward Dimendberg has put it, the city is according to noir formula "a highly rationalized and alienating system of exploitative drudgery permitting few possibilities of escape," a space in which men like Marlowe operate as "as a kind of mobile perceptual center," our Virgil in the Inferno.[16]

As "Night Tide" continues, we follow Marlowe's efforts to protect his client Basso from a former employee recently out of prison—"a hothead punk by the name of Johnnie Dyke"—whom Basso has previously accused of stealing. Soon Basso washes up dead under a pier, and Marlowe sets out to solve the crime, interviewing ladies of the night, breaking into darkened offices, manhandling Dyke's best friend Ed Giles, and demanding "a double order of plain facts" from Dyke's wife, a waitress named Christine. Marlowe discovers a conspiracy between Christine and Giles, who framed Johnnie and burglarized Basso in the first place. The plot contains a list of generic items identical to what we find in crime film of the 1940s: a disloyal and avaricious wife, mistaken identities, fistfights in the dark, "a big guy with a Latin jaw" named "Sharky," bumbling police, and the gradual alignment of the detective with Dyke, the type of idealized brother-figure that critic Sean McCann has identified strongly with Chandler's narratives.[17] In the end, Marlowe exonerates Johnnie, who decides to push out to sea. Marlowe is alone in the gloom:

> I watched him walk away, until he'd gone the length of the empty pier and was swallowed up by the night. Then I turned back to the shallow black water beneath me, which, where the sea and the land were close to meeting, was coated thick with oil, and dirty, almost stagnant. [Sighs] And I thought a lot about Johnnie, the people like Chris and Giles he mixed with and trusted. I felt sorry for him. But then . . . then, I looked up a little, away from the water at the pier, and out toward the open sea, where it was deeper and cleaner. The further I looked, the cleaner it seemed to be. Then I remembered that that was where Johnnie Dyke was heading, and I felt better.

The Janus-faced city is beyond repair; the only solution is escape. In this episode, Marlowe has not only succeeded in de-alienating Dyke, but also given impetus for the idealized sap to abscond the duplicitous city, to go out where the water runs clear, rid of encrusted scum and two-faced men. But the very tincture of regularization inherent to serial narrative also undermines this statement by promising that any imposition of order will only last until next week's episode.

THE LONESOMEST MILE IN THE WORLD

The conjunction of features that coalesce in *Marlowe* began to pass out of popularity as the genre evolved into the 1950s. One interesting transitional case is *Broadway Is My Beat*, a dramatic series helmed by such broadcasters as Elliot Lewis, a veteran of *The Columbia Workshop*, and John Dietz, who had run sound effects for Orson Welles in the 1930s and staged "The Hitch-Hiker" in 1942. A stylized procedural, *Broadway* resembled social documentary crime tales inasmuch as it used a murder to introduce us to a city.[18] The intro ran, "Broadway is my beat, from Times Square to Columbus Circle—the gaudiest, the most violent, the lonesomest mile in the world!" Although following a policeman, the program already sounds like *Marlowe*. Here is Larry Thor playing Detective Danny Clover, opening up an early episode entitled "The Julie Dixon Case":

> Broadway is a place that can fool you. It walks by the lost and the broken and the dying without batting an eye. But when one of its own lies dead, Broadway tears its collective breast, dons the sack cloth and ashes, and sends up a shrieking lament that can be heard round the world. For a little while you believe it. You believe Broadway is heartbroken. Because death came on a man who called himself Max Magnificent and stuck a knife in his back... Then you take a good look at Broadway and you know you're out of your mind.

The prose is the same, and the city just as duplicitous. There are also plenty of innocents to be rescued, often members of subordinate ethnic groups. In episodes that ran from 1949 to 1954, Clover turns up the bodies of reformed Harlem delinquents, Chinese professors from Mott Street who are killed over precious idols, and young Italians gunned down after talking the rap for bosses. But no idealized brother-figure is available, as the detective's affection is channeled into a cavity left behind by a character dead at the outset of the play. As a result, even when the culprit is caught, the drama lacks the sense of release in "Night Tide." Clover may feel for the victims of crime, but he cannot free them from Broadway, because they have already left this Hades for another.

This lack of release is just the first step in a gradual undoing of the *Marlowe* formula. *Broadway* became more engrossed in procedures than in victims, and the more that detection seemed like science, the more the detective himself began to unravel. Here is the introduction to "Ruth Larson" from the summer of 1951, a few months after "Julie Dixon":

> The nighttime starts at the river before it closes over Broadway. A wind drifts in with the moistened shadows, flings them into the street, flattens them against the gutter, picks a man waiting for a bus and wraps darkness around him. And a light comes on, and another, and down the street, there where the crowd is gathered against the traffic signal, high above them a neon sputters, flames, the spectaculars dance, somebody runs into the street and yells "come on" and everybody does. Night has come to Broadway! And where I was there was a wind. A built-in wind, a thing composed of poor ventilation. Tears, shed and unshed and bottled chemicals. It was the basic ingredient of the city morgue, though not to be found on blueprints or bills of specifications, it was something new to the man walking beside me.

Rather than prowling the lonesomest mile in the world, Clover prowls headquarters alongside Dr. Zinsky, an increasingly significant figure who solves chemical riddles and diagnoses "paranoids" who commit crime. *Broadway* was adopting a type of crime fighting associated with the machinery of modern policing, including laboratories, reports, and clerk work, changes that represented the professionalization of law enforcement.[19] As this procedural activity expanded, Clover's narration began to change. This pattern is clear in "Joe Blair" just a few months on:

> The sounds you hear on Broadway are fragments. Words, broken off and windblown that drift your way, the swift dart of subway noises, and a horn, and a whistle, and footsteps. The brief wild sob of the faraway river. You've got to listen close so you'll know if *that* sound began with laughter or with despair. The difference it makes, not much. Broadway reacts to clowns and death in nearly the same way. The blonde who had a little accident on the street corner, or the dead man you saw propped against a fence in an alley. Something to tell your family about—how the policeman pushed you back so you couldn't see how it all ended. But I saw. I had to stay to the end. Till Tommy Camp was lifted down and shrouded and taken away. Until Tommy Camp was made a matter of official concern. Then I left. Go to a place, back to headquarters. Write it down to be transcribed later by a stenographer. To be dated by a dater, stamped by a stamper, to be put in a file by a clerk.

As stenographers, daters, stampers, and filers take over work once performed or eschewed by the lone detective, he suddenly begins to anguish over the fragmentary nature of his existence, a fact reflected at the sentence level. In the passage above, we start out with the second-person subject pronoun used in the sense of "one" ("You've got to listen close"), which is associated with falsehood and is contrasted with the first person ("But I saw"). Toward the end both "you" and "I" are confused with the imperative ("Go to a place, back to headquarters, write it down") in which there is uncertainty about whether these actions are meant for the second-person subject or the narrator. Clover speaks to himself as if he were another individual, then speaks as himself, finally self-disassociating and telling himself to carry out actions. Just a year earlier, prefatory narration had been used to give us a sense of the duplicitous and polarized city in which savvy detectives encounter alienated men and women, but now these same addresses are being used to prove that the detective himself is alienated. By late 1952, there is little doubt that this identity crisis has been produced by a rise of scientific policing. In "Kenny Purdue," Clover is called to a party where a drunk finds a dead body in his bed. Clover describes the arrival of crime scene specialists:

> The men whose fate it was to walk through strange doorways made it to the entrance. The men from technical. Standard operating procedure, varied as to the size of the room and the location of the body. Death measured on a steel tape, noted as a symbol, photographed with the best of equipment. Result? Kenny Purdue's dying recorded precisely in space. So leave there. Up the flight of stairs again, and home again. Bed again. Three hours of it. Then up again, and coffee against the rest of the sleep that didn't happen. Call headquarters, give my routine for the morning. And go now to the address of the dead man.

Here we have missing subject nouns, absent verbs, and sloppy syntax. Even the tense of narration is confused. Marlowe seemed beset by his responsibility to free the righteous from the shadowy city, but Clover is psychically disintegrated by the very policing structure of which he is a component. The larger the whole, the more splintered the part.

As 1950s dramatists drew "the authorities" into the noir world, it seems, the lone-wolf persona dissolved. This development is partly the result of long-standing efforts to rebrand law enforcement, but it should also be understood in the context of postwar backlash against the large organizational forms associated with the New Deal. Richard Hofstadter has pointed out that among the underpinnings of the paranoid style of the 1950s was distaste for Depression-era reforms and their emphasis on the social; historian Mark Fenster has suggested that 1950s culture confronted the old paradox of imperiled individualism in the face of systemic culture.[20] This very struggle is reproduced in *Broadway*, as Clover is driven to self-alienation by systemic machinery exemplified by "the men from technical." Perhaps surprisingly, J. Edgar Hoover had long favored the latter—representations of the FBI that feature a scientific model of crime fighting.[21] According to Richard Gid Powers, Hoover disdained many of the G-man films and radio shows of the 1930s and '40s, many of which feature apparently loyal Bureau men who nevertheless work outside the law.[22] This "loner" G-man diminished in the 1950s. As historian J. Fred MacDonald explains, during the 1950s, crime radio set out to prove that the incorruptible FBI was the pinnacle of progress, "the most efficient, the most scientific law enforcement organization in the world."[23] Mindful of this image, writers provided heroes that no longer intimated feelings, beat up suspects, or empathized with saps, but rather conducted themselves as professional men dispassionately obeying protocol and gathering facts for use at trial. Feds typed out reports rather than stalking mean streets. On *Dragnet*, we learn nothing about Joe Friday's inner life, aside from tiny hints: Friday lives with his mother; he has an uncle in Washington state; he has spent time in Phoenix; he has a pet pug. Friday goes whole seasons without ever drawing his gun, and his dialogue is closer to "police blotterese" than "Mickey Spillanish" to borrow Nachman's terms.[24] On *This Is Your FBI*, Special Agent Jim Taylor never has a love interest, seldom appears in a private setting, rarely finds himself in mortal jeopardy, and avoids decisions about how to pursue an investigation. Instead, he is forever returning to headquarters to file a report, following the next step in the chain of command, relying on experts, and obediently waiting for test results. "The G-Man was different from the other detective heroes," Gerald Nachman explains, "because he spent less time with his girl and more time in the classroom, the laboratory, and on the phone to Washington. . . . He had the whole F.B.I. organization backing him up: files, labs, machinegun ranges, and step-by-step instructions from the Director."[25] The story of 1950s law enforcement is the Sisyphean routine of chasing down endless background checks and making long lists of leads. Although Hoover was the very symbol of postwar anticommunism, stoking the paranoid style, he nevertheless endorsed a New Deal–style of crime fighting on the airwaves, at the expense of characters such as Clover.

DIRECTING ADDRESS

To get an idea of the sort of depiction of law enforcement that Hoover preferred, and of how it embraced systemic models that the paranoid style ostensibly abhorred, I would like to take a deeper look at Jerry Devine's *This Is Your FBI*, a show endorsed by Hoover, with whom Devine built a friendship.[26] In its mature form of the 1950s, *Your FBI* crystallizes the conventions of the late procedural and represents a stunning stylistic break with *Marlowe* and *Broadway*, particularly in how it addresses members of the audience. At the top of the hour, we hear a billboard issued in the voice of the Bureau: "This is your FBI, the official broadcast from your Federal Bureau of Investigation. Tonight's FBI File: Citizen Caldwell!" Before a syllable of story material is uttered, announcer Dean Carleton begins to declaim:

> In times of stress, times like today, ordinary words take on different meanings to different people. Take the word "loyalty investigation" for example. There are some who regard the current government employee investigations as witch hunts. Some feel that they do not go far enough. Others are not clear how the FBI goes about such investigations. Tonight's program will answer those and other questions about loyalty investigations. Of approximately 4 million government employees whose records have been checked by the Federal Bureau of Investigation, in about 18,000 instances information was discovered that warranted a full field examination. This is the case history of one such investigation.

The tale is not a melodrama, but purports to be a true story. It is not offered through the consciousness of one of its participants or even from a speaking character, but as if a federal institution is speaking to us as real citizens. The introduction is not designed to set out a scene in which acts take place, but to serve a tutelary function, providing us with public information. In the 1930s, as Kathleen Battles has shown, police figures had served as hosts in crime docudramas on the radio in a way that confirmed the authority of real police to understand and make sense of events.[27] Carleton's diatribe takes this to the next level—not merely narrating events but delivering polemical commentary that makes sense of the world outside of the fiction. The show may not even be crime "fiction" at all. Indeed, because the play addresses us as the people we really are, it is closer to a public affairs commentary than it is to a film noir. When Clover says "you," he employs a figure of speech, but when Carleton uses it, he intends the second person literally. Just two minutes into the broadcast, we are already far from the lone-wolf era, and one thing drawing us out of it is direct address (speech to an audience *as* an audience), which had not occupied such a large part of the conventional dramaturgy of the crime show since before World War II.

As the broadcast continues, we segue to a Chicago apartment, where Thomas Caldwell is chatting with his wife Clara about changing careers, contemplating entering the civil service so that his war record will count toward retirement. No sooner does the idea occur to Tom than we segue to an FBI office in which G-men are looking over "Caldwell's file," discovering that although he has a Purple Heart and the Bronze Star, Caldwell has at least once entered Communist party headquarters. Is he *really* loyal? Our external narrator returns:

> No FBI investigation is ever a one-man job. By its very nature a full field examination, like this one is, can only be operated by a tightly knit, smooth operating law enforcement organization. For example, wires were sent to three other field offices, offices in Boston, Milwaukee and Cincinnati, where Caldwell had previously resided. In Washington, FBI notified the civil service administration that an investigation was being conducted. An investigation that was digging deeper, ever deeper into the background of one Thomas Caldwell, an investigation for one thing: the truth.

Agent Jim Taylor discovers that Caldwell has no police record and is part of a machinists union. We segue to Taylor interviewing Caldwell's best friend, who does not hesitate to provide details about Caldwell's voting pattern, his belief in civil liberties, and his support of an antilynching petition. So far Caldwell is the model of a postwar middle-class man. We segue back to Carleton:

> Special agents of the FBI don't spend their days behind mahogany desks. They investigate. They develop new leads. They start with a slip of paper containing a man's name: Thomas Caldwell. Soon it's the Caldwell File, as agents talk to his minister, to his neighbors, to a man with whom he once worked, to members of the veteran's group he joined. Reports are written after each interview, whether they seem important at the time or not. For the agent's job is not to judge, but to collect information. To collect the facts. As many facts as possible. And the search for facts concerning Thomas Caldwell takes Taylor to meet the head of the Machinery Worker's Union . . .

The union boss is immediately suspicious of Taylor, insisting that he is a Gestapo-type and refusing to answer questions unless under subpoena. Unfazed, Taylor types up the report, as another agent enters to advise him that an undercover operative might have some information. We segue back to Caldwell telling his wife about the ongoing investigation of which he has apparently been apprised. Clara does not understand the fuss, explaining that she does not see anything wrong with government investigations: "So that we can be sure no Communists are working for them." Then Tom confesses: he joined the Party back in 1946! In a (presumably unintentional) comic segue, the program then cuts to an advertisement for life insurance.

As the next act opens, Carleton is proud to deliver a message to the audience that supposedly comes from Hoover himself, a device that the program employed on several occasions:

> The Communist Party in this country has adopted the technique of the big lie. They try to smear everyone who disagrees with them by labeling those people fascists. They look upon the real liberals of the nation as their sworn enemies and at the same time seek to corrupt liberalism in every possible way. Other people, equally mistaken, sometimes attempt to smear all liberals by labeling them as Communists. Some even set themselves up as self-appointed investigator of their fellow citizens' political beliefs. Don't make that mistake yourself. Vigilante action and witch-hunts are never any help. They only contribute to the Communist cause by pitting loyal Americans against other loyal citizens. We need no secret weapons or vigilantes to defeat Communism in this country. Your civil liberties are at stake in this fight to keep America free of Communism. Contribute to the protection of those liberties. Don't call names, and don't investigate on your own. If you suspect anyone of sabotage, espionage or subversive activities, look for the number of your local FBI field office in your local telephone directory, and call it.

It will turn out that this is good advice. As the story resumes, Taylor uncovers that Caldwell is part of a local cell of the Party, but word comes down from the judicious Mr. Hoover that they need proof beyond accusations. Taylor arranges for an undercover man to drop handwriting evidence from Party headquarters inside a magazine at the public library. Comparing Caldwell's samples with Party documents, it seems that all is lost for Caldwell. But at the last moment, the union boss from the first act of the program returns, now willing to testify. "Caldwell never really was a Communist," he explains "for almost three years he tipped me off as to what they were up to, and at the same time used his influence to get them to move slow." We segue to the joyous Caldwell family, celebrating Tom's approval by the Loyalty Board of the Department of Supply. Before sitting down to write a thank-you letter to Hoover, Caldwell muses, "You know, Clara, you never really appreciate democracy until you see it work. Until you see it do *this* kind of a job." In a conclusion that is intended to be reassuring, the announcer explains that this story is just one of thousands of ongoing cases and promises that the agency is "a true believer in the Biblical prophecy that the truth will set you free."

The play's bounteous mystification can receive only a fraction of the study that it deserves here. To begin with, note that the drama has absorbed the normative vision of duplicity from the noir cycle. *Your FBI* continues to profit from the idea that the world has an underside that exists throughout the nation, though it is now made up of organized Communists rather than petty hoods. In a key difference from the 1930s, we are presented with an ideological underworld, not a social one. Combating such a conspiracy is a task not for a lone detective or even a police force, but for a large institution, as the play describes a threat whose scale seems to justify the Bureau's self-designating as the only thin blue line capable of cordoning it off.[28] Also, unlike the worlds of earlier crime radio, this one seems eminently remediable. Many episodes conclude like "Caldwell," with an unambiguously happy ending, contrasting the melancholy of *Marlowe* and *Broadway*. This resolution arrives thanks to the notion that testifying before the authorities is a civic obligation. Once provided with information by witnesses via agents, federal institutions can be counted on to provide justice. The system always *works* in these shows. To empower this ideology, *Your FBI* developed a dramaturgy that is aesthetically dominated by a series of direct addresses to the audience and thematically engrossed with the production of a file replete with testimony, two sorts of communication that had not been as crucial in previous eras nor in many other prominent forms of crime fiction. Direct address is one key to this fantasy, since it dispenses with the personality of the first-person narrator, a necessary move if we are to view facts as facts and not as information about the individual relating them to us.[29] Because he is a "Voice of the FBI" rather than a character, Dean Carleton is not liable to use points of information to express himself, so we know that direct address is offered as information about an objective world. Indeed the narrative arc of "Citizen Caldwell" contains little else besides a supposedly faithful outline of FBI activities to produce a complete file, a set of actions interspersed with—and brokered by—informative commentary. Caldwell the man is utterly tangential to this drama, and he is only liberated from suspicion because Caldwell the file has been ideally completed with all relevant testimony, since one's virtue as a citizen is knowable only through "the facts." Many plays of the 1950s are described as official "files" or "confidential reports" that have been long hidden but are now revealed for our edification. In the same years as radio stars were cited on secret lists of subversives, radio glorified those lists and the machinery that produced them.

TESTIMONY, CONTAMINANT, CRISIS

If direct address is one convention of the "factual aesthetics" of 1950s crime stories, another is testimony.[30] Of course, the act of extracting a confession had long been among the most vivid activities in radio. But when Marlowe beats the truth out of Ed Giles under a misty dock, he is not collecting facts for a transcript. In the old noir city, the stories of detective work wash away, leaving traces only in personal memory. For a confession to really be *testimony* it has to be dated by a dater and stamped by a stamper, as Clover might put it, or entered into a file, like the facts provided by the union boss that closes the case on Caldwell. Both of these acts of communication are tendered to a higher power whose present authority turns confession from moral catharsis into juridical act. Contrast "Caldwell" with the retrospective confessions common in shocker drama. In *Suspense* dramas, characters often speak to us as if the listener is the intimate chosen confessor of the protagonist, such as a lover, a psychiatrist, a reporter, or a representative of posterity. We are addressed as if the character speaking to us has mistaken us for someone else in the fiction. The confession is often a piece of writing, as in "Death on Highway 99," in which a man confesses murder in a letter to a high school crush whom he has never had the courage to approach. On *Quiet, Please*, dramas often center on an act of confessing to crime, and the twist at the end occurs when the teller learns the identity of the person to whom he or she has just related the tale. In "I Always Marry Juliet," for instance, a stage director admits that he has married and murdered three of his Juliets before he learns that he is dead and confessing to the shade of the Bard. In these cases, confession has a therapeutic dimension, the feeling of release of interior pressure through verbal expurgation. Like Ronald Adams in "The Hitch-Hiker," these character-narrators believe that their telling will serve a palliative function. But in few of these confessions is the character testifying for an official purpose. By introducing an authority figure into the genre, as Jerry Devine and others did, the context of speech acts is altered, transforming intimate confession into formal testimony. In this way, "secret file" dramas like "Caldwell" justified the subversive lists plaguing radio personnel, since these dramas lionized the men and women who testified before federal authorities (and *not* before union bosses) at just the historical moment at which public figures were being urged to name names.

Despite the extremism of *Your FBI*, it would be incorrect to characterize all dramas of address, fact, and testimony as inherently conservative. In fact, Devine's model of drama is resisted by *Dragnet*, the signature program of the 1950s, and one whose reputation depended on procedural "realism" that drifted into grim nihilism. Star Jack Webb was famous for attending police classes, riding along with real cops, and representing the Los Angeles Police Department headquarters so scrupulously that characters walked up the same number of stairs as there were at the real building. These choices were intended to make the program seem "as real as a guy pouring a cup of coffee" as Webb explained to *Time*.[31] On the strength of this approach, along with Webb's deadpan voice, *Dragnet* has been hailed as a "stylistic breakthrough" and has been considered as exciting an "on-air laboratory" as *The Columbia Workshop*.[32] Media scholar Jacob Smith has characterized Webb's delivery (known as "underplaying") as a reaction to the overplaying found in preceding microphone technique.[33] For a long time, Smith explains, many radio actors had played parts close to the microphone and filled their portrayals with excesses of emotion as a way of making up for the absence of reverse shots and mise-en-scène. Webb reversed the convention, providing less emotion and longer silences, a blankness that gave criminals' confessions spotlight. While crime shows from *Mr. District Attorney* to *Boston Blackie* usually ended with an apprehension, arrest, or sentencing, *Dragnet* almost always concluded in an interrogation room at the moment that the suspect agreed to talk. Just as Marlowe's special aptitude was to de-alienate saps, Friday's talent was to bully, finesse, or coax testimony from suspect and witness alike. "Whenever you guys show up I make up my mind you're not going to get any answers from me, I'm not even going to give you the time of day," a bartender quips in one play, "and then what happens? I end up telling you everything you want to know."

But is this the kind of truth that "sets you free"? In "The Big Knife" in 1950, Joe Friday and his partner Ben Romero are called to a school where girls have been slashed by someone bearing a razor, crimes that are described with enough ambiguity to allow listeners to read the attacks as rape. After asking victims to describe their experiences, an overachieving boy named Jim takes charge, leading the detectives around campus, and bragging that he has organized a "Boy's Watch" to apprehend the culprit. Fascinated with crime and eager to become a police officer, Jim explains that he did a class project on lie detectors and enjoys books on the criminal mind. Jim even speculates that "whoever the perp is" he is probably mentally ill and does not know it. Of course, Jim is quickly the prime suspect, and Friday and Romero methodically build a case by interviewing his girlfriend Barbara and taking samples from his hands, claiming that they have a machine to detect what he has recently held. In the end, the gambit works and Jim confesses. In his testimony to a stenographer, he claims, "I couldn't study unless I did it"; he even thanks Friday and Romero for apprehending him: "It was getting worse. Barbara . . .

after the prom tonight ... I was going to kill her." In this case, the culprit was already anxious to explain his crime, and Friday merely had to contrive the circumstances to allow this urge to play out. But as years went on, episodes of *Dragnet* were increasingly structured as little else besides confessions. "The Big Phone Call" takes place almost entirely in an interrogation room, as Friday and Frank Smith pile up evidence—secret wiretaps, spending reports, and eyewitnesses—in front of a suspect, asking the same questions over and over again until he caves in. In "The Big Parrot" a man is interrogated for two-thirds of the broadcast, although when he finally admits to murdering his landlord and burning down his house the only motivation he can offer is "I dunno." "The Big Cast" starts with an arrest after the investigation is long over, and the episode is nothing but a dialogue of Friday slowly getting more and more details from a serial killer as he dines on a plate of liver and onions, at last admitting his horrible deeds. Yet when it comes to a motive, all he can say is that he did it for "no big reason."

None of these dramas are really mysteries. The story is not propelled by evidence or by lists of suspects. There is almost no uncertainty involved. Instead, and perhaps more than any other program, the action of *Dragnet* is really structured to "produce" a piece of testimony, as the lack of it drives decisions made by the characters, and the final attaining of it brings a closure. Yet in the broadcasts that exemplify this impulse most directly, testimony curiously fails to produce an understanding of motive. Even in the case of the boy slasher, surely some psychiatric explanation could have been concocted, particularly for the listening public of the 1950s, which would have been savvy about such an account. Yet nothing is offered explicitly. In fact, the program proves that no matter how effective interrogation is, facts are still "just facts" and rarely illuminating of the interior of individuals relating them. This represents a challenge to the ideology of *Your FBI*, which seems to suggest that testimony is the only way to ascertain as privately held a "truth" as one's fidelity to country. For Devine and his FBI patron, the whole purpose of extracting testimony is to get to the bottom of belief. But Webb's program exhibited consistent incredulity with this scenario. With all the facts in the world, a character from *Dragnet* would never be as knowable as Caldwell. To put it in legal terms, *Your FBI* is about testimony as a way to access *mens rea*, the guilty mind, but *Dragnet* is about testimony that can only ascertain *actus reus*, the guilty act. Rarely in postwar crime radio do these two components seem to coexist.

Dragnet also challenges the affirmative tone of the postwar procedural with a mood of sheer futility that endured from the lone-wolf plays. Something of Marlowe remains in Friday, famous for his taciturn nature and pithy double entendres that lament the ceaseless tide of crime. Underplaying makes it difficult to know Friday very well, but there is at least one play in which his grim quiet is breached. In "The Big Tear," Friday's partner Frank Smith and his family receive threats from a thug. With both cops under strain, Friday launches into one of the longest monologues in the program's run, a rich anomaly essential to any understanding of him:

> All right look, Frank, I could try to spell it for you a hundred ways and it still wouldn't come out with any kind of a total. I know what you must be going through and I'm sorry it's got to be you. I'm sorry it's got to be any cop ... but it's squeezing me just as hard as it is you. You and I both knew when we filled out those application blanks what we were taking on. It's the job, Frank. Maybe I never told you before. Maybe you been through this too, I don't know. There isn't a day that goes by that I'm on the job here that I don't run into something that breaks my heart. Why did Ed Wilson have to get shot when he and his partner tried to take that guy out of that rooming house over on 8th Street? Why did Olson and his partner have to end up in the P & F ward last week? Why do you have to run the gamut of every human emotion and every class of people on earth? Poor to rich, it's all the same. Somebody picks up a gun to knock over the corner gas station, everything hangs in the balance. The minute that thief slips a cartridge into the gun, the gas station attendant, his family, his kids and every relative he's got slides right off the short end. Everybody seems to think that wherever crime is concerned, they never get any closer to it than the front page of the morning paper. I know you know this, Frank, for every crime I don't know how many innocent people are thrown into the balance even before the crime's been committed. That's the way it is, Frank. I'm not trying to tell you anything you don't know, but the minute you passed that exam, you had to make book that something like this could contaminate you just as surely as if you weren't a cop. I guess it's almost a legacy. Nobody's found a vaccine for crime, nobody's immune. It's worse with you and me because we know better than to think we can't be touched. We *know* we can every time we roll on a call, every time we pick somebody up, every time a guy gets out of line. It's the job, Frank.

Friday focuses on contingency and the incomprehensibility of fate in the work that he has chosen. Note the rhetorical questions, the repetition of "don't know," the confused metaphor of calculation (Friday cannot "spell out" a "total") and the use of conditionals such as "could" and "maybe." Friday also associates the theme of meaningless contingency with the argot of infection, explaining there is no "vaccine" or "immunity" and that he feels "contaminated." These attributes and themes suggest that testimony does not make the world easy to understand. Actually, testimony seems to make the man extracting it ever more susceptible to the capricious hand of fate. Neither the criminal nor the investigator can withstand the process of official investigation.

In this sense, although *Dragnet* is in many ways a deeply conservative program, it also prefigures a set of critiques of postwar life that were beginning to find articulation in the writing of David Riesman, William H. Whyte, and others—the worry that Americans had lost meaning in their lives as a result of organizations, "other-directedness," and "yesmanship."[34] It is an inversion of the themes in 1930s political culture. "What the writers of the 1930s called 'community,' the postwar intelligentsia labeled 'conformity,'" historian Richard Pells observes. "Social consciousness had turned into 'groupism'; solidarity with others implied an invasion of privacy; 'collectivism' ushered in a 'mass society.'"[35] In the *Dragnet* march, which thundered with an unmitigated obedience toward the policemen patrolling social mores, there is a note portending critique of the meaningless conformity that it modeled and enforced. In this way, the "lessons" of the postwar procedural phase of crime radio and its "factual aesthetic" tend to thwart one another. Embracing testimony as a form of communication to the exclusion of the richer field of communicative acts of the early 1940s, the crime serial theorized that the mind is wholly knowable to authority through divulged facts. Yet, at the same time, programs such as *Dragnet* cast a jaundiced eye on this procedure. One can testify and provide ever more factual quanta to arrive at an account of a series of events, but in the process both the confessor and the interrogator become contaminated with a terrible nihilism. In one way, this account shows that experiments with forms of communication continued to drive the agenda of radio aesthetics. In another way, it shows that radio still negotiated collective understanding, mulling over contradictory positions on pressing postwar cultural issues with remarkable articulateness. But the stark contradictions between the models proposed by these programs also reveal that the intuitive connection between communication and the mind that was established in wartime radio—the sense that these two ideas implied one another—was reaching a moment of crisis.

CHAPTER 10

In Trials

In February of 1950, a Pittsburgh clerk named Matt Cvetic achieved minor fame after his testimony at the House Un-American Activities Committee (HUAC), the first of over sixty appearances by the erstwhile FBI "undercover man."[1] Such figures were hot news that year thanks to a string of espionage scandals. In 1945, Soviet spy Igor Gouzenko had defected, while Elizabeth Bentley exposed Soviet informers at the Departments of Treasury and State. Writer Whittaker Chambers soon fingered State Department official Alger Hiss, leading to a 1948 show trial. The next year, the Congress of Industrial Organizations purged all members associated with the Communist Party, losing nearly a million members. Abroad, Berlin was blockaded and China was "lost" ahead of the outbreak of the Korean War, while security lapses surfaced at home: weeks after the test of a Soviet atomic weapon in 1949, an investigation opened on Manhattan Project mole Klaus Fuchs. And the Red Scare would reach fever pitch in 1950, as Julius Rosenberg was arrested and Senator Joseph McCarthy began his notorious crusade. Because Cvetic appeared before HUAC in the same week as McCarthy's infamous Wheeling speech, his photograph was attached to a *New York Times* article about the Wisconsin senator's numerically ambiguous allegations.[2] Stocky, wide-jawed, and balding, Cvetic could even be mistaken for McCarthy, and he proved nearly as melodramatic, accusing "front organizations" such as the American Slav Congress of plotting the overthrow of the US government.[3] According to his own account, Cvetic agreed to testify to relieve the strain and pressure of years posing as a Communist, a burden for which he blamed the death of his mother and the failure of his marriage. This account proved bogus: Cvetic had been fired by the FBI after incessant demands for more money. Jilted by the Bureau, he became a professional witness, nicknamed "Mr. Communism" by derisive attorneys.[4] Cvetic's diatribes on the witness stand soon devolved into inconsistency, as a record of alcoholism and mental illness came to light. Appearing before one hearing, he brought a dossier of evidence of Red infiltration of Congress, a file that was mysteriously "mislaid" during a lunch break.[5] In the end, Cvetic was an embarrassment, and by 1955 he retired to give lectures to groups such as the John Birch Society. Irate at Cvetic's apostasy, J. Edgar Hoover issued a release stating that whatever his delusions of grandeur, "Mr. Communism" did not in any way "represent the FBI."[6]

Yet he did "represent" the FBI regularly. In the summer of 1950, Cvetic was the subject of a *Saturday Evening Post* series entitled "I Posed as a Communist for the FBI," which told the tale of how he obtained as many as fifty thousand pages of documents implicating clergymen, union leaders, scientists, educators, businessmen, and lawyers.[7] In 1951 this inflated tale became *I Was a Communist for the FBI*, a Warner Brothers feature starring Frank Lovejoy, part of a cycle of paranoia movies of the 1950s, including *The Red Menace* (1949) and *Red Snow* (1952), films of humorless Commie thugs passing off microfilm and their nymphomaniac protégées luring otherwise upright men into treason.[8] Appearing just after the Hollywood Ten trial introduced blacklisting to the entertainment industry, this film cycle plainly set out to appease Red-baiters and calm the prosecutorial atmosphere. Scholar Thomas Doherty has described such films as "protection payments in 35 mm."[9] Seizing the opportunity, Cvetic sold the radio rights to his story for $10,000 to producer Frederic W. Ziv, a distributor of *The Cisco Kid* and *Boston Blackie*. Ziv also acquired the rights to the tale of Herbert Philbrick, whose best selling *I Led 3 Lives* (1952) told a similar undercover story. Ziv was no zealot. Actually, he was known to hire blacklisted writers to develop Red Scare kitsch, knowing that a fast buck had to be out there for shows with a little more dash than *This Is Your FBI*.[10] As he rewrote the Philbrick tale for TV, Ziv produced *I Was a Communist for the FBI* for radio, recruiting Dana Andrews to portray Cvetic. Not only did Andrews have plenty of radio experience, but he had also portrayed Igor Gouzenko in 1948's *The Iron Curtain*, while achieving critical success as working-class airman Fred Derry in 1946's definitive film of postwar manhood, *The Best Years of Our Lives*. In December of 1951, the first of seventy-eight episodes of *I Was a Communist* aired out of WIP Philadelphia, syndicated by transcription to stations nationwide.

THE ROMANCE OF THE UNDERCOVER MAN

The Cvetic show is irresistible to students of the Cold War. Over time, it has become perhaps the most listened-to radio program of the 1950s, a catalog of fantasies of Red mischief, few of which had much to do with revolution. Reds rob banks in "The Rat Race," scam money from the families of fallen Korean War soldiers in "Pennies from the Dead," and defame upstanding judges in "A Suit for the Party." Operatives forge banknotes, blackmail former members, hold up trains, and counterfeit a Rembrandt. In "My Friend the Enemy," Party boss Comrade Revchenko plots to annex the Pittsburgh rackets, while in "The Red Snow" Cvetic infiltrates a community of drug users and obtains as much morphine, heroin, and cocaine as possible, to force desperate junkies into the Party. In light of these activities, it is no surprise that the Cvetic tale appears in historian Ellen Schrecker's account of the postwar effort to use the media to transform Communism from a political philosophy into a crime; historian Richard Gid Powers has also studied the program, astutely noting that Dana Andrews delivers dialogue more slowly while under Communist cover than he does to the audience, as if "adapting to the cumbersome mental habits of his comrades"; and writer David Everitt has called attention to the show's stylization, suggesting that its soundscapes "could only be imagined by a cinematographer with an expressionistic bent."[11] These attributes feed the show into a mythology of the undercover federal agent, one of the key figures in the postwar imagination. Historian Michael Kackman has shown that portrayals of undercover men like Cvetic helped to construct postwar models of "acceptable citizenship."[12] In this way, *I Was a Communist* emblematizes a cultural atmosphere in which average citizens were entreated to do the righteous underground work of purging Reds from positions in government, entertainment, and education. That is why the Cvetic program is typically considered to be a drama about postwar American political identity, an exhibit of rabid loyalty during a period in which it was easily questioned.

Yet the themes perceived by these authors are poorly contextualized, a problem that relegates the show to the dustbin of Cold War oddities a little too hastily. Wider study of contemporaneous programming reveals that the common reading of *I Was a Communist* deserves to be revisited. For one thing, the show is in many ways no different from other 1950s plays about heroes "going undercover," whether to infiltrate fellow travelers or to break up cartels. On *Dragnet*, Joe Friday went undercover into narcotics organizations more than a dozen times in the first three years of the program. *Suspense* aired all sorts of undercover man stories: in "Allen in Wonderland," a lawyer overhears spies on a train, and goes undercover to foil an assassination plot; in "The Girl in Car 32," a cop insinuates himself into a group of underworld figures to foil a jewel fence. On *Counterspy*, special agent Harry Peters poses as a sympathizer to stop a fascistic cabal infiltrating Civil Defense Leagues in "Double-crossing Defender," and he assumes an identity to bust a smuggler in Mexico who is using flying saucers to ferry cocaine over the border in "Soaring Saucer." But even this farfetched yarn cannot top "Little White Pill" on *Dangerous Assignment*, in which Steve Mitchell goes undercover as a correspondent touring a Russian gulag and enlists the help of a female photographer (who is also undercover) to foil a kangaroo court and administer a mind-control antidote pill to an American businessman (who has been in Russia undercover) and rescue him from a jailor (who is himself an undercover operative for the West).

With so many programs dealing with disguise and undercover work by both authorities and civilians—many even politicizing this theme to the same degree as did the Cvetic program—it seems clear that the overmeaning of *I Was a Communist* must reside in a romance with the undercover life that underlies its efforts to model civic duty. With this in mind, consider the opening of "The Wrong Dream," the first episode of the Cvetic program. Amid thumping drums, an announcer begins:

ANNOUNCER: Many of the incidents in the story you are about to hear are based on the actual incidents and authentic experiences of Matt Cvetic, who for nine fantastic years lived as a Communist for the FBI. Here is our star, Dana Andrews as Matt Cvetic.
CVETIC: Externally you look the same, nine years later. A little more settled-looking, a little more tolerant of things . . . a little tired. But that's all—they think. Inside, are the ruins. Inside are the wreckage of loneliness and compromise, disappointment . . . and fear. That's the price you pay for being, for nine years, a Communist for the FBI, like me.
ANNOUNCER: In just a moment listen for Dana Andrews as Matt Cvetic, Un-der-coverman! [*Commercial break*] Now here is Dana Andrews as Matt Cvetic, undercover man. This story from the confidential file is marked "The Wrong Dream"!
CVETIC: I stumble out of Comrade Revchenko's office feeling like a guy who's just caught a live grenade and can't let go. All I can think about is getting to a pay station and call my FBI contact and maybe he can get the stinger out. We arrange to meet at "rendezvous green," currently a vacant garage away

from things on the North side. It's a place where a guy with a case of jitters can smoke and pace the concrete and flip his lid a little until the mimis go away.

The passage resembles most of the undercover man stories cited above, as well as many of the noir crime programs discussed in chapter 9, complete with the intimation that the drama is a "secret file." But there is also an interesting new theme: "the ruins inside," "the wreckage of loneliness," "the jitters," and "the mimis." This is the argot of neurosis, and nothing like it had aired explicitly among the heroes of shockers, police procedurals, or detective programs of the past. Yet the terminology of inner torment matches perfectly the "double voices" that Powers has perceived and also Everitt's idea of expressionism, both of which are dramatic conventions that gel with Cvetic's "psychologized" sense of his own situation. The consensus on the Cvetic program has been to consider its hero to be an embodiment of geopolitical conflict, a situation that engenders paranoid schizophrenia in which the undercover man must be two beings at once. In this way, *I Was a Communist* feeds into a critical belief that the Cold War colonized the American sense of the psyche. It is on this basis that critic Alan Nadel argues that the Red Scare helped to "institutionalize the traits of the asylum" into everyday life, for instance.[13] But there remains a whole underexplored side to this argument. In what anthropologist Catherine Lutz calls "the marriage of the spy and the psychoanalyst," not only did Cold War ideas influence psychology, but concepts of psychology also influenced how the Cold War was imagined.[14] This is just what is going on in *I Was a Communist*, a program whose dramatic conceit is to psychologize the Cold War, a task befitting a theater of the mind, one practiced at promulgating models of psychic life and putting them in extreme situations of assessment.

I make this claim in order to introduce this final chapter with a sense of how dominant "psychological tests" became in postwar radio, subsuming every other type of subject matter to a dramatic agenda bent on depicting minds in states of ordeal. Even in *I Was a Communist*, a series pushing all the hot buttons of 1950s politics, the intrigue is in many ways only accidentally involved in the Red Scare. The real focus is on pure mental tribulation. In 1950s radio plays, the prominence of psychology *as* psychology exceeds even that found in the works of Lucille Fletcher, Wyllis Cooper, or Arch Oboler; late radio is a theater descended directly from "Donovan's Brain," "The Horla," and "Green Light." But this continuity is limited. Dramas such as *I Was a Communist* are also deeply at odds with the sender/receiver dynamic that had long prevailed in psychological plays, as described in chapter 6. World War II radio characters are chiefly coded based on how they receive or issue messages relative to an outside world. These dramas are like radio systems, in which every mind is but a node in the larger architecture, one that either transmits or receives in order to reveal one's innermost "programming." The characteristic tests depicted in wartime radio have little to do with "the mimis." Instead a character produces or responds to some terrifically important signal or transmission that circulates through and defines the world of the play. This model remains remarkably consistent across a number of genres of the 1940s. No matter what else happens in a drama, at the level of motivation, we can count on the fact that the characters that we meet will be revealed by a message of some kind. This is one of the compositional benefits of using acts of communication to shape the idea of consciousness—it makes the illustration of interiority attributable to dramatic events that have to do with external connectedness.

But the situation is very different in the scene above. The fact that Cvetic gets orders and transmits them as intelligence is not nearly as revelatory or even as interesting as the fact that he must disguise the fact that he is doing so. In this way, the scene is symptomatic of the postwar period, during which radio depicted minds structured by a panoply of new formulas, including schizoid personality, madness, repression, tests of conscience and brainwashing, a series of psychological narratives that are difficult to reconcile into as coherent a schema as the sender/receiver binary had been in the 1940s. Dramatists were grasping for novel models of psychology in order to rationalize a new approach to portraying the mind in aural form. The only consensus among these dramas is something that the Cvetic program hints at when it depicts a mind attempting to protect inner thought from externalization and to thereby avoid the outer communicative relationship by which interiority had once been made knowable. In radio plays of the 1950s, that is, a situation of mental duress usually involves a character who struggles to maintain a boundary between the self and the outside world, to prevent inner voices from erupting, even if this means suffering the "jitters" in order to stop "inner wreckage" from showing in the exterior façade. By convention, characters came to be illustrated by lives lived behind a scrim raised between an inner voice and the society surrounding it. In general terms, if Depression radio pretended that interiority did not exist, and wartime radio considered it to be a permeable membrane, 1950s radio often seemed obsessed with hardening, managing, and disciplining the relationship between inside and outside. Of course, this is a very simplistic account of the human mind, one that could sustain a number of different interpretations and models. And although a great many radio dramas of the postwar era seem driven to investigate excesses of 1950s culture, these themes are only pretexts for tales that are more strongly driven by states of psychic distress. Many dramatists were performing "trial runs," attempting to find a powerful new account of the mind to replace the one built around models inherited from theorists such as Paul Lazarsfeld, but they did not settle on any one satisfactory version of the psyche. There is no unitary theory of mind on the airwaves in the postwar years. Yet the frenetic attempt to find such a model successfully raised a trenchant question about how the mind would be configured after the radio age, investing this issue with a degree of consequence that helped to make the relationship of the mind to the media continuously captivating for decades to come.

So what does this experimentation entail? In what ways did late radio "think about" the mind? To find an answer, it will help to look in greater detail at *I Was a Communist*, which I consider to be an exercise in the engendering of schizoid personality, a drama that tests the sturdiness of psychic structures that prevent externalization of speech. According to a fairly routine formula, each play in this series opens after a Party meeting, as Cvetic is asked to stay and receive a special assignment. A Party leader reveals that he has received urgent instructions from the Politburo, or even "Uncle Joe Stalin" himself. Based on Cvetic's excellent Party record, the cell leader chooses him for a task—picking up secret material, fomenting a riot, accompanying a visiting apparatchik, or committing a criminal act to "weaken" America—an assignment that is never what it seems. Dispatched alongside Cvetic is someone to watch his every move, and the first act of the drama finds the undercover man struggling to discover the "real" purpose of his assignment and to communicate it to the FBI without arousing suspicion from his Party shadow, lest Cvetic fall prey to the "Commie knuckle-boys," or worse, the "control commission." The protagonist has two tasks, at odds with one another: Cvetic needs to find out what the Reds are up to and to stop it by way of a message; at the same time he must also keep his real agenda a mystery to those watching over him. Action is bound by the effort to both divulge and seem to protect, to both reveal and seem to hide, and all the while Cvetic must expertly manipulate the perceptions that others have of his behavior in what he often calls his "shadow dance." With this tension in place, the first act of the program ends as something happens to make the minder suspicious, and a commercial is timed to a cliffhanger in which Cvetic's hidden task to inform the FBI is in danger of being discovered as his enthusiasm for doing Party bidding is cast under suspicion. When we return, Cvetic has somehow barely survived blowing his cover. Eventually, he finds some way to reveal the plot to the FBI in the nick of time, covering his tracks by setting up another conspirator as the more likely informant.

The absurdity of the scenario masks a complex series of challenges, as the narrative centers on a character with two sets of motives that twist him to the point that the play takes the form of a ritualized ordeal. In the dramas about infiltrating organizations, Cvetic joins one of the groups that zealots claimed needed investigation, such as an arts guild ("Draw the Red Curtain"), the Boy Scouts ("Hate Song"), the City Housing Authority ("Home Improvement"), the National Employment Office ("Traitors for Hire"), or a liberal magazine ("Rich Man, Poor Man"). In the course of these episodes, Cvetic encounters an old union man, a naïve immigrant, a charming fisherman, a young child, or a poor family—someone whom he "pretends" to befriend in order to pervert the agenda of the group to which they belong, according to the plan of Comrade Revchenko. Of course, as we learn from Cvetic's constant narration, he really does have affection for these characters, expressing remorse at the fact that they have become targets and rage at the Communist plans to exploit the innocent through him. But the sentiments that we hear in direct-address narration must never issue into the world of the fiction, lest the undercover work be exposed. Indeed, any inner affection must be purposefully miscommunicated in order to successfully perform it. As a result, when the objects of affection realize the "truth" that Cvetic's friendship was but a ruse to achieve an act of subversion that has mysteriously gone awry, they turn their backs on him in scenes of public shaming. Over the two-year program run, this is the effect that Cvetic mentions most, being "friends with people you despised, and being despised by people you would have liked to have had as your friends."

In the same way that Cvetic becomes decoupled from his public name and his own emotions, he also suffers interruptions of his yearning for fulfilling domestic life. In "No Second Chance," for example, Cvetic is ordered to Los Angeles to cooperate with the Soviet secret police in a matter of "termite exterminating." On the train he meets Helen Enright, with whom he develops a romance, before learning that his mission is really to infiltrate Helen's household to discover if her brother Jack, a high-ranking Party official, is planning to defect to the FBI. By the end of the episode, Cvetic must seem to betray Helen in front of her brother, in order to save Jack from assassination and to help him maintain his cover as an informant. But Cvetic loses Helen forever. "Never mix romance and undercover work," he notes dejectedly on the train back east. To remain an asset, all personal feeling must be cloaked, and because Cvetic divulges everything to the FBI and nothing to anyone else, the Bureau takes the place of his inner self, a transposition that is mentally ruinous. As Cvetic describes it in "Where Red Men Roam":

Your friends avoid you. Your enemies smirk and say they always knew Matt Cvetic was a wrong number. Your mother accepts you as you are, with a reproach in her eyes, breaking your soul. Your dad gives you ten dollars to change your name so you won't disgrace the name of Cvetic. Nine years of it, and you're talking to yourself, like I'm doing now. It's a brilliant morning with 24-karat sunshine, and the sky diamond bright, but I'm seething. All morning I've been pushing doorbells, getting housewives to sign Commie-inspired "peace" petitions. Yesterday, I picketed an aircraft factory. The day before that, I dated the secretary to a War Department official. Object? Sedition. And the day before that, I solicited ads for "The Worker's Daily." And the day before that I—ah, never mind. All I know is that I'm sick of it, work, the Party, discipline, the revolution, until you drop in your tracks, Cvetic. Until you wind up in the booby hatch, boob.

Cvetic recalls interrogations by Party officials, who can spend whole episodes quizzing him on his apparent acts of disloyalty, only to reveal at the end that another Commie in the room has been the real object of inquisition. He is tested on principles of Party orthodoxy, often forced to humiliate others in sequences during which only the strictest mental discipline allows him to withstand the "poison" of Communist doctrine, afterward learning that all along most of the participants in the kangaroo court were FBI agents just playing along. Cvetic cites nervous sweating, sudden weight loss, the whitening of hair, and insomnia. Men like him are "walking alone," "walking the tightrope"; they are living "a lie," "a double life," or "a triple life"; they await "the crack-up" and muse over the ignorance of the public for whom they toil—*if they only knew*. *I Was a Communist* asks us to what degree one can withhold expressing feelings without actually becoming two people, and in this way the drama provides the listener with evidence that repression leads to insanity, an experiment in which the fight against Communism is but a catalyst.

DISCIPLINING THE INNER VOICE

The Cvetic program was one of many groping toward a new model of the psyche in the postwar era. Other postwar dramatists began their search by forswearing the languages promulgated by mass psychology theorists such as Paul Lazarsfeld in favor of a lexicon of psychobabble derived from behavioral psychology, criminology, and Freudian psychoanalysis, all by way of Hollywood. The war had brought millions of Americans in touch with psychiatry for the first time; afterward its terms were popularized, stereotyped, and vulgarized (particularly during the years surrounding Freud's centenary in 1956), as the media not only turned to models derived from psychoanalysis, psychiatry, and psychotherapy but also tended to confuse them.[15] This was certainly the case on radio. In "The Frightened Fugitive" on *This Is Your FBI*, the introductory diatribe explains that the FBI has learned that honest children become criminals as a result of too much "eggo." "Phobias" define characters often on *Radio City Playhouse*, and on *Inner Sanctum Mysteries*, obsessions are described as "psychoses" and false statements are the result of "vivid hallucinosis." On *Broadway Is My Beat*, culprits are "paranoids" while *Dragnet* has "psychos," "maniacs," and men with "compulsion complexes" in grisly murder plays. On *Suspense* plays such as "Drive-in" and "To Find Help," men with "something wrong with their mind" take innocent waitresses and old ladies hostage. Alongside the mentally ill, psychological professionals also appeared on the airwaves with new frequency. We hear taped sessions with a psychiatrist in *Mr. District Attorney*'s "Blackmail Murder" and meet psychiatrists who organize elaborate scenarios to rid war vets of their neuroses in "Mission Accomplished" on *Suspense*, and experimental techniques save women from "guilt complexes" in "Local Storm" on *Radio City Playhouse*. In "Bad Medicine" on *X Minus One*, comedy erupts when a man purchases a robot therapist but accidentally gets a model calibrated to treat only Martian psychology.

In some of these usages, the trappings of psychology are not as full with meaning as they appear. Psychobabble is often deployed in lieu of revelations about inner life. This was often the case in *Dragnet*, in which the designation "maniac" indicated that detectives need not trifle with chasing down leads to determine a motive in a crime or the traumatic experience from which mental illness sprang. Instead, Detectives Friday and Smith concentrate on apprehension, because in the understandings held by characters in the program, mania is incomprehensible. As a consequence, the appearance of psychological language is ironically used as a reason to neglect a perpetrator's actual interior life. On the other hand, several dramas costume themselves in psychoanalytic regalia for the opposite reason, employing it to make behavior decipherable to a postwar audience familiar with cues weighted with psychological meaning. Consider "Elwood" on *Suspense*. In this 1947 play, Eddie Bracken plays a postpubescent gas station attendant. The play relates to us Elwood's unhappy life at home ("every time I got near that place I hated it more"), his fraught memory of his stepfather (whose photograph he surreptitiously tears to pieces), his fetishization of knickknacks (a flyer's pin, a gold fountain pen, which he "tries out in front of the mirror"), and his close relationship with a former schoolteacher, who stands in for lover and surrogate mother. This material would have been transparent to its audience, as the importance of childhood was one of the most successfully popularized of all psychoanalytic tenets in the period.[16] Now that family psychodynamics are at the very center of the drama—and after we have heard Elwood detail his interest in the works of Freud and Adler—we learn that Elwood himself is a serial attacker, as he begins to "feel strange" and get a headache just before savage beatings take place in the small town. In this sense, the play relates violence as an externalization of the "inner wreckage" derived from the vague neuroses alluded to at the outset. One mark of being a maniac is the troubled home life from which one springs, but another lies in one's manifest inability to prevent the divulging of dark impulse. While Cvetic's derangement is concomitant with the prevention of externalizing inner truth, Elwood's maladjustment is the cause of just such an externalization.

"Elwood" aired in 1947, at a moment in which the narrative of psychological repression and displacement was as yet not firm in popular understanding. But as this account of the mind became more widely available, the repression model soon appeared explicitly among more mainstream sorts of characters, particularly in domestic settings. *Suspense* aired several stories along these lines, perhaps never more clearly than in "I Saw Myself Running," a 1955 play starring Charlotte Lawrence. The play begins at the breakfast table of a suburban home, where husband Freddy offers his wife Susan the newspaper—she refuses "the women's page," foreshadowing ensuing developments—and complains of her recent spate of bad dreams. In these dreams, Susan finds herself hovering outside of her body as it flees monsters and falls from airplanes without a parachute. "Too much brandy," harrumphs Freddy. Later that night the same dream recurs, and the heroine relates it in richly illustrated sequences that showcase all of the detail that transistor radio promised to amplify. Vertiginous spaces have vertiginous underscores and long corridors have overlong reverb. As Susan's next dream begins, we meet "Sue," Susan's uncanny second self, who lives in these dreams all of the time, always running from horrors while Susan remains awake. Sue is afflicted with the fear of things that Susan has grown out of, such as faceless men at the top of the stairs, wriggling caterpillars, and other signs suggesting abuse. "I mind everything you think you've forgotten," Sue explains, self-designating as the return of the repressed. Susan wakes up screaming from this encounter with her double, and she is so shaken that she tries to give up sleeping altogether. All that a doctor can offer is the unhelpful observation that Susan is just tired, and he prescribes a soporific drug. The next day, with her husband away, Susan falls asleep on a couch. In the dream world, Sue waits, ready to run the same sequence of chases that it is her lot to undergo over and over again. Once more the two women are chased by darkness at the top of the stairs, and soon Susan hears both her husband and the doctor calling her to awaken. But Sue has the idea of leaving the dream world herself, trading places with Susan in the exterior world. Suddenly two male figures appear by the door that leads out from the dream. Freddy and the doctor confer with Sue and vote that Susan ought to remain in the dream world alone. To her horror, Susan has lost her ability to wake up, and she plummets into repetitive sequences of childish terror. When we return to the exterior waking world, Freddy wakens Susan on the couch to find that her voice is changed entirely. Alas, the play ends before we learn whether or not "Sue" will "take the women's page" at breakfast and thereby assume the uxorial subjectivity whose configuration the play has called into question.

As in "Elwood," a second inner voice has erupted through the hardened perimeter of the mind, bringing disaster. But in this case the source of the problem is less mental illness and more misogynistic social norms, which produce pathology. If "Elwood" is an ancestor of Hitchcock's *Psycho* (1960), "I Saw Myself Running" is a descendant of Charlotte Perkins Gilman's "The Yellow Wallpaper" (1892). On the one hand, this play uses a "psychological" narrative to warn against the psychic effect of patriarchal domesticity. On the other hand, during a period in which, as Elaine Tyler May has explained, many Americans came to cherish the ideal of traditional heteronormative marriage, the use of the language of psychology through radio plays could be less of a critique of that system and more a means of perpetuating it, giving American men and women "a vocabulary with which to tame and manage their frustrations."[17] In any case, the key to "I Saw Myself" is that a repressed inner voice has become an arena of trial, a medium in which the tension over postwar married life would be negotiated, just as repression has become a code in which tests of mental fortitude might be written. In the same way that the Cold War offers a pretext to explore Cvetic's schizoid attempt to remain silent to the ears of others, "I Saw Myself" uses the tensions of the postwar family utopia to test the degree of stability needed for the mind to successfully inhabit it. According to this view, both efforts are experiments in duress and ordeal before they are critiques of their respective cultural excesses.

But not all radio plays about resisting externalization are so easy to thread into broader cultural narratives. There is also a tradition of radio dramas that aired around this time centered on trials of conscience. In "Hit and Run" on *Radio City Playhouse*, a salesman is driving home after a long trip and hits a child on the way, fleeing the scene. The remainder of the play concerns little else but the man's efforts to build the courage needed to turn himself in. In "Pass to Berlin" on *Escape*, a military attaché in the decimated city murders a woman of the night and spends the entire drama being baited into confessing by characters who seem to be manifestations of his own death instinct. *Diary of Fate* often aired tales of weak underclass men who "can't get ahead" and are tempted to take an easy road to social mobility through crime: a chemist is presented with a chance to take credit for a dead colleague's discovery; a diplomat recovers jewels stolen by the Nazis and conspires to keep them; a simpering male secretary reaches his wit's end with his boss and seizes an opportunity to kill him and steal his identity. In each of these cases, we meet a character who has a dark inner purpose and a separate voice to relate it to us, and the drama becomes an account of how these characters try to keep this desire at bay (or fail to) while also keeping the voice that represents it from issuing into the public world of the drama. When a character triumphs, it is likely that a second voice has been successfully restrained. When a character fails, it is likely that this voice has been accidentally released. If 1940s dramas had been about the careers of external signals weaving in and out of people, 1950s dramas were about the careers of private thoughts kept out of worldly circulation. In this way, a golden era of government surveillance was also an era that coveted privacy, and a society obsessed with secrets was also a society obsessed with secretiveness.

Suspense's "Subway" is an early example of the postwar plays of conscientious ordeal. At the outset of the program, we meet an aspiring young actress named Paula in a New York subway station, contemplating throwing herself onto the tracks. Boarding an uptown train, she runs into a dizzy rival named Ruth, whom she has not seen since their time together at an acting academy. Ruth has had a choice Broadway role handed to her, while Paula's own career is in the doldrums. The two women chat, to all appearances a warm conversation, as Ruth offers to help Paula obtain an understudy position, should her producer decide to move Ruth to a new show. Our heroine's mind swirls with jealousy in passages that are given a sense of boundary from the external world not only through close miking, but also as a result of the use of the noise of the train to break passages of external dialogue away from passages of internal monologue. The scene becomes tenser as the train fills. Paula realizes that she has a pair of scissors in her pocket, an intended loan to her mother. As she keeps up her jovial banter, her inner voice churns in its own train of thought, plotting to lure Ruth to an empty spot and stab her in order to take her role in the hit play:

Ruth standing between me and the break I'd dreamed of: understudy in a hit show. But she had said "Don't build up your hopes, honey. There isn't the slightest chance." But wasn't there the slightest chance? The slightest chance of something happening to her. The train started up again, jolted me so that I was thrown up sickeningly against Ruth. My fingers were testing the points of the scissors in my bag. No one could see me; we were packed in too solidly. The scissors were sharp, cold and long. Yes, they seemed long enough. I kept my eyes on the dim light and the dirty tiles of the stations as the train throbbed along uptown. I was holding the scissors as if they were a weapon. I was sure that at some time or another scissors had been used as a weapon. The scissors in my bag seemed to grow bigger with the idea. Idea and scissors. Scissors and idea. They were increasing in size. The ache in my throat had gone up into my ears. Throbbing, keeping time with the throbbing of the subway.

The music and the sound of the train swell, but afterward the dialogue continues in the "external" voice that Paula has maintained despite her impulse:

PAULA: Ruth, you play on Sundays, don't you?
RUTH: Why, yes.
PAULA: No show tonight, then. Ruth, why don't you come home and eat with me? There's just mother and me. We'd be all alone.
RUTH: Oh, I'd love to, Paula! Frankly, I didn't have anything to look forward to but a boring evening. I'm so glad now that I couldn't find a taxi!
PAULA: Mother will be very glad to see you!
RUTH: That's very sweet of you to ask me.

After undergoing a masculinized arousal in her inner monologue, Paula invites Ruth into a kind of courtship. In this way, the sequence features an illicit urge generated in an inner voice, one that is in danger of becoming sexualized murder, as a process of exteriorization is one and the same with the moral temptation that Paula is being presented with, be it murderous, erotic, or both. As the play proceeds, Paula begins to justify her desire for this outcome as a simple "hit and run." But then, quite suddenly, the whole situation transforms. "Then I looked at her," Paula says. "I realized she was a human being." Visual recognition instantly recodes a play hitherto dominated by the sound of a rocketing train, a change that is underscored when Paula asks Ruth to come with her "into the light" at the end of the line, as if out of a world of hidden inner motives and back into a world where these urges remain in apparent subterranean abeyance. In the moral transformation of light, Paula's inward desires have not leached out; her external persona has overcome them, for now.

PLAYING PEEK-A-BOO

In the ordeal narratives discussed above, I have argued that radio plays of the 1950s use the voice to test a diverse range of ideas about how minds work. Cvetic, Elwood, Susan, and Paula each grapple with the fact that they have two voices, a doubleness that cannot be escaped, only disciplined. Meanwhile, schizoid personality, madness, repression, and tests of conscience are all strung together as if they were each part of a broad research program that deals with inner voices suffering from fraught relationships with external worlds. There remains a type of psychological test that is yet to be discussed, one that I will argue cannot quite be mediated by sound and was also associated with the "light" that seems to dispel temptation in "Subway": brainwashing, a phenomenon that came to public attention during the Korean War and then enjoyed a long and complex life in the popular imaginary.[18] Here I would like to propose that brainwashing is associated with visual experience in a way that aural narrative could not satisfactorily approximate. In some of the last really innovative sequences in radio's theater of the mind, American radio drama seemed to be suggesting that as the televisual age dawned, it opened the way for a model of *total* mental programming far more powerful than what had been available to the aural media, contemplating brainwashing as a fusion of medium and mind that was prefigured by the radio age but unachievable within its signature genre.

The play in which this idea finds its best articulation is "Zero Hour" on *Suspense*, perhaps the last major radio play to air in the network age, and surely one of the final great dramas in the most creative phase of the program. We know a little about the provenance of this play. At the end of 1954, veteran director Norman MacDonnell left *Suspense* to concentrate on *Gunsmoke*, and CBS's flagship thriller program was left in the hands of Antony Ellis, an actor and writer who had been working on the program for several years while he also put in time on *Pursuit* and *Escape*. Ellis looked to the fiction of Ray Bradbury, who had been an occasional writer of crime thrillers for *The Molle Mystery Theater* and *Radio City Playhouse* in the late 1940s. By the end of that decade he became something of a celebrity following his collections *The Martian Chronicles* (1950) and *The Illustrated Man* (1951). In 1950, *Dimension X* aired a half dozen of Bradbury's stories, and it is to one of these that Ellis looked to find "Zero Hour," a tale that depicts a revolution in which groups of children all around the United States murder their parents at the behest of an alien being, who has cunningly tricked them into playing a game that he calls "Invasion." After a version of the play aired on *Dimension X*, Ellis directed a rewritten script for *Escape* in 1953, then he did it again twice for *Suspense* in 1955, after which the network received thousands of letters, both of praise and protest.

Ellis's play begins with a score that quotes from "Twinkle, Twinkle, Little Star," here played *adagietto* and a little off-key, to come across as uncanny. This is coupled with the sound of children laughing with abandon. "What a game!" exclaims a narrator external to the drama, "Kids laughing with excitement they hadn't known in years." We are presented with a scene of youngsters talking to an imaginary creature that lives in a rosebush. A door opens and closes as a young girl named Mink runs into her house, "all dirt and sweat," and begins to fuss with the pots and pans near her mother, Mary Morris, who is peeling vegetables. Mrs. Morris asks about the commotion outside, learning that the kids in town are all playing "invasion." Mary thinks little of it at first, and soon we are back in the garden as Mink and her friend Anna assemble objects, repeating instructions (inaudible to us) that issue from an invisible being guiding the construction of some kind of portal out of household materials and reminding everyone that "Zero Hour" is at five o'clock. Soon it is lunchtime, and Mink comes into her house through the front door once again. Mary quizzes her daughter on the premise behind the new game. Mink explains that the neighborhood kids are getting together to overthrow the grown-ups at the behest of a being named "Drill." "They couldn't find a way to attack Earth," she explains. "They couldn't get in or something. And Drill says they have to do it by surprise, and even get help from your enemy. One day they thought of children." Mary plays along, applauding this "fifth column" strategy. Soon Mink is ready to return to her game, exclaiming, "They're going to let us run the world when they get in, all us kids, and I might even be Queen." Mrs. Morris is charmed by the story, until Mink once more runs to the doorway and looks back to call out almost sadly, "Mom, we'll have to get rid of you after the invasion, but I'll be sure it won't hurt very much." Mrs. Morris is taken aback, but later learns from an old school friend who calls on the telephone that children all over the country are playing invasion, and the two speculate about childish word of mouth, calling "Drill" the new "password."

Already this drama is a composite of several of the dramatic conventions that I have discussed in part 3. A "Destiny-machine" is constantly at work in the background, as the steady approach of Zero Hour is softly underscored by the chiming of a clock in Mrs. Morris's living room. The drama also makes fascinating use of radial levels of action around an immobile center. Although we remain inside the kitchen and living room with Mrs. Morris for most of the play, almost every time that Mink appears in the house, she pauses at the front door in an aural background beyond our principal audioposition, in order to intimate that ominous events will come after Zero Hour. In this way, our attention is called to a perimeter in the world of the drama from which Mink disappears and reappears aurally to us and visually to her mother. Our heroine is even curiously entrapped in that closed world. Despite the fact that Mary has worries that the game out by the rosebush may be more than just a game, she never actually exits the house to look for herself, and she only once approaches the doorway. Mrs. Morris is the archetypal repressed housewife, one whose thoughts are often revealed to us not through an inner voice but through a "voice of authority" narrator whose godlike omniscience is no different in weight and timbre from that exhibited by the external announcer of *Your FBI*'s direct addresses. It is during Mary's conversation with her friend that these inner thoughts begin to trouble her subconscious, alienating Mary from the chatter in which she is engaged. The narrator tells us of Mrs. Morris's wandering mind, explaining that as she went about her routine tasks, "Mary was thinking of other things. Rosebushes, dimensions. Things she had forgotten about being a child." Mink is the one that is being brainwashed by a powerful external "invisible friend," but the language of wandering minds and repression hints that Mrs. Morris has a repressed child somewhere inside her as well. Perhaps she once "played Invasion."

As hours "drowse by" in the complacent world of the home, there soon appear warning signs over its defenses, as we hear of "an occasional hum in the interior of the house as a car drove by," foreshadowing other penetrations soon to assail the suburban utopia. "All over the neighborhood children playing games," our narrator explains, "talking to trees and shrubs and rosebushes. Even children in apartment houses high in the air conversing with potted plants." He also tells us more about Mary's inner worries. "She didn't know why," he explains, "but there was something about parents shutting ears and eyes to what was happening. And because she was disturbed, she did something that she did not usually do. She called her husband at the office." In the conversation, Mary seems on the verge of exteriorizing her worry, but at the last moment she recoils from an admission that may conjure just what it hopes to avoid, and Mary says goodbye to her husband Henry without explaining much at all. Hours pass until it is half past four o'clock, as "shadows lengthen on the green lawn" and Mink and her friend fidget with their many implements, while Mary drinks coffee and turns over her thoughts in the first true soliloquy to give us access to her inner feeling:

> Children, children. Love and hate side by side. Sometimes children love and hate you all in half a second. Strange, children. Do they ever forget the whippings and the hard, strict words of command? I wonder, I wonder. How can you forget or forgive those over and above you. Those tall, silly dictators, those parents ...

Suddenly, Mink appears at the threshold of the house, asking for a lead pipe and a hammer, as if answering the secret question that Mary has just let loose. Mrs. Morris also fails to read signs that Mink is going through a kind of motherhood: "Drill's half way through," she explains. "If we could get him all the way through it would be easier. Then all the others could come through after him." Once again brainwashed Mink is being associated with the category of the visual. She passes in and out of the visible world of the house; she draws something out from a hidden sonorous dimension into the bright world of sight. Feeling a chill, Mrs. Morris closes the window, blinding herself to the fate that she knows to be coming.

Meanwhile the Destiny-machine of the drama continues unabated, as silence falls across the house and chimes tell us that five o'clock has at last arrived. Instead of sighing with relief, Mary calls the operator to get the correct time—another slip of the anxious inner voice—and a machine tells her that five o'clock and Zero Hour are as yet still five minutes away. A car arrives and Mrs. Morris hears Henry calling to the children on his way in. Henry kisses Mary and begins to mix cocktails. Despite everything that has transpired, Mrs. Morris persists in repressing her worry, relating details of her daily tasks, clinging to the façade that gives her life coherence. When Henry perceives something amiss, his wife will only admit that she is "a little tired, upset." As if on cue, a hum penetrates the interior of the house from outside, and all of Mrs. Morris's gradually accumulating fears are realized in an instant. Mary's external voice transforms, in the last stage of a process that has been going on since the outset of the play. At first, Mary's inner feeling has been vocalized by the external narrator, then it turns into a soliloquy in her own voice, but now Mary's voice drops an octave—the same slippage that Paula avoided, and to which Susan succumbed—and her breath comes in clipped animalistic bursts just as the dramatic world fills with the sound of explosions and brainwashed children rejoicing as they slaughter their elders.

On a pretext, Mary convinces Henry to run upstairs to the attic with her, locking the door and throwing away the key, explaining, "We've got to stay here. Be quiet. Please be quiet." Soon, the hum from the outside world has penetrated the interior of the house and begins to move up the stairs toward Mary and Henry, a scene that quotes "Sorry, Wrong Number." "Hush, please be quiet," Mary pleads. "They might go away. Please . . . please . . . please . . ." Cowering in the dark, Henry and Mary hear hard steps coming up the stairs. Once more approaching from the outer edge of the audible, Mink calls out, "Moooommy, Daaaaddy!" The two parents perceive an "alien sound of eagerness" in her voice, intuiting the presence of Drill behind her words. A cold light seeps under the door and Mink at last arrives, completing the implosion of the home and filling out the primal scene of a child catching her parents upstairs together, and in the dark. The parents cower before the brainwashed girl with the "bright little eyes" who can transform sounds from a rosebush into fearsome blue light, a creature of the televisual age under the control of "wavering blue shadows" that hover in the doorway behind her. We hear no attack. Instead, at the head of the stairs, Mink calls out, "Peek-a-boo"—three syllables that perpetrate an act of vision on her parents, making them utterly *seen*, a violence from which no amount of quiet can protect. A series of psychological ordeals has arrived at a kind of test that the equipment of the radio drama could conceptualize but not adequately meet, an imaginary mythological encounter between radio and television that ends not with a whimper but a scream.

TABLEAUX

It seems that this history of the "theater of the mind" has circled away from one invasion allegory only to arrive at another. My exploration began with "The Fall of the City," a play in which aural tensions set the stakes of heeding orators, and it concludes at "Zero Hour," a tale of aural self-control that becomes a rampageous allegory of visual peril. Each of these plays thematizes issues of vivacity to the unique historical moment in which it aired, and together they index a series of social transformations. A public square has become a nuclear home, suggesting changes in how broadcasters conceptualized the worlds in which didactic allegories are appositely set, as well as in the audience's idea of the places that are meaningfully imperiled. The first play is directed at a society in the midst of reconceptualizing its public space in the age of radio, while the second responds to a society with deeply troubled private space in the age of television. A correspondent has been replaced by an omniscient narrator, a change that reflects an authoritative and even authoritarian feel that became attached to radio style as part of its multidimensional relations with the conservative impulses of the 1950s. And the culminating moments of these plays also bring broad social developments down to a microdramatic level. Nineteen-thirties America is presented with the tableau of a panicking mob unwilling to look upon a hollow conqueror; 1950s America is presented with the tableau of a child hollowed out by a conqueror, turning weaponized eyes at her parents, a scene that is a fitting endpoint for this study and the phenomenon it chronicles.

Indeed, on one level this book has been the story of the expansion and collapse of *scene* itself—as a dramatic convention on which all others are built—in the radio field. There is an expansive sweep to setting in prominent radio plays of the late 1930s, a fascination with ballooning external space that gradually diminished as many dramatists narrowed their work to focus on smaller spaces and internal states over the course of the 1940s. Perhaps the plays of late network radio represent a last stage of that implosion, and the scream of a hitherto sublimated inner voice in "Zero Hour" signals that interiority has caved in as well, leaving no spot for the microphone to be, no point from which the representation of action can begin and end. That is one way to think of the golden age of radio from the perspective of media aesthetics. What died in the 1950s wasn't a medium (music and talk formats boomed for decades), but radio as a place in which to "try out" habits of visualization. Radio is certainly still used to tell stories in a variety of ways, but its "imaging mode" resembles that of reading more than that of watching because it is no longer customarily framed by proscenia past which there lie spatialized scenes in which things happen as if spread before a spectator-listener. In other words, radio stories are no longer *dramas*. We do not experience them as one of what Roland Barthes called the dioptric arts: those mediations that exist beyond a cutout rectangle, the arts that (unlike music) can convey tableaux.[19] In that sense, the rise and fall of the American radio play is the story of the flirtation of aural representation with the logic of visual exhibit, of the construction and destruction of a beguiling fantasy of listening as through a window. As a cultural-historical matter, how should we assess that practice? If an aural medium could masquerade as a dioptric art during the middle decades of the twentieth century, then there must have been a habit of transposition between the senses in this period, a protean fluctuation in the very process by which modern feeling was structured. Historian Warren Susman wrote that the Depression ushered in a "culture of sound and sight."[20] Perhaps we can also say that this culture was synaesthetic in quality, a theater of the mind in which sound *was* sight, or dreamed itself to be. That dream ended with the implosion of aural "scene" and the radio platform retreated from its area of overlap with the codes of the visual. If the game of the radio play is to bridge the world of sound and the world of picture, "Zero Hour" mocks that idea with a cleavage between those worlds, marking a new commitment to sensorial antinomy.

But just as dreams follow us in waking hours, the dizzying outpouring of radio plays that filled the airspace of the twentieth century left a residue on modern understanding. For one thing, it made the golden age of radio fully imaginable and worthy of a degree of nostalgia that approaches mawkishness. The era could now seem to promise everything that our own experiential paradigm of mediation does not, an idealized fantasy that the radio receiver is more freeing than the screen. Also, ever since the radio age, we remain convinced that the mind and the media coevolve. This notion may have been born in the nineteenth century, but it became widely available during the network era, whose hundreds of thousands of hours of entertainment from "Fall of the City" to "Zero Hour" were enamored by the conceit. Indeed, if the premise of *Theater of the Mind* is correct, and classic radio plays involved an aesthetic mode whose historical circumstances and core expressive challenges made it into a theater that takes place in the mind, about the mind, for the mind, then it follows that the radio age played a part in producing the publics that today resolutely believe that televisions and computers (and the modes of address that they reify) shape or even ordain interiority. To dismiss this notion as a myth would be futile. As "Zero Hour" suggests, an attempt to expel one media fantasy tends to produce another. The issue is not how to break this "unnatural" fusion of mind and medium so as to replace it with some "purer" idea of inner life that is somehow free of technological inscription. Rather, the issue is how to historicize this connection in a way that enables its demystification. There is no single apt and unproblematic way of regarding the defining relationship between media and those of us situated within it. There is only the possibility of grasping that relation in a fully revealed manner, so as to more presently perceive the state of perpetual reconfiguration in which both experience and its categorization are held.

CODA

Instruction and Excavation

Today there remain a few Americans for whom the golden age of radio is not a historical enigma but a live memory. Until his recent passing, one of them was Norman Corwin, the singular radio dramatist of his era, who died a national treasure at over one hundred years of age in 2011. Corwin outlived every accuser who expelled him from the industry in the 1940s, and he once again became synonymous with the glorious era of broadcasting late in life. Like old radio, Corwin was always being rediscovered. In the 1970s and '80s, he worked in film and television while writing social criticism. Later he wrote for National Public Radio and remade his 1941 play "We Hold These Truths" to celebrate the two-hundredth anniversary of the Bill of Rights in 1991. In 2005, Corwin was the subject of an Academy Award–winning short film.[1] I interviewed him on a number of occasions for this book. In one of those conversations, I outlined the objectives of *Theater of the Mind* and asked him what he believed was lost when the golden age of radio ended. Corwin replied that audiences today lack a sense of the ear as a passageway, of the imagination as a living medium in its own right:

> In that time, you see, my job was in essence to create an instructive experience for the ear.... The ear is the most poetic organ of the body in that it absorbs music, which is abstract, and the finest of writing, which is Shakespeare. Any organ through which we perceive a high level of instruction, and find beauty and instruction in it, that's the prized receptor of the body. Today we don't give ourselves enough credit for the native gifts of hearing and digesting and creating. Great credit has to be given to human imagination. It's a great organ on which to play, and I've always been conscious of that.[2]

Even as Corwin credited the creativity of audition, he emphasized the broadcaster's role in instruction. In his model, the "theater of the mind" was a mode of poetic reciprocity, in which a creative dramatist orchestrated an equally creative process taking place inside a listener stationed at the far end of a relay that ran from microphone to transmitter, from radio to eardrum, finally terminating in the absorbing "organ" of the imagination.

That reciprocity is not at the forefront of our repertoire of listening habits now. Listeners today are surely no less aurally capable than their counterparts of the 1940s in absolute terms, but we do not highly credit the abstract listening that Corwin describes. That discrepancy explains why the genre seems to have achieved finalization generations ago. Because competence in the mode of receptivity that we call the theater of the mind has atrophied, we fixate only on what we "fill in" as we listen, discrediting all instruction and sentimentalizing an absent experience of absence itself. Perhaps our nostalgia for radio plays is just that: evidence of our deep expectation of license to breathe our own life into the mediated narratives we encounter. If so, then the aesthetics of radio listening are little more than a rallying metonym for the feeling of being smothered by contemporary mass media. To exalt old radio is to demand more hermeneutic generosity than we feel we are allowed, but it is also to ignore the way that radio can limit our unbridled creativity with its own instructions, and to suppress our awareness that there is always someone on the other side of reciprocity, be it auditory or otherwise.

And even if we were able to quash nostalgia and reorient listeners to credit the plenitude of the play over its scarcity, could we really cast off the sense of absence that has become such a part of how we now think about old radio recordings? Today we call classic plays "old-time radio," an idiom that itself seems part of a bygone era. Radio fans seldom listen to dramas that aired after 1960, as if to associate *The Mercury Theater on the Air* with newer performances would disturb a pleasing aura of finality. They have a point. While powerful dramas still air around the globe, none have the feel of *Cavalcade of America* or *Dragnet*, in part because classic plays presume a pact between listener and broadcaster that had been guaranteed by a historically unusual system of national-scale distribution. Airing an old radio play today is like staging a show devised for a coliseum in a café. The result may be rich with insight, but it also points to the distance of a buried past. New plays have a quality peculiar to all "revivals" in that they provide irrefutable evidence of death. But my purpose is not to scold postwar culture for leaving *The Shadow* in the dark. Rather, I want to emphasize that things like the scale of the audience and the nature of the media platform are not just lovable features of the "voice" of classic radio but the very precondition of its aesthetics. To truly rebroadcast "The Fall of the City," "The Hitch-Hiker," or "Zero Hour" we would have to re-create the network system and listen over the course of many years as part of a large public, so that our art of listening could fit the art of sound that it engages, and absorption become one with instruction. As *Theater of the Mind* has argued from the outset, the activity of radio lies in the many interpretive errands that it prompts us to perform, using normative conventions crafted to suggest certain "pictures in the mind" to certain publics in a certain time. If that is so, then we will surely be hard-pressed to fulfill those errands as time goes on and it becomes increasingly difficult to relate to the experience of encountering a mass media system like the one that shaped practically everything aired by Corwin and his contemporaries.

On the other hand, *can't* we appreciate exactly the reciprocity Corwin describes? When the radio play went (almost) extinct, it also entered an unusual afterlife. Many of the plays discussed in this book have remained constantly available at the periphery of American culture in a variety of repackaged popular nostalgia products. I researched *Theater of the Mind* using recordings from online vendors whose tracks were digitized from sources that run the gamut of modern sound technology: compact discs, cassettes, records, magnetic tape reels, and acetates, all duplicated and preserved over decades by radio buffs. As a critical matter, how should we grasp the persistent availability that these recordings provide? What does it mean to borrow a sense experience devised for a body that is now dust and offer it to one now living? At airtime, Corwin's words played to the prized receptors of millions, but they also spoke to the unborn somehow. What is the vital status of the classic radio play right now, as it radiates more through time than space?

Is it even *radio* anymore? Maybe not. What we call old-time radio is scrawled with marks left by many generations of devices, some of which are themselves obsolete. Once the antithesis of the inscriptive media, radio drama now catalogs them, as the hiss of the cassette tape and the compression artifact of the mp3 scumble the broadcast as such and work against the searching ear. Of course, recording had always been a resource for making radio. During the 1940s and 1950s, plenty of broadcasts were transcribed before airtime, and most plays incorporated a recorded effect or tune. The line between inscriptive media and amplifying media was never a bright one. But now the supplemental relation of recording to broadcasting is reversed. Broadcasts persist *only* because they were transformed before dispersing in space, plucked from the air and mineralized. Like fossils, radio recordings preserve the pattern by gradually filling it in with something else, so that when we study old radio today, our ears dig into sedimentary strata that have not just hidden what broadcasters aired, but replaced it. That changes the rules. In Corwin's time, the essence of radio rested in its capacity to produce "mental pictures," but today his plays seem to have another ontology because our experience of them is inseparable from an ever-increasing number of sticky tangibilities, remainders interposed in the relay between the broadcast and our ears. If the dream of the radio play was to deny its aesthetic machinery and become a pure "sonorous object"—alive, untethered, in the "out there" part of our inner auditorium—it has belatedly become a sonorous *thing*, dirtied, pockmarked, and brought down to earth.[3]

That is a blessing in disguise. By mineralizing radio, we congealed the instructions it set forth, petrifying its expectation of reciprocity in the moment of formation. So the patterns of listening of the mid-twentieth century are not really there for us as they were, but neither are they truly lost. That presents an opportunity. Thanks to it, when we listen to these recordings, we "do" a sort of listening that *seeks* rather than *projects*, sorting through sounds and interrogating how they were crafted, asking digging questions that are impossible to ask of any pure sonorous object. That practice of "excavational listening" serendipitously helps us to understand the phenomenon of projection more clearly. Rather than treating a transcribed radio broadcast as a container of old attitudes, excavational listening uses the recording of a broadcast as evidence of the activity of taking in that broadcast. Perhaps doing media history is never more than that. Theorist Friedrich Kittler once provocatively remarked that in a certain sense Marshall McLuhan's idea of "understanding media" is out of the question.[4] By determining our situation, media control both understanding and illusion, framing all possible knowledge of the past. Media even condition any resource that we might use to theorize their own pervasiveness. As a result, all history must be "media history," and yet no history can be *about* media exactly. Does it need to be? By extracting perceptions from the crust of recordings, we may not learn the "ultimate" meaning of radio—is there such a thing?—but we might find an encounter between listeners and their world that cannot be otherwise got. An effective "media archaeology" provides for a paleontology of experience, and that is a marvel on its own.[5]

So the trick is not to listen like Americans of the 1940s. That's impossible. It is even against the spirit of radio, which was always reinventing how listening is done. The trick is to honor Corwin's words by repurposing them, crediting our own "native gifts" and making a mode of historical listening that uses recordings to learn how they expected to be apprehended, communing with what we might call the listening consciousness of an age. After all, what are the techniques described in this book but amplified hypotheses about the common sensory field of a public, expressive conjectures about sensation that were conventionalized into the kaleidosonic style, keynotes, signals, radial immobility, inner voices, and so forth? Making sense of that is not a matter of drawing aesthetics into cultural history so much as it is a matter of realizing that anything we call culture happens in an aesthetic sphere already, one animated by energies of apperception and affect, of experience and memory—what Miriam Hansen has called "the political ecology of the senses."[6] To do media history is to accept the errand of learning how beauty and instruction reach through that ecology to the living human interior, our first recording mechanism and our last.

GUIDE TO RADIO PROGRAMS

For this project, I selected approximately six thousand broadcasts to study. My selections fell into two general categories. The first consisted of seventy-four "primary" programs that I studied in great detail, using a sample of between 4 and 650 recordings, spread evenly over the years covered in this book, as available. The second category consisted of up to three sample broadcasts selected from about ninety-four programs. All recordings were obtained commercially from the Old Time Radio Catalog (otrcat.com) between 2002 and 2010. Below, I list the programs studied in each category, and provide a guide for readers interested in listening to the examples discussed in each chapter. A more complete listing is available in Neil Verma, "Theater of the Mind: The American Radio Play and Its Golden Age, 1934–1956" (PhD diss., University of Chicago, 2008), 384–441.

PRIMARY PROGRAMS

Adventures by Morse (52 broadcasts, 1944)
The Adventures of Philip Marlowe (103 broadcasts, 1947–51)
An American in England (6 broadcasts, 1942)
Amos 'n' Andy (128 broadcasts, 1929–47)
Arch Oboler's Plays (36 broadcasts, 1939–45)
Battle Stations (4 broadcasts, 1943)
Boston Blackie (102 broadcasts, 1944–47)
Broadway Is My Beat (136 broadcasts, 1949–54)
Calling All Cars (276 broadcasts, 1933–1939)
The Campbell Soup Playhouse (46 broadcasts, 1938–40)
Cavalcade of America (368 broadcasts, 1935–53)
Ceiling Unlimited (5 broadcasts, 1943–44)
Chandu, the Magician (166 broadcasts, 1948–49)
Columbia Presents Corwin and other Corwin specials (11 broadcasts, 1941–45)
The Columbia Workshop (85 broadcasts, 1936–51)
Dangerous Assignment (76 broadcasts, 1949–53)
Dark Fantasy (32 broadcasts, 1941–42)
David Harding, Counterspy (59 broadcasts, 1942–53)

The Diary of Fate (24 broadcasts, 1948)
Dimension X (45 broadcasts, circa 1950–54)
Dragnet (192 broadcasts, 1949–53)
Edgar Bergen and Charlie McCarthy (40 broadcasts, 1937–44)
Escape (121 broadcasts, 1947–50)
Father Coughlin (24 broadcasts, 1937–39)
The FBI in Peace and War (47 broadcasts, 1950–57)
Fibber McGee and Molly (97 broadcasts, 1935–41)
The Fifth Horseman (8 broadcasts, 1946)
Fireside Chats and other FDR recordings (20 broadcasts, 1936–45)
The First Nighter Program (22 broadcasts, 1939–52)
The Ford Theater (17 broadcasts, 1947–49)
The Fred Allen Show (51 broadcasts, 1932–42)
Front Page Drama (131 broadcasts, 1939–52)
George Burns and Gracie Allen (33 broadcasts, 1934–40)
Hermit's Cave (30 broadcasts, circa 1930s–40s)
Hop Harrigan (117 broadcasts, 1942–48)
I Love a Mystery (154 broadcasts, 1939–52)
Inner Sanctum Mysteries (101 broadcasts, 1941–49)
I Sustain the Wings (4 broadcasts, 1943–44)
I Was a Communist for the FBI (73 broadcasts, 1952–53)
The Jack Benny Show (184 broadcasts, 1933–55)
Jergen's Journal with Walter Winchell (11 broadcasts, 1941–49)
Lady Esther Presents Orson Welles (7 broadcasts, 1941–42)
Lights Out! (82 broadcasts, 1936–47)
The Lux Radio Theater (435 broadcasts, 1937–55)

The Man behind the Gun (10 broadcasts, 1943–44)
The March of Time (7 broadcasts, 1931–42)
The Mayor of the Town (25 broadcasts, 1942–48)
The Mercury Theater on the Air / The Mercury Summer Theater (32 broadcasts, 1938–46)
Les Misérables (7 broadcasts, 1937)
The Molle Mystery Theater (55 broadcasts, 1944–48)
Mr. District Attorney (73 broadcasts, 1939–57)
Mystery in the Air (8 broadcasts, 1947)
Quiet, Please (89 broadcasts, 1947–49)
Radio City Playhouse (61 broadcasts, 1948–50)
The Sealed Book (26 broadcasts, 1945)
The Shadow (47 broadcasts, 1937–48)
Soldiers of the Press (10 broadcasts, circa 1942–45)
The Strange Dr. Weird (28 broadcasts, 1944–45)
Studio One (48 broadcasts, 1947–48)
Superman (302 broadcasts, 1941–1946)
Suspense (647 broadcasts, 1940–56)
The Theater Guild on the Air (48 broadcasts, 1945–52)
This Is Your FBI (73 broadcasts, 1945–51)
26 by Corwin (4 broadcasts, 1941)
2000 Plus (13 broadcasts, 1950–51)
The Weird Circle (78 broadcasts, circa 1940s)
The Whistler (90 broadcasts, 1942–45)
Wings to Victory (14 broadcasts 1940–43)
The Witch's Tale (41 broadcasts, 1931–38)
Words at War (89 broadcasts, 1943–45)
You Are There (27 broadcasts, 1947–49)
You Can't Do Business with Hitler (12 broadcasts, 1941)
Your Army Air Force (13 broadcasts, 1944–45)
X Minus One (73 broadcasts, 1955–56)

SECONDARY PROGRAMS (1-3 BROADCASTS EACH, 1935-60)

American Theater of Radio
A Salute to the Law
Adventures in Research
The Adventures of Sam Spade
The Air Adventures of Jimmy Allen
The American Trail
Archie Andrews
Author's Playhouse
The Babe Ruth Story
Baby Snooks and Daddy
Best Plays
Beyond Tomorrow
Big Town
The Black Museum
Black Stone Detective
The Bob Hope Show
Bold Venture
Box 13
Bring 'em Back Alive
Captain Midnight
Casey, Crime Photographer
Charlie Chan
Chicago Theater of the Air
The Cisco Kid
Command Performance
Corliss Archer
Dick Tracy
Dr. Christian
Dr. Six Gun
Easy Aces
Everything for the Boys
Exploring Tomorrow
The Falcon
Family Theater
Favorite Story
Front Page Farrell
Frontier Gentleman
Fu Manchu
The Globe Theater
The Great Gildersleeve
The Green Hornet
The Guiding Light

Gunsmoke
Hallmark Playhouse
Harry Lime
Have Gun Will Travel
Hilltop House
Hopalong Cassidy
I Sustain the Wings
Journey into Space
The Jimmy Durante Show
Let George Do It
Life of Riley
Little Orphan Annie
The Lone Ranger
Luke Slaughter
Mail Call
The Man Called X
The Milton Berle Show
My Favorite Husband
Mysterious Traveler
Nero Wolfe
Nick Carter
Nightbeat
One Man's Family
On Stage
Origin of Superstition
Our Miss Brooks
Out of the Deep
Out of This World
Ozzie and Harriet
Pat Novak for Hire
The Phillip Morris Playhouse
Philo Vance
The Red Skelton Show
Richard Diamond
Rin Tin Tin
Romance of the Rancho
The Saint
Screen Director's Playhouse
Screen Guild Theater
Sgt. Preston of the Yukon
Sleep No More
Space Patrol

Strange Wills
Tales of the Foreign Service
Tales of the Texas Rangers
Tales of Tomorrow
Tarzan
Theater of Romance

This Is War
Tom Mix
You Bet Your Life
Young Widder Brown
Yours Truly, Johnny Dollar

PRINCIPAL BROADCASTS DISCUSSED, BY CHAPTER

Chapter 1. Most of the plays discussed in chapter 1 come from *The Columbia Workshop*: "Broadway Evening" (07/25/36), "Cartwheel" (08/01/36), "The San Quentin Breakout" (09/05/36), "Voyage to Brobdingnag" (09/12/36), "Pitch Perception" (02/13/37), "The Fall of the City" (04/11/37), and "Air Raid" (10/27/38). To get a sense of how exploratory cues became important, try *First Nighter*, whose introductions "set the scene." Examples include "A Symphony with Your Spaghetti" (10/13/39) and "Love and Gazooza" (03/25/41). To compare styles between the 1920s and the 1930s, listen to *Amos 'n' Andy* episodes like "Ruby's Father Sees Amos Talking to Another Girl" (04/22/29), "The Marriage of Andrew H. Brown" (04/03/39), and "Turkey Trouble" (11/12/43). For a sense of the work of dramatists who came into network radio around the time of "Fall of the City," try listening to *Cavalcade of America*, which featured a biographical sketch of Joel Chandler Harris by Arthur Miller (06/23/41) and one of Emily Dickinson by Norman Rosten (01/22/41). Norman Corwin's first major play for *The Workshop* was "Poetic License" (11/03/38). Also try the following printed anthologies: Erik Barnouw, *Radio Drama in Action: Twenty-Five Plays of a Changing World* (1945), Norman Corwin, *Thirteen by Corwin: Radio Dramas* (1943), and Douglas Coulter, ed., *Columbia Workshop Plays: Fourteen Radio Dramas* (1939). Orson Welles's first major project as a radio director was a seven-week adaptation of Victor Hugo's *Les Misérables*, beginning with "The Bishop" (07/23/37).

Chapter 2. In the second chapter, I note Orson Welles's penchant for "sham resurrections" in the episode "The Grave" from *Les Misérables* (08/07/37) and *The Mercury Theater on the Air*'s "Count of Monte Cristo" (08/29/38). I also cite Carleton E. Morse's sequences, some of which are only available in re-creations. Try listening to *I Love a Mystery*'s "Pirate Loot of Island of Skulls" (4 parts, incomplete, 04/20/42 to 06/15/42), "Snake with the Diamond Eyes" (3 parts, incomplete circa 1940), "Temple of the Vampires" (20 parts, 01/02/50 to 01/27/50), and "The Thing That Cries in the Night" (15 parts, 10/31/49 to 11/18/49); and, on *Adventures by Morse*, "City of the Dead" (10 parts, 01/08/44 to 03/11/44) and "Cobra King Strikes Back" (10 parts, 04/08/44 to 06/10/44). To hear how audioposition structured sequences well into the 1950s, listen to *Suspense* plays such as "The Most Dangerous Game" (09/23/43), "The Butcher's Wife" (02/09/50), and "How Long Is the Night?" (10/13/52). To hear Irving Reis use a parabolic microphone, listen to *The Columbia Workshop*'s "The Finger of God" (07/18/36). To hear how long it takes to get to hell according to Norman Corwin, listen to *The Workshop*'s "Plot to Overthrow Christmas" (12/22/40). To hear the best basket to catch decapitated heads, listen to *The Mercury Theater*'s "A Tale of Two Cities" (07/25/38). Additional stylized uses of audioposition occur in "Incident in the Pacific" (02/19/44) on *The Man Behind the Gun* and in "The Burning Court" (06/17/42) on *Suspense*. A simpler version of positioning can be found in the opening segments of *Calling All Cars* (e.g., "The Blonde Menace," 04/09/35) and *The Lone Ranger* (e.g., "The Origin of Tonto," 12/07/38). For two acoustically differing explanations of the attack on Pearl Harbor, compare *The Jergen's Journal with Walter Winchell* broadcast (12/07/41) with Franklin Roosevelt's well-known "Day of Infamy" speech (12/08/41). This chapter concludes with an analysis of *The Workshop's* "The Fall of the City" (April 11, 1937); readers interested in a contrasting sound design might also listen to a second production of this play performed on *The Workshop* (09/28/39), which aired from a stadium at the University of Southern California.

Chapter 3. To get a sense of how *The Columbia Workshop* used "intimate" audioposition, try listening to "A Trip to Czardis" (12/15/38), "Nine Prisoners" (02/20/39), and "The Story in Dogtown Common" (12/14/39). It takes extensive listening to appreciate how *The Shadow* used intimacy to construct the invisible presence of its protagonist. I direct the reader to the following episodes: "The Temple Bells of Neban" (10/24/37), "Triangle of Death" (12/12/37), "The League of Terror" (01/09/38), "The Firebug" (06/19/38), and "The Creeper" (06/07/38). For examples of how audioposition was thematized by programs into the 1940s, try "Kidnapping" (02/02/45) and "Murder in the Jury Box" (11/29/46) on *This Is Your FBI*. Orson Welles's *Mercury Theater on the Air* experimented with audioposition extensively. Try listening to "Dracula" (07/11/38), "Julius Caesar" (09/11/38), "Hell on Ice" (10/09/38), and especially "A Passenger to Bali" (11/13/38). Other instances of intimate audioposition include "July Fourth Is a Radio Car" (07/04/34) on *Calling All Cars*, "The Odyssey Of Runyon Jones" (06/08/41) on *26 by Corwin*, "Ecce Homo" (05/21/38) on *The Columbia Workshop*, "I Sing a New World" (03/19/41) on *Cavalcade of America*, and "My Chicago" (07/26/45) on *Arch Oboler's Plays*. Some wartime versions of the same phenomenon include "Cromer" (12/01/42) and "Women of Britain" (08/24/42) on *An American in England*, as well as "Assignment U.S.A." (02/22/44), "Last Days of Sevastopol" (07/17/43), and "Firm Hands, Silent People" (07/31/43) on *Words at War*. Some examples of the kaleidosonic style include *The Mercury Theater on the Air*'s "The War of the Worlds" (10/30/38), *The March of Time* (10/5/34), *This Is War*'s "Navy" (02/28/42), *The Columbia Workshop*'s "Daybreak" (07/10/45) and "Air Raid" (10/27/38), and "Night" (06/07/45) on *Arch Oboler's Plays*. Additional plays of this type are cited in chapter 4.

Chapter 4. Two Norman Corwin plays that exemplify the impermanence of both intimate and kaleidosonic radio are "Clipper Home" (12/22/42) from *An American in England* and "Unity Fair" (07/03/45) from *Columbia Presents Corwin*. The broadcasts of *Cavalcade of America* discussed in chapter 4 include "The Will to Rebuild" (10/30/35) and "The Bridge of Builders" (02/19/36). Interested readers should also try "Defiance of Nature" (12/18/35), "Songs That Inspired a Nation" (03/04/36), "Showmanship" (09/30/36), "Tillers of The Soil" (05/13/36), "Hardness" (05/20/36), and "Songs of the Sea" (01/20/37). For a contrast between the *Cavalcade* of the early 1930s with its postwar programming, listen to the account of the life of Mark Twain in "Fame in Literature" (04/01/36) and the one related in "Life on the Mississippi" (02/24/53). The centerpiece of this chapter is Norman Corwin's special "We Hold These Truths" (12/15/41). Those interested in this broadcast will also want to consult "On a Note of Triumph" (05/13/45), which marked the end of the war in Europe, as well as two other landmark broadcasts on *Columbia Presents Corwin*, "Untitled" (05/30/44) and "Fourteen August" (08/14/45). In part 2 I go into greater detail on wartime programs cited in this chapter, but readers may want to consider a few specimens. On the *Inner Sanctum Mysteries* episode "Island of

Death" (12/07/41), there is little mention of the attack that day, except a note at the outset of the broadcast. *Superman* was also caught off guard. When Pearl Harbor was attacked, the program was in the midst of a two-week sequence in which Clark Kent set out to rescue missing American engineers kidnapped by Incas—"The Pan American Highway" (15 parts, 11/14/41 to 12/17/41). By the end of the story, the titular highway was a link in a defense chain and Kent was a special operative under sealed orders by the government. Students interested in Kent's wartime exploits should also consult "Lita, the Leopard Woman" (12 parts, 01/12/42 to 02/06/42) and "Mystery in Arabia" (7 parts, 09/30/42 to 10/09/42). On *Amos 'n' Andy,* there is evidence of the effect of the onset of war in "Impersonating an Officer" (05/12/44) and "Nazi Spy" (06/02/44). Enemy spies were a major theme in every genre throughout the war: "Enemy Agents" (12/09/42) on *The Mayor of the Town,* "Menace in Wax" (11/17/42) on *Suspense,* "The Man Called X" (circa 1944) on *The Globe Theater,* and "The Secret Room" (02/13/45) on *The Strange Dr. Weird.* Listeners are asked to turn in waste kitchen fats for pharmaceutical manufacturing in "Little Old Lady" (05/25/43) on *Lights Out!;* while in "Mirage" (10/03/43) on *The Whistler,* spokespeople assure listeners that Signal Oil's "go-farther gasoline" will keep a car running for the duration. To get a sense of how Pearl Harbor affected the content of antifascism, try listening to *You Can't Do Business with Hitler,* which was tame while managed by the Office of Emergency Management (e.g., "Heads They Win, Tails We Lose," circa early 1942), but became far more polemical after being taken over by the Office of War Information (e.g., "The Third Horseman," circa late 1942).

Chapter 5. Lucille Fletcher's break as a writer came with the scenario for "My Client Curly," which was scripted by Norman Corwin and aired on *The Columbia Workshop* (03/7/40). Some plays that exhibit the 1940s interest in the "psychological" include "Johnny Got His Gun" (03/09/40) on *Arch Oboler's Plays,* "The Apple Tree" (01/12/42) on *Lady Esther Presents Orson Welles,* "Under the Volcano" (04/29/47) on *Studio One,* "The Path and the Door" (08/25/46) on *The Columbia Workshop,* "Ten Men in a Flying Fortress" (01/28/43) on *Wings to Victory,* "The Fine Art of Murder" (12/22/46) and "The Shadow's Revenge" (05/11/47) on *The Shadow,* "Prep Joe" (02/12/44) on *The Man behind the Gun,* "Strange Interlude" (04/07/46) on *The Theater Guild on the Air,* and "Elementals" (02/14/49) on *Radio City Playhouse.* The principal play discussed in this chapter is *Suspense*'s "The Hitch-Hiker" (09/02/42), which is not to be confused with a later version on *The Mercury Theater Summer Theater* (06/21/46). In the *Suspense* broadcast, Orson Welles claims to have produced the play for his *Mercury Theater on the Air* in the past, but I have no recording of that broadcast. The "shocker" programs referred to in chapter 5 include *Suspense, Lights Out!, The Whistler, The Sealed Book, Inner Sanctum Mysteries, Mysterious Traveler, Mystery in the Air,* and *The Molle Mystery Theater.* For a longer look at blood-and-thunder radio, listen to versions of Edgar Allan Poe's "The Tell-Tale Heart," which aired over the years on *The Fleishmann's Yeast Hour* (04/04/35), *The Columbia Workshop* (07/11/37), *Inner Sanctum Mysteries* (08/03/41), *The Mercury Summer Theater* (08/23/46), *The Weird Circle* (circa 1940s), and *The Hall of Fantasy* (06/01/53). Homage to Poe can also be heard in "The Black Cat" (08/22/39) on *Calling All Cars* and "Big Crazy" (08/30/51) on *Dragnet.* For uses of the voices of the dead that are more rhetorical than gothic, also consider *Studio One*'s "Thunder Rock" (09/02/47) and Norman Corwin's "On a Note of Triumph" (05/08/45). No study of supernatural radio would be complete without Alonzo Deen Cole's *The Witch's Tale,* a pivotal program of the 1930s. For an introduction, try listening to such plays as "Graveyard Mansion" (03/06/33), "The Flying Dutchman" (04/36/34), "Bronze Venus" (08/21/35), and "Haunted Crossroads" (02/12/37). Some examples of what I call "signal-based" dramas include "Tension in Room # 643" (08/01/49) on *Radio City Playhouse*, "Voices of Destruction" (03/14/41) on *Front Page Drama*, "Continue Unloading" (10/04/43) on *Cavalcade of America*, "Condition Red" (01/25/44) on *Words at War*, "The Panther Story" (02/09/46) on *Out of the Deep*, and "The Beckoning Fair One" (06/05/45) on *The Molle Mystery Theater*. Signal-based dramas that aired on *Suspense* include "Lord of the Witch Doctors" (10/27/42), "Sorry, Wrong Number" (05/25/43), "The White Rose Murders" (07/06/43), "Donovan's Brain" (2 parts, 05/18/44 and 05/25/44), "History of Edgar Lowndes" (06/08/44), "Search for Henri LeFevre" (07/06/44), "Heart's Desire" (03/22/45), "August Heat" (05/31/45), and "The Thirteenth Sound" (02/13/47).

Chapter 6. I begin chapter 6 with the original broadcast of Lucille Fletcher's *Suspense* play "Sorry, Wrong Number" (05/25/43). The final line of this broadcast was flubbed; readers may also want to listen to a later repeat on the same program (08/21/43). Some examples of dramas that involve feats of aural perception include "Footfalls" (07/12/45) on *Suspense* and "The King of the World" (03/25/45) on *The Sealed Book*. Here are a few stories in which characters discover that they are being manipulated by a hidden radio: *Inner Sanctum*'s "Voice on the Wire" (11/29/44), *The Lux Radio Theater*'s "Dark Waters" (11/27/44), and *The Whistler*'s "The Body Wouldn't Stay in the Bay" (01/08/45). Some plays about garrulous talkers and acts of sheer vocal prowess include "Mr. Ten Percent" (05/17/45) on *Arch Oboler's Plays*, "Poison Peddler" (02/24/49) on *Counterspy*, and "Voice of Death" (03/06/45) on *The Strange Dr. Weird*. Several plays involve a scene in which a man alone in the dark hears the voice of a hidden being who is trying to influence him. Some instances include "The Pit and the Pendulum" (01/12/43) on *Suspense*, "Kill" (04/20/43) on *Lights Out!*, and "The Horla" (08/21/47) on *Mystery in the Air*. On *The Whistler*, *The Diary of Fate*, and *The Sealed Book*, there are many episodes in which the host of the program speaks to a character in the play he is narrating—for example, in *The Whistler*'s "Final Decree" (04/29/43). Plays about "transmitter" characters include *The Columbia Workshop*'s "The Pied Piper of Hamelin" (07/21/46), *Suspense*'s "The Angel of Death" (01/03/46) and "The Doctor Prescribed Death" (02/02/43), *The Molle Mystery Theater*'s "Talk Them to Death" (06/27/47), *Studio One*'s "An Enemy of the People" (05/13/47), *The Mayor of the Town*'s "Tom Williams Wants to Enlist" (09/06/42), *Counterspy*'s "Murmured Millions" (08/11/49), and *Mr. District Attorney*'s "Labor Pirates" (08/19/42). "Receiver" characters seem to become "transmitters" in "War Orphan" (10/07/42) on *The Mayor of the Town*, "The Man Who Wanted to Be Edward G. Robinson" (10/17/46) on *Suspense*, and "Gumpert" (08/21/45) on *Columbia Presents Corwin*. Here are two plays in which "transmitter" characters seem to become "receivers": "It's a Gift" (03/21/48) on *The Ford Theater* and "Anonymous" (09/19/48) on *Quiet, Please*. Chapter 6 concludes with "The Tell-Tale Heart" (08/03/41) on *Inner Sanctum Mysteries* and "The Man Who Liked Dickens" (10/09/47) on *Suspense*.

Chapter 7. In the section on eavesdropping in chapter 7, I discuss plays about espionage, smuggled documents, and spy networks. Here are a few examples: "Five Graves to Cairo" (12/13/43) on *The Lux Radio Theater*, "The Secret Room" (02/13/45) on *The Strange Dr. Weird*, "Washington Woman Spy" (06/08/42) on *David Harding, Counterspy*, "Last Days of Sevastopol" (07/17/43) and "Prisoner of Japs" (08/07/43) on *Words at War*, *Hop Harrigan* (10/08/42),

"The Man Called X" (circa 1944) on *The Globe Theater*, "Menace in Wax" (11/17/42) on *Suspense*, and "Enemy Agents" (12/09/42) on *The Mayor of the Town*. The passage on overhearing proceeds by describing techniques employed to signify eavesdropping. In sequence, these examples are "The Voice Machine" (16 parts, 07/25/41 to 08/29/41) on *Superman*, "Lloyd Mawson" (05/18/48) and "Nelson Walker" (06/15/48) on *The Diary of Fate*, "The Search for Robert Regent" (68 parts, 06/28/48 to 09/30/48) and "The Return of Roxor" (50 parts, 12/08/48 to 04/28/49) on *Chandu, the Magician*, "Unknown Source" (07/10/48) on *Mr. District Attorney*, *Captain Midnight* (11/23/39), "The Walkie-Talkie Stick-Up" (07/19/46) on *This Is Your FBI*, "Death Bound" (02/03/47) on *Inner Sanctum Mysteries*, "Jealousy" (09/27/42) on *The Whistler*, "The Man Who Talked" (09/29/38) on *Calling All Cars*, "Death Talks Out of Turn" (06/20/44) on *The Molle Mystery Theater*, and "Secret Agent 23" (circa 1944) on *Ceiling Unlimited*. Readers interested in ventriloquism should begin with Edgar Bergen's broadcasts, for instance, "Mae West" (12/12/37), "Charles Boyer" (05/23/43), and "Charlie Answers Ad for Edgar" (12/10/44). Here are a few instances in which ventriloquist characters are featured as murder suspects: "Jane Arnold" (05/19/50) on *Broadway Is My Beat*, "Dead of Night" (03/24/47) on *Out of This World*, "Deadly Dummy" (01/24/49) on *Inner Sanctum Mysteries*, "The Marvelous Barastro" (08/07/47) on *Mystery in the Air*, and "Ria Bouchinska" (11/13/47) on *Suspense*. Plays about quasi-ventriloquial relationships include "Count of Monte Cristo" (08/29/38) on *Mercury Theater on the Air*; "The Black Curtain" (12/02/43) on *Suspense*; "Cyrano de Bergerac" (10/12/47) on *The Theater Guild on the Air*; "Naughty Marietta" (06/12/44) and "Break of Hearts" (09/11/44) on *The Lux Radio Theater*; "Help Wanted Female" (01/08/48), "A Writer in the Family" (01/29/48), and "Love Is Stranger Than Fiction" (02/12/48) on *The First Nighter*; "A Bell for Adano" (03/28/44) and "Rainbow" (03/06/45) on *Words at War*; and "Killer, Come Back to Me" (05/17/46) on *The Molle Mystery Theater*. The final third of chapter 7 concerns characters involved in conveying signals, such as "K9 Corps" (06/06/43) on *The Man behind the Gun*. For plays that show the importance of "nerve centers," listen to "Vicious Visitor" (09/29/49) and "Curious Conspiracy" (10/20/50) on *David Harding, Counterspy*, "The Big Man" (01/12/50) and "The Big Couple" (51/02/22) on *Dragnet*, "The Scientific Touch" (10/07/55) on *The FBI in Peace and War*, "Richard McMillan" (circa 1942–45) and "William Tyree" (circa 1942–45) on *Soldiers of the Press*, "The Battle of the Atlantic" (08/05/43) on *Battle Stations*, "To All Hands" (11/09/43) on *Words at War*, and "Submarine Astern" (03/29/43) on *Cavalcade of America*. For a contrasting example of how journalists were depicted prior to the Second World War, listen to *Big Town*'s "Deep Death" (40/01/01). The signalman broadcasts discussed in depth at the end of this chapter are "Meridian 71212" (08/24/39) on *The Columbia Workshop*, "Weather Ahead" (03/14/49) and "Correction" (01/10/49) on *Radio City Playhouse*, "Twelve to Five" (04/12/48) and "Green Light" (01/26/48) on *Quiet, Please*, and "The Signal Man" (08/24/46) on *Lights Out!*

Chapter 8. I begin chapter 8 by citing comments from writer-director Arch Oboler. To get a sense of his work, readers should listen to "This Precious Freedom" (12/30/39) and "The Family Nagachi" (09/27/45) on *Arch Oboler's Plays*, as well as "Sakhalin" (03/03/37), "Murder Castle" (02/16/38), "Valse Trieste" (03/30/38), "Murder in the Script Department" (05/11/43), and "Bathysphere" (06/29/43) on *Lights Out!* One noteworthy stylistic change in postwar radio is its use of illustrative sound effects, an example of which is *Escape*'s "The Fall of the House of Usher" (10/22/47). I also call attention to the postwar fascination with technology: *Counterspy* uses a punch-card computer to track down a missing biologist in "Captured Contact" (01/14/51), and an "IBM machine" is used to track a cross-dressing killer in "Big Girl" (02/09/51) on *Dragnet* and to gamble on horse races in "Big Brain" (03/14/50) on *Mysterious Traveler*. "The Outer Limit" represents a common story of alien visitors, airing on *Escape* (02/07/50), *Suspense* (02/15/54), and *Dimension X* (11/16/55). Nineteen-fifties radio also aired many tales about nuclear disaster, including "The Rocket Ship" (07/28/45) on *Lights Out!* and "Adam and the Darkest Day" (11/07/48) on *Quiet, Please*. Several programs experimented with new genres. On *Suspense*, for instance, the 1950s witnessed experiments with westerns ("A Killing in Abilene," 12/14/50), true crime ("The McKay College Basketball Scandal," 09/24/51), and programs based on songs ("St. James Infirmary," 02/23/53). Cold War adventures from this period include "Relief Supplies" (07/01/49) on *Dangerous Assignment*, "Fabulous Formula" (04/18/50) on *Counterspy*, "Listen, Young Lovers" (05/31/54) on *Suspense*, and "Rim of Terror" (12/02/56) on *Escape*. I argue in chapter 8 that postwar "closed world" plays feature immobile characters trapped in an aural space characterized by radial balance. Some examples of this include *Escape*'s "Three Skeleton Key" (11/15/49), *Suspense*'s "Pigeon in a Cage" (05/25/53), and *Dimension X*'s "Hello Tomorrow" (11/03/55). I also make the case that many postwar radio plays are filled with a sense of time running out. For examples, see "Corpus Delecti" (01/17/47) on *Molle Mystery Theater*, "Marvin Thomas" (06/08/48) and "Victor Wakeman" (06/29/38) on *The Diary of Fate*, "Long Distance" (07/03/48) on *Radio City Playhouse*, "Later Than You Think" (08/02/48) on *Quiet, Please*, "Big Bomb" (07/13/50) on *Dragnet*, "Sabotage in Paris" (05/06/50) on *Dangerous Assignment*, "Statement of Fact" (05/14/53) on *On Stage*, and "Vial of Death" (05/18/53) on *Escape*.

Chapter 9. To hear the lurid tone of postwar crime programming, readers should listen to *Suspense*'s "Can't We be Friends?" (07/25/46) and "Too Hot to Live" (10/26/50), *Dragnet*'s "Child Killer" (02/02/50), *Mr. District Attorney*'s "Hot Rod Killer" (circa 1952), or *This Is Your FBI*'s "The Baby Peddlers" (11/28/52). In chapter 9, I describe a strategy of depicting small and large spaces that made *Dragnet* a program about containing threats. Some episodes that exhibit these properties include "The Big Actor" (08/10/50), "The Big Betty" (11/23/50), "The Big Saint" (04/26/51), and "The Big Set-Up" (07/12/51). For a brief sample to discover how major crime authors were adapted for radio anthologies, consult "The Glass Key" (03/10/39) on *The Mercury Theater on the Air*, "Murder in City Hall" (09/19/44) and "Two Men in a Furnished Room" (09/27/46) on *Molle Mystery Theater*, and "Love's Lovely Counterfeit" on *Suspense* (03/08/45). The majority of chapter 9 focuses on the following radio plays, in order of mention in the text: "Night Tide" (05/21/49) on *The Adventures of Phillip Marlowe*; "Julie Dixon" (02/10/50), "Ruth Larson" (06/23/51), "Joe Blair" (11/03/51), and "Kenny Purdue" (11/15/52) on *Broadway Is My Beat*; "Citizen Caldwell" (10/26/51) on *This is Your FBI*; "Death on Highway 99" (10/04/45) on *Suspense*; and "I Always Marry Juliet" (04/05/48) on *Quiet, Please*. At the end of the chapter, I cite the following four episodes of *Dragnet*: "The Big Knife" (05/11/50), "The Big Phone Call" (02/14/52), "The Big Cast" (02/08/51), and "The Big Tear" (09/11/52).

Chapter 10. I begin chapter 10 with some episodes of *I Was a Communist for the FBI*: "The Rat Race" (09/17/52), "A Suit for the Party" (10/08/52), "My Friend the Enemy" (05/13/53), and "The Red Snow" (08/05/53). I also note contemporaneous plays about undercover men:

Dragnet's "The Big Meet" (10/26/50), *Suspense*'s "Allen in Wonderland" (10/27/52) and "The Girl in Car 32" (03/15/54), *Counterspy*'s "Double-crossing Defender" (01/21/50) and "Soaring Saucer" (05/02/50), and *Dangerous Assignment*'s "Little White Pill" (05/30/50). Remaining *I Was a Communist* episodes cited in this chapter include "The Wrong Dream" (04/01/53), "Red Red Herring" (05/14/52), "Hate Song" (12/31/52), "Traitors for Hire" (05/28/52), "Rich Man, Poor Man" (07/16/52), "No Second Chance" (11/05/52), and "Where Red Men Roam" (07/02/52). The next section explores the use of psychoanalysis, psychiatry, and psychotherapy. Some plays that illustrate this phenomenon are "The Frightened Fugitive" (11/08/46) on *This is Your FBI*; "Of Unsound Mind" (07/17/48) on *Radio City Playhouse*; "Death Out of Mind" (12/29/47) on *Inner Sanctum Mysteries*; "Ruth Larson" (06/23/51) on *Broadway Is My Beat*; "Big Compulsion" (04/12/53) on *Dragnet*; "Drive-in" (01/11/45), "To Find Help," (01/18/45), and "Mission Accomplished" on *Suspense* (12/01/49); "Blackmail Murder" (circa 1950s) on *Mr. District Attorney*; and "Bad Medicine" (07/10/56) on *X Minus One*. The two plays that I read closely on this subject are from *Suspense*: "Elwood" (03/06/47) and "I Saw Myself Running" (05/24/55). The next section of the chapter deals with plays about trials of conscience. Some examples of this type of drama include *Radio City Playhouse*'s "Hit and Run" (08/14/48), *Escape*'s "Pass to Berlin" (05/19/50), *Diary of Fate*'s "Walter Vincent" (05/25/48), and *Suspense*'s "Subway" (10/30/47). The final radio play discussed in this chapter is *Suspense*'s "Zero Hour" (04/05/55).

NOTES

INTRODUCTION

1. Walter Ames, "Radio Still Here and Even Growing," *Los Angeles Times,* March 11, 1956, D4. There are similar anecdotes in many commentaries about radio. See, e.g., Gerald Nachman, *Raised on Radio* (Berkeley: University of California Press, 1998), 7.

2. "NBC Video Series to Depict Psychological Motivations in Human Behavior," *New York Times*, July 7, 1949, 50.

3. Eugene E. Brussell, ed., *Webster's New World Dictionary of Quotable Definitions* (Englewood Cliffs, NJ: Prentice Hall, 1988), 474; James Beasley Simpson, *Simpson's Contemporary Quotations* (Boston: Houghton Mifflin, 1997), 554.

4. Christopher H. Sterling and Michael C. Keith, "Where Have All the Historians Gone? A Challenge to Researchers," *Journal of Broadcasting and Electronic Media* 50 (2006): 351; Matthew Murray, "'The Tendency to Deprave and Corrupt Morals': Regulation and Irregular Sexuality in Golden Age Radio Comedy," in *The Radio Reader: Essays in the Cultural History of Radio*, ed. Michele Hilmes and Jason Loviglio (New York: Routledge, 2002), 141.

5. Joseph Julian, *This Was Radio: A Personal Memoir* (New York: Viking, 1975), 235.

6. W. J. T. Mitchell, "Image," in *Critical Terms for Media Studies*, ed. W. J. T. Mitchell and Mark B. N. Hansen (Chicago: University of Chicago Press, 2010), 41. See also W. J. T. Mitchell, *Iconology: Image, Text, Ideology* (Chicago: University of Chicago Press, 1986), 7–46.

7. David Hume, *A Treatise of Human Nature*, ed. L. A. Selby-Bigge (New York: Oxford University Press, 1978), 253.

8. Mitchell, *Iconology*, 15.

9. Stanley High, "The Not-So-Free Air," *Saturday Evening Post*, February 11, 1939, 8. These figures must be an exaggeration. My guess is that networked broadcasts were tallied as if they were unique on each station that aired them, which seems unfair. Nevertheless, whether it was a hundred thousand shows each year or a million, the output of broadcasting was impressively vast.

10. At the end of this book, I provide a chapter-by-chapter guide to broadcasts discussed. A list of nearly all plays studied in my research can be found in the appendix to my dissertation:

Neil Verma, "Theater of the Mind: The American Radio Play and Its Golden Age, 1934–1956" (PhD diss., University of Chicago, 2008), 384–441. All recordings were obtained from the Old Time Radio Catalog (otrcat.com) between 2002 and 2010. I used the following reference books in framing my sample: John Dunning, *The Encyclopedia of Old Time Radio* (New York: Oxford University Press, 1998), J. David Goldin, *Radio Yesteryear Presents the Golden Age of Radio* (New York: Yesteryear, 1998), Ron Lackman, *The Encyclopedia of American Radio* (New York: Checkmark Books, 2000), Mitchell E. Shapiro, *Radio Network Prime Time Programming, 1926–1967* (Jefferson, NC: McFarland, 2002), and Thousand Oaks Library, American Radio Archives, *Radio Series Scripts, 1930–2001*, compiled by Jeanette M. Berard and Klaudia Englund (Jefferson, NC: McFarland), 2006.

11. Classic discourses on drama tend to concern structure rather than technology. See, e.g., John Dryden, "An Essay of Dramatic Poesy," in *English Critical Essays: Sixteenth, Seventeenth and Eighteenth Centuries*, ed. Edmund D. Jones (New York: Oxford University Press, 1952), 107–74, and G. E. Lessing, *Hamburg Dramaturgy*, trans. Helen Simmern (New York: Dover, 1962).

12. Aristotle, *Poetics*, trans. Michael Heath (New York, Penguin, 1996), 10–13. Of course, radio cannot provide another key tragic element: "spectacle."

13. Roland Barthes, "Diderot, Brecht, Eisenstein," in *Image-Music-Text*, trans. Stephen Heath (New York: Noonday, 1992), 70.

14. Lessing, *Hamburg Dramaturgy*, 141–42. "Drama" derives from the Greek word for "action." See also Aristotle, *Poetics*, 5–6, 10–11, and Jean-Paul Sartre, "Beyond Bourgeois Theater," in *Theatre in the Twentieth Century*, ed. Robert W. Corrigan (New York: Grove, 1963). Lessing's characterization ought to be viewed in the context of his framework for distinguishing time-based and space-based media. See G. E. Lessing, *Laocoon: An Essay upon the Limits of Painting and Poetry*, trans. Ellen Frothingham (Boston: Roberts Brothers, 1883).

15. Jindřich Honzl, "Dynamics of the Sign in the Theater," in *Modern Theories of Drama: A Selection of Writings on Drama and Theater, 1840–1990*, ed. by George W. Brandt (New York: Oxford University Press, 1998), 270.

16. Raymond Williams, *Drama from Ibsen to Brecht* (New York: Oxford University Press, 1969), 12–16. See also Raymond Williams, *Marxism and Literature* (New York: Oxford University Press, 1977), 173–79.

17. See Raymond Williams, *Television: Technology and Cultural Form* (Middletown, CT: Wesleyan University Press, 1993), 17–23. In broadcasting, Williams argues, the invention of transmission preceded the invention of content. The big developments in the 1920s had to do with technology; those of 1930s broadcasting had to do with using it.

18. See, e.g., Carolyn Marvin, *When Old Technologies Were New: Thinking about Electric Communication in the Late Nineteenth Century* (New York: Oxford University Press, 1988); Jeffrey Sconce, *Haunted Media: Electronic Presence from Telegraphy to Television* (Durham, NC: Duke University Press, 2000); and Paul Young, *The Cinema Dreams Its Rivals: Media Fantasy Films from Radio to the Internet* (Minneapolis: University of Minnesota Press, 2006).

19. Sconce, *Haunted Media*, 7.

20. I have in mind Jacques Ellul, *The Technological Society* (New York: Knopf, 1964); Marshall McLuhan, *The Gutenberg Galaxy* (Toronto: University of Toronto Press, 1962); and Alvin Toffler, *Future Shock* (New York: Random House, 1970). In the last decade, scholars have re-

turned to the relationship between mind and medium to develop approaches to literature. See, e.g., Lisa Zunshine, *Why We Read Fiction: Theory of Mind and the Novel* (Columbus: Ohio State University Press, 2006).

21. See Susan J. Douglas, *Inventing American Broadcasting, 1899-1922* (Baltimore: Johns Hopkins University Press, 1987); Robert McChesney, *Telecommunications, Mass Media, and Democracy: The Battle for the Control of U.S. Broadcasting, 1928-1935* (New York: Oxford University Press, 1993); and Susan Smulyan, *Selling Radio: The Commercialization of American Broadcasting, 1920-1934* (Washington, DC: Smithsonian Institution Press, 1994). See also Paul Starr, *The Creation of the Media: Political Origins of Modern Communications* (New York: Basic Books, 2004), 327-384, and Steve J. Wurtzler, *Electric Sounds: Technological Change and the Rise of Corporate Mass Media* (New York: Columbia University Press, 2007). These authors follow in the footsteps of Erik Barnouw, the first major scholar of broadcasting. See the first two parts of his three-volume study: *A Tower in Babel: A History of Broadcasting in the United States to 1933* (New York: Oxford University Press, 1966) and *The Golden Web: A History of Broadcasting in the United States, 1933-1953* (New York: Oxford University Press, 1968).

22. Touchstone works in the "social construction of technology" include Wiebe Bijker, Thomas Hughes, and Trevor Pinch, eds., *The Social Construction of Technological Systems* (Cambridge: MIT Press, 1990); Thomas Kuhn, *The Structure of Scientific Revolutions* (Chicago: University of Chicago Press, 1962); Bruno Latour, *Science in Action: How to Follow Scientists and Engineers through Society* (Cambridge: Harvard University Press, 1987); and Donald MacKenzie and Judy Wacjman, eds., *The Social Shaping of Technology: How the Refrigerator Got Its Hum* (Philadelphia: Open University Press, 1985).

23. See, e.g., Douglas B. Craig, *Fireside Politics: Radio and Political Culture in the United States, 1920-1940* (Baltimore: Johns Hopkins University Press, 2000); Susan J. Douglas, *Listening In: Radio and the American Imagination* (Minneapolis: University of Minnesota Press, 1999); Melvin Patrick Ely, *The Adventures of Amos 'n' Andy: A Social History of an American Phenomenon* (New York: Free Press, 1991); Nathan Godfried, *WCFL, Chicago's Voice of Labor, 1926-1978* (Urbana: University of Illinois Press, 1997); Michele Hilmes, *Radio Voices: American Broadcasting, 1922-1952* (Minneapolis: University of Minnesota Press, 1999); Gerd Horten, *Radio Goes to War: The Cultural Politics of Propaganda during World War II* (Berkeley: University of California Press, 2002); Michael C. Keith, *Sounds in the Dark: All-Night Radio in American Life* (Ames: Iowa State University Press, 2001); Kristine M. McCusker, *Lonesome Cowgirls and Honky-Tonk Angels: The Women of Barn Dance Radio* (Chicago: University of Illinois Press, 2008); Philip Rosen, *The Modern Stentors: Radio Broadcasters and the Federal Government, 1920-1934* (Westport, CT: Greenwood, 1980); Alexander Russo, *Points on the Dial: Golden Age Radio beyond the Networks* (Durham, NC: Duke University Press, 2010); Barbara Dianne Savage, *Broadcasting Freedom: Radio, War, and the Politics of Race, 1938-1948* (Chapel Hill: University of North Carolina Press, 1999); Holly Cowan Shulman, *Voice of America: Propaganda and Democracy, 1941-1945* (Madison: University of Wisconsin Press, 1990); Hugh Slotten, *Radio's Hidden Voice: The Origins of Public Broadcasting in the United State*s (Urbana: University of Illinois Press, 2009); and Michael Stamm, *Sound Business: Newspapers, Radio and the Politics of New Media* (Philadelphia: University of Pennsylvania Press, 2011).

24. Influences include Warren Susman and Lizabeth Cohen, both of whom also write directly about radio: Lizabeth Cohen, *Making a New Deal: Industrial Workers in Chicago,*

1919–1939 (New York: Cambridge University Press, 1990), 99–158, 291–322; Warren Susman, *Culture as History: The Transformation of American Society in the Twentieth Century* (Washington, DC: Smithsonian Institution Press, 2003), 150–83.

25. I have in mind such works as Kathleen Battles, *Calling All Cars: Radio Dragnets and the Technology of Policing* (Minneapolis: University of Minnesota Press, 2010); Howard Blue, *Words at War: World War II Era Radio Drama and the Postwar Broadcasting Industry Blacklist* (Lanham, MD: Scarecrow, 2002); Mary Desjardins and Mark Williams, "Are You Lonesome Tonight?": Gendered Address in *The Lonesome Gal* and *The Continental*," in *Communities of the Air: Radio Century, Radio Culture*, ed. Susan Merrill Squier (Durham, NC: Duke University Press, 2003), 251–74; Douglas Gomery, *A History of Broadcasting in the United States* (Malden, MA: Blackwell, 2008), 71–103; Tona J. Hangen, *Redeeming the Dial: Religion and Popular Culture in America* (Chapel Hill: University of North Carolina Press, 2002); Bruce Lenthall, *Radio's America: The Great Depression and the Rise of Modern Mass Media* (Chicago: University of Chicago Press, 2007); Jason Loviglio, *Radio's Intimate Public: Network Broadcasting and Mass-Mediated Democracy* (Minneapolis: University of Minnesota Press 2005); J. Fred MacDonald, *Don't Touch That Dial! Radio Programming in American Life from 1920 to 1960* (Chicago: Nelson-Hall, 1991); Allison McCracken, "Scary Women and Scarred Men: Suspense, Gender Trouble and Postwar Change, 1942–1950," in Hilmes and Loviglio, *The Radio Reader*, 188–92; Kathy M. Newman, *Radio Active: Advertising and Consumer Activism, 1935–1947* (Berkeley: University of California Press, 2004); and David Suisman and Susan Strasser, eds., *Sound in the Age of Mechanical Reproduction* (Philadelphia: University of Pennsylvania Press, 2010).

26. On the dearth of work on radio writers, see Sterling and Keith, "Where Have All the Historians Gone?" Important humanistic work on radio programs includes Rudolf Arnheim, *Radio* (Salem, NH: Ayer Company, 1986); Timothy Campbell, *Wireless Writing in the Age of Marconi* (Minneapolis: University of Minnesota Press, 2006); Debra Rae Cohen, Michael Coyle, and Jane Lewty, eds., *Broadcasting Modernism* (Gainesville: University Press of Florida, 2009); Andrew Crisell, *Understanding Radio* (New York: Routledge, 1994); Tim Crook, *Radio Drama: Theory and Practice* (New York: Routledge, 1999); Margaret Fisher, *Ezra Pound's Radio Operas: The BBC Experiments, 1931–1933* (Cambridge: MIT Press, 1993); Elissa Guralnick, *Sight Unseen: Beckett, Pinter, Stoppard, and Other Contemporary Dramatists on Radio* (Athens: Ohio University Press, 1995); Richard Hand, *Terror on the Air! Horror Radio in America, 1931–1952* (Jefferson, NC: McFarland, 2006); and Paul Heyer, *The Medium and the Magician: Orson Welles, the Radio Years, 1943–1952* (New York: Rowman and Littlefield, 2005).

Historians have also opened up the production process tremendously. See, e.g., Elena Razlogova, "True Crime Radio and Listener Disenchantment with Network Broadcasting, 1935–1946," *American Quarterly* 58 (March 2006): 137–58; Alexander Russo, "Roots of Radio's Rebirth: Audiences, Aesthetics, Economics and Technologies of American Broadcasting, 1926–1951" (PhD diss., Brown University, 2004); Jacob Smith, *Vocal Tracks: Performance and Sound Media* (Berkeley: University of California Press, 2008); and Shawn VanCour, "The Sounds of 'Radio': Aesthetic Formations of 1920's American Broadcasting" (PhD diss., University of Wisconsin–Madison, 2008). Students should also consult the following books, which were slated to be published beyond the production deadline for *Theater of the Mind*: Michele Hilmes, *Network Nations: A Transnational History of British and American Broadcasting* (New York:

Routledge, 2011), and Elena Razlogova, *The Listener's Voice: Early Radio and the American Public* (Philadelphia: University of Pennsylvania Press, 2011).

27. Rick Altman, "Deep Focus Sound: Citizen Kane and the Radio Aesthetic," in *Perspectives on Citizen Kane*, ed. Ronald Gottesman (New York: G. K. Hall, 1996), 100.

28. Paul Fussell, *Wartime: Understanding and Behavior in the Second World War* (New York: Oxford University Press, 1989), 181; Marshall McLuhan, *Understanding Media: The Extensions of Man* (Cambridge: MIT Press, 2002), 22–32. Like Fussell, most writers consider radio "cool" in contrast to "hot" television, but McLuhan himself considered radio to be a "hot medium" because he compared it to telephony, concluding that the former provides richer detail than the latter. My disagreement is more with the metric than the measurement: so long as we ask how much "more" or "less" radio "gives," our approach can only address a narrow set of problems.

29. Several other writers have identified the "blind medium" fallacy. See, e.g., Clive Cazeaux, "Phenomenology and Radio Drama," *British Journal of Aesthetics*, 45, no. 2 (April 2005): 157–74; Crisell, *Understanding Radio*, 143–63; Guralnick, *Sight Unseen*, xvi.

30. Russo, *Roots of Radio's Rebirth*, 244.

31. On "distant reading," see Franco Moretti, *Graphs, Maps, Trees: Abstract Models for a Literary History* (New York: Verso, 2005).

32. See Aristotle, *De Anima*, trans. R. D. Hicks (Buffalo, NY: Prometheus Books, 1991), 37–71; Walter Benjamin, "The Work of Art in the Age of Mechanical Reproducibility," 2nd rev., in *Selected Writings*, vol. 3, *1935–1938*, trans. Edmund Jephcott, Howard Eiland et al., ed. Howard Eiland and Michael W. Jennings (Cambridge, MA: Belknap, 2002), 101–33; Susan Buck-Morss, "Aesthetics and Anaesthetics: Walter Benjamin's Artwork Essay Reconsidered," *October* 62 (Fall 1992): 3–41; and Miriam Bratu Hansen, "Benjamin and Cinema, Not a One-Way Street" *Critical Inquiry* 25 (Winter 1999): 306–43. I should specify that, while my work never fully leaves the area of philosophy associated with Alexander Baumgarten, Immanuel Kant, David Hume, and others concerned with art and beauty, this is not directly a book about an aesthetics of judgment and taste in the way that these philosophers intend.

33. Miriam Bratu Hansen, "The Mass Production of the Senses: Classical Cinema as Vernacular Modernism," *Modernism/Modernity* 6 (April 1999): 59–77.

34. The German title of Arnheim's manuscript was *Rundfunk als Hörkunst*. English editions either reduce this title to "Radio" or translate it as "Radio, an Art of Sound."

35. For touchstone texts in culture-as-ideology, see Louis Althusser, "Ideology and Ideological State Apparatuses," in *Lenin and Philosophy and Other Essays* (London: New Left Books, 1971), 127–86; Karl Marx and Frederick Engels, *The German Ideology* (New York: International, 1989), 64–81. See also Terry Eagleton, *Ideology* (New York: Verso, 1991), 63–124; Max Horkheimer and Theodor Adorno, "The Culture Industry: Enlightenment as Mass Deception," *Dialectic of Enlightenment* (New York: Continuum, 1994), 120–67; and Henri Lefebvre, *The Sociology of Marx* (New York: Columbia University Press, 1982).

36. Theodor Adorno, "Free Time," in *The Culture Industry: Selected Essays on Mass Culture*, ed. J. M. Bernstein (New York: Routledge, 2003), 196, 197. My reading of this essay is inspired by James W. Cook, "The Return of the Culture Industry," in *The Cultural Turn in U.S. History*, ed. James W. Cook, Lawrence B. Glickman, and Michael O'Malley (Chicago: University of Chicago Press, 2008), 291–317.

37. See Richard Schechner, *Performance Theory* (New York: Routledge, 1988), 190.

38. Williams, *Drama from Ibsen to Brecht*, 17. See, also: Williams, *Marxism and Literature*, 127–35.

39. Williams, *Marxism and Literature*, 132.

40. Jonathan Sterne, *The Audible Past: Cultural Origins of Sound Reproduction* (Durham, NC: Duke University Press, 2003), 19.

41. Lisa Gitelman, *Scripts, Grooves, and Writing Machines: Representing Technology in the Edison Era* (Stanford, CA: Stanford University Press, 1999) 227.

CHAPTER ONE

1. The Columbia Broadcasting System, *Radio Alphabet* (New York: Hastings House, 1946), 30.

2. Orrin E. Dunlap Jr., "A Year's Stardust," *New York Times*, January 2, 1938, 134; *Newsweek* quoted in John Dunning, *The Encyclopedia of Old Time Radio* (New York: Oxford University Press, 1998), 171; Wolfe Kaufman, review of "The Fall of the City," *Variety*, April 14, 1937, 38. See also Earle McGill, *Radio Directing* (New York: McGraw-Hill, 1940), 21–22.

3. Douglas Coulter, ed., *Columbia Workshop Plays: Fourteen Radio Dramas* (New York: McGraw-Hill, 1939), 349. As of this writing, the other two plays in the registry are *The Mercury Theater*'s "The War of the Worlds" (October 31, 1938) and Norman Corwin's "We Hold These Truths" (December 15, 1941). Individual episodes of *Gangbusters* and *The Lone Ranger* are also listed.

4. See Michael Kammen, *The Mystic Chords of Memory: The Transformation of Tradition in American Culture* (New York: Vintage, 1991), 419.

5. See Coulter, *Columbia Workshop Plays*, 169–71; Benjamin L. Alpers, *Dictators, Democracy, and American Public Culture: Envisioning the Totalitarian Enemy, 1920s–1950s* (Chapel Hill: University of North Carolina Press, 2003), 91–93; Erik Barnouw, *The Golden Web: A History of Broadcasting in the United States, 1933–1953* (New York: Oxford University Press, 1968), 66–68; Howard Blue, *Words at War: World War II Era Radio Drama and the Postwar Broadcasting Industry Blacklist* (Lanham, MD: Scarecrow, 2002), 26–27, 81–82; Daniel J. Czitrom, *Media and the American Mind, from Morse to McLuhan* (Chapel Hill: University of North Carolina Press, 1982), 86; Michael Denning, *The Cultural Front: The Laboring of American Culture in the Twentieth Century* (New York: Verso, 1998), 382–84; Judith E. Smith, "Radio's Cultural Front," in *The Radio Reader: Essays in the Cultural History of Radio*, ed. Michele Hilmes and Jason Loviglio (New York: Routledge, 2002), 209–230.

6. Kaufman, review of "The Fall of the City," 38.

7. See Denning, *The Cultural Front*, 382–84. For an overview of the aims of radio writers in this period, see Bruce Lenthall, *Radio's America: The Great Depression and the Rise of Modern Mass Media* (Chicago: University of Chicago Press, 2007), 175–206.

8. See Robert McChesney, *Telecommunications, Mass Media, and Democracy: The Battle for the Control of U.S. Broadcasting, 1928–1935* (New York: Oxford University Press, 1993); Kathy M. Newman, *Radio Active: Advertising and Consumer Activism, 1935–1947* (Berkeley: University of California Press, 2004); Hugh Slotten, *Radio's Hidden Voice: The Origins of Public Broadcasting in the United States* (Urbana: University of Illinois Press, 2009); and Susan Smulyan,

Selling Radio: The Commercialization of American Broadcasting, 1920–1934 (Washington, DC: Smithsonian Institution Press, 1994).

9. MacLeish is quoted in Orrin E. Dunlap Jr., "The Verse Play: 'The Fall of the City' Reveals Possibilities for Theatre of the Air," *New York Times*, April 18, 1937, 178. See also Lenthall, *Radio's America*, 183–93.

10. Eric Bentley, *The Playwright as Thinker: A Study of Drama in Modern Times* (New York: Harcourt, Brace, 1967), 232–46.

11. See Tim Crook, *Radio Drama: Theory and Practice* (New York: Routledge, 1999), 3–41; Shawn VanCour, "The Sounds of 'Radio': Aesthetic Formations of 1920's American Broadcasting" (PhD diss., University of Wisconsin–Madison, 2008), 228–32.

12. See Joan Shelley Rubin, *The Making of Middlebrow Culture* (Chapel Hill: University of North Carolina Press, 1992), 266–329.

13. "CBS' Dramatic Repertory," *Variety*, April 14, 1937, 35; Chris Vials, *Realism for the Masses: Aesthetics, Popular Front Pluralism, and U.S. Culture, 1935–1947* (Jackson: University Press of Mississippi, 2009), 40–41.

14. "Class Cycle" *Variety*, June 30, 43; "Great Plays" *Time*, April 13, 1942, 43–44.

15. Malcolm Goldstein, *The Political Stage: American Drama and Theater of the Great Depression* (New York: Oxford University Press, 1974), 250.

16. Zora Neale Hurston, *A Life in Letters*, ed. Carla Kaplan (New York: Anchor Books, 2002), 421.

17. "BSH Secrecy Oath," *Variety*, May 4, 1938, 29. Radio writers and directors are listed in sidebars of several issues of this publication in 1942 and 1943, see, e.g., *Variety*, November 4, 1942, 30.

18. See Michael J. Socolow, "'Always in Friendly Competition': NBC and CBS in the First Decade of National Broadcasting," in *NBC: America's Network*, ed. Michele Hilmes (Berkeley: University of California Press, 2007), 25–43.

19. Bob Landry, "Radio Program Factory," *Variety*, December 30, 1936, 33. See also Michele Hilmes, "NBC and the Network Idea: Defining the 'American System,'" in *NBC: America's Network*, ed. Michele Hilmes (Berkeley: University of California Press, 2007) 17; Alexander Russo, *Points on the Dial: Golden Age Radio beyond the Networks* (Durham, NC: Duke University Press, 2010), 126–27.

20. McChesney, *Telecommunications, Mass Media, and Democracy*, 115–16. See also Steve J. Wurtzler, *Electric Sounds: Technological Change and the Rise of Corporate Mass Media* (New York: Columbia University Press, 2007), 195–96.

21. "New CBS Copy Rules Ban Laxatives and 'Bloody' Juveniles," *Advertising Age*, May 20, 1935, 46. See also Sally Bedell Smith, *In All His Glory: The Life of William S. Paley, the Legendary Tycoon and His Brilliant Circle* (New York: Simon and Schuster, 1990), 138–43.

22. For a more nuanced view on how advertisers considered radio and their role in modernizing society, see Roland Marchand, *Advertising the American Dream: Making Way for Modernity, 1920–1940* (Berkeley: University of California Press, 1986), 1–94.

23. "Radio I: A $140,000,000 Art," *Fortune*, May, 1938, 114.

24. Remy Brunel, "Radio Magic Expected to Emerge from Columbia Workshop Experimentation," *Washington Post*, September 6, 1936, AA5.

25. Robert Reinhart, "Drama in Control Room Overshadows Irving Reis' CBS Lads and Lassies," *Variety*, October 27, 1937, 38.

26. Denning, *The Cultural Front*, 381.

27. On Lewis, see R. LeRoy Bannerman, *Norman Corwin and Radio* (Tuscaloosa: University of Alabama Press, 1986), 39; and Barnouw, *The Golden Web*, 64–65.

28. On local content, see Robert L. Hilliard and Michael C. Keith, *The Quieted Voice: The Rise and Demise of Localism in American Radio* (Carbondale: Southern Illinois University Press, 2005), 1–68.

29. Smulyan, *Selling Radio*, 62. See also Wurtzler, *Electric Sounds*, 19–69.

30. In recent work, a much more complex picture of media ownership in this period is emerging. See Alexander Russo, *Points on the Dial*, 47–76; Michael Stamm, *Sound Business: Newspapers, Radio and the Politics of New Media* (Philadelphia: University of Pennsylvania Press, 2011).

31. "NBC in 1935 Paid 2,400,000," *Variety*, January 29, 1936, 37; "Affiliates Outgross Indies," *Variety*, June 15, 1938, 32; Robert J. Landry, "Comment," *Variety*, August 26, 1936, 45.

32. "78,000,000 Radio Payroll; 60,000,000 for Talent," *Variety*, July 22, 1936, 1–36.

33. On September 5, 1936, *The Workshop* broadcast a "re-creation of the re-creation" that took listeners behind the scenes of the *Calling All Cars* broadcast.

34. Rice is quoted in Martin Grams Jr., *Suspense: Twenty Years of Thrills and Chills* (Kearney, NE: Morris, 1997), 123.

35. "Radio's Sharecroppers," *Variety*, April 21, 1937, 31; "Jell-o Signs Benny," *Advertising Age*, March 1, 1937, 8.

36. "Radio's Rapid Growth," *Variety*, March 23, 1938, 1; "Sustaining Time Outweighs Sponsored Hours," *Advertising Age*, April 17, 1939, 26.

37. John K. Hutchens, "Drama on the Air," *New York Times*, November 2, 1941, X12; Susan J. Douglas, *Listening In: Radio and the American Imagination* (Minneapolis: University of Minnesota Press, 1999), 196.

38. Denning, *The Cultural Front*, 42. See Blue, *Words at War*, 167; Melvin Patrick Ely, *The Adventures of Amos 'n' Andy: A Social History of an American Phenomenon* (New York: Free Press, 1991), 26–96.

39. Robert J. Landry, "Take Off Those Handcuffs," *Variety*, January 8, 1935, 85; Eddie Cantor, "Radio Needs Showmen," *Printer's Ink*, October 25, 1934, 19; Mary Pickford, "Radio as Mary Sees It," *Printer's Ink*, December 20, 1934, 57.

40. Martin Heidegger, *Being and Time*, trans. John Macquarrie and Edward Robinson (San Francisco: Harper, 1962), 140.

41. John Dos Passos, *U.S.A.* (New York: Library of America, 1996), 260; F. T. Marinetti, "La Radia," in *Wireless Imagination: Sound, Radio and the Avant Garde*, ed. Douglas Kahn and Gregory Whitehead (Cambridge: MIT Press, 1992), 267. See also Timothy Campbell, *Wireless Writing in the Age of Marconi* (Minneapolis: University of Minnesota Press, 2006), 67–96.

42. Douglas Kahn, *Noise, Water, Meat: A History of Sound in the Arts* (Cambridge: MIT Press, 1999), 16.

43. Smith, *In All His Glory*, 139.

44. Robert Lynd and Helen Lynd, *Middletown* (New York: Harcourt, Brace, 1929), 269n;

Peter Simonson, *Refiguring Mass Communication: A History* (Urbana: University of Illinois Press, 2010), 1–28.

45. Rudolf Arnheim, *Radio* (Salem, NH: Ayer, 1986), 238; Lewis Mumford, *Technics and Civilization* (New York: Harcourt, Brace, 1934), 241.

46. Robert Lynd and Helen Lynd, *Middletown in Transition* (New York: Harcourt, Brace, 1937), 264.

47. The President's Research Committee on Social Trends, *Recent Social Trends in the United States* (Washington, DC: GPO, 1933), 153–56.

48. Hadley Cantril and Gordon Allport, *The Psychology of Radio* (New York: Harper and Brothers, 1935), 20.

49. E. B. White, *One Man's Meat* (New York: Harper and Brothers, 1944), 35.

50. See, e.g., Alan Brinkley, *Voices of Protest: Huey Long, Father Coughlin and the Great Depression* (New York: Vintage, 1983), 156; Lizabeth Cohen, *Making a New Deal: Industrial Workers in Chicago, 1919–1939* (New York: Cambridge University Press, 1990), 99–158, 323–33; Neal Gabler, *Winchell: Gossip, Power, and the Culture of Celebrity* (New York: Knopf, 1995), xiii; David M. Kennedy, *Freedom from Fear: The American People in the Great Depression* (New York: Oxford University Press, 1999), 227; Ronald Tobey, *Technology as Freedom: The New Deal and the Electrical Modernization of the American Home* (Berkeley: University of California Press, 1996), 159. The homogenization of radio listening is intimately related to electrification. See Ronald S. Kline, *Consumers in the Country: Technology and Social Change in America* (Baltimore: Johns Hopkins University Press, 2002), 113–78, and David Nye, *Electrifying America: Social Meanings of a New Technology, 1880–1940* (Cambridge: MIT Press, 1990).

51. See, e.g., Susan Merrill Squier, "Communities of the Air: Introducing the Radio World," in *Communities of the Air: Radio Century, Radio Culture*, ed. Susan Merrill Squier (Durham, NC: Duke University Press, 2003), 1–35; Douglas, *Listening In*, 24; Michele Hilmes, *Radio Voices: American Broadcasting, 1922–1952* (Minneapolis: University of Minnesota Press, 1999), xvii–20; Kate Lacey, *Feminine Frequencies: Gender, German Radio, and the Public Sphere, 1923–1945* (Ann Arbor: University of Michigan Press, 1996), 221–43; Jason Loviglio, *Radio's Intimate Public: Network Broadcasting and Mass-Mediated Democracy* (Minneapolis: University of Minnesota Press, 2005), xvi–xxv; and Timothy D. Taylor, "Music and the Rise of Radio in 1920s America: Technological Imperialism, Socialization, and the Transformation of Intimacy," *Historical Journal of Film, Radio, and Television* 22 (October 2002): 425–43. The use of these terms is interpretive. Anderson associates his ideas with print culture, and Habermas rejects the notion that mass media offer a true public sphere. See Benedict Anderson, *Imagined Communities: Reflections on the Origin and Spread of Nationalism* (London: Verso, 1983), 37–46, and Jürgen Habermas, *The Structural Transformation of the Public Sphere: An Inquiry into a Category of Bourgeois Society* (Cambridge: MIT Press, 1989), 162.

52. See, e.g., Lenthall, *Radio's America*; Paul Young, *The Cinema Dreams Its Rivals: Media Fantasy Films from Radio to the Internet* (Minneapolis: University of Minnesota Press, 2006); Russo, *Points on the Dial*; and Bill Kirkpatrick, "Sounds Local: The Competition for Space and Place in Early U.S. Radio," in *Sound in the Age of Mechanical Reproduction*, ed. David Suisman and Susan Strasser (Philadelphia: University of Pennsylvania Press, 2010), 47–68.

53. Wurtzler, *Electric Sounds*, 170. For examples of earlier uses of "space-binding" language,

see e.g., Menachem Blondheim, *News over the Wires: The Telegraph and the Flow of Public Information in America, 1844–1897* (Cambridge: Harvard University Press, 1994), 190–91, and Carolyn Marvin, *When Old Technologies Were New: Thinking about Electric Communication in the Late Nineteenth Century*, (New York: Oxford University Press, 1988), 191–231.

54. Larry Wolters, "Chicago Leads Nation in Radio Dramatic Shows," *Chicago Tribune*, December 8, 1935, SW6; Larry Wolters, "New Gold Rush! 50 Chicagoans to Hollywood," *Chicago Tribune*, February 14, 1937, S11.

55. "Air's Million-or-More List," *Variety*, December 30, 1936, 31; "Standard Brands," *Fortune*, January, 1938, 78.

56. "National Brands Gain," *Printer's Ink*, April 8, 1937, 66–70.

57. See VanCour, "The Sounds of 'Radio,'" 224–25.

58. Ibid., 252; Smulyan, *Selling Radio*, 93–117.

59. "Broadcast Review of the Month," *Advertising Age*, March 23, 1935, 34; Michael C. Keith, *Sounds in the Dark: All Night Radio in American Life* (Ames: Iowa State University Press, 2001), 32–33. For a firsthand account, see Gertrude Berg, *Molly and Me: An Autobiography* (New York: McGraw-Hill, 1961), 193–205.

60. This phenomenon is the basis for the theorization of broadcasting put forward by Raymond Williams. See his *Television: Technology and Cultural Form* (Middletown, CT: Wesleyan University Press, 1993).

61. VanCour, "The Sounds of 'Radio,'" 340–45.

62. Alec Nisbett, *The Technique of the Sound Studio* (New York: Hastings House, 1962), 154.

63. Norman Corwin, *Thirteen by Corwin: Radio Dramas* (New York: H. Holt, 1943), 83.

64. Larry Wolters, "Amos 'n' Andy Start a Trend: Longer Shows," *Chicago Tribune*, November 21, 1943, NW6.

65. Leon Meadow, "But Nothing Ever Happens," *Hollywood Quarterly* 2 (January 1947): 139.

66. See Andrew Tolson, *Media Talk: Spoken Discourse on TV and Radio* (Edinburgh: University of Edinburgh Press, 2006), 9–10, 26–36.

67. See Robert L. Mott, *Sound Effects: Radio, TV, and Film* (Boston: Focal Press, 1990), 38–44.

68. Earle McGill, *Radio Directing* (New York: McGraw-Hill, 1940), 38.

69. Shortwave transmissions bounce off the earth's ionosphere and arrive on the other side of the globe in less than an eighth of a second. By contrast, in wired network radio, signals were sent from a New York station to a Chicago affiliate by way of a wire before going on the airwaves. The resistance in the wire resulted in a lag slightly greater than that of the shortwave.

70. Max Wylie, *Radio Writing* (New York: Farrar and Reinhart, 1939), 16.

71. This statement is cited very often. See, e.g., Douglas, *Listening In*, 190.

72. This is transcribed in A. M. Sperber, *Murrow, His Life and Times* (New York: Freundlich Books, 1986), 174.

73. See Philip Seig, *Broadcasts from the Blitz: How Edward R. Murrow Helped Lead America into War* (Washington, DC: Potomac Books, 2006), 81.

74. See Tom Gunning, *D. W. Griffith and the Origins of American Narrative Film: The Early Years at Biograph* (Urbana: University of Illinois Press, 1991), 105–6.

CHAPTER TWO

1. Douglas Coulter, introduction to *Columbia Workshop Plays: Fourteen Radio Dramas* (New York: McGraw-Hill, 1939), xiv; "Drama and Sound," *New York Times*, August 9, 1936, X10.

2. Pierson is quoted in Remy Brunel, "'March of Time' Scenarists and Actors Trained to Swift Changes to Keep Up with News," *Washington Post*, September 13, 1936, AA5.

3. R. Murray Schafer, *The Soundscape: Our Sonic Environment and the Tuning of the World* (Rochester, VT: Destiny Books, 1992), 272. Social researchers found these strategies effective. See, e.g., Tor Hollonquist and Edward A. Suchman, "Listening to the Listener," in *Radio Research, 1942–43*, ed. Paul Lazarsfeld and Frank Stanton (New York: Duell, Sloan and Pearce, 1944), 294. See also Michel Chion, *Audio-Vision: Sound on Screen*, trans. Claudia Gorbman (New York: Columbia University Press, 1994), 55.

4. W. J. T. Mitchell, *Picture Theory: Essays on Verbal and Visual Representation* (Chicago: University of Chicago Press, 1994), 153n.

5. Brown is quoted in Ira Skutch, *Five Directors: The Golden Years of Radio* (Los Angeles: Directors Guild of America, 1998), 29.

6. Roland Barthes, "Diderot, Brecht, Eisenstein," in *Image-Music-Text*, trans. Stephen Heath (New York: Noonday, 1992), 69.

7. Earle McGill, *Radio Directing* (McGraw-Hill: New York, 1940), 130.

8. Brecht argued that radio drama ought to be informed by his concepts of alienation in the theater, but that it should also outgrow the structures of his epic theater. See Bertolt Brecht, *Brecht on Theatre*, ed. and trans. John Willett (New York: Hill and Wang, 1964), 91, and *Brecht on Film and Radio*, ed. and trans. Marc Silberman (London: Methuen, 2000), 44–45.

9. Of course, many scholars have already noted how "perspective" is used in plays at the level of dialogue. See, e.g., Michele Hilmes, *Radio Voices: American Broadcasting, 1922–1952* (Minneapolis: University of Minnesota Press, 1999), 221; Elena Razlogova, "True Crime Radio and Listener Disenchantment with Network Broadcasting, 1935–1946," *American Quarterly* 58 (March 2006): 144; and Joan Shelley Rubin, *The Making of Middlebrow Culture* (Chapel Hill: University of North Carolina Press, 1992), 295. Bruce Lenthall has a passage on the use of sound to convey distance, as well. See his *Radio's America: The Great Depression and the Rise of Modern Mass Media* (Chicago: University of Chicago Press, 2007), 199–201.

10. See Rick Altman, "Sound Space," in *Sound Theory, Sound Practice*, ed. Rick Altman (New York: Routledge, 1992), 46–64; Michel Chion, *Film, a Sound Art*, trans. Claudia Gorbman (New York: Columbia University Press, 2009), 485–86; Mary Ann Doane, "Ideology and the Practice of Sound Editing and Mixing," in *Film Sound: Theory and Practice*, ed. Elisabeth Weis and John Belton (New York: Columbia University Press, 1985), 54–62; James Lastra, *Sound Technology and the American Cinema: Perception, Representation, Modernity* (New York: Columbia University Press, 2000), 132–42 and 195–98.

11. See Chion, *Audio-Vision*, 88–93, 97.

12. Welles is quoted in John S. Hays and Horace J. Gardner, *Both Sides of the Microphone: Training for the Radio* (New York: Lippincott, 1948), 122. See also Shawn VanCour, "The Sounds of 'Radio': Aesthetic Formations of 1920's American Broadcasting" (PhD diss., University of Wisconsin–Madison, 2008), 348–51.

13. See Max Wylie, *Radio Writing* (New York: Farrar and Reinhart, 1939), 179–220. Sometimes losing track of characters "on stage" was a good thing. Jack Benny's wife, Mary Livingston, was famous for the "disappearing character" gag. The idea is that if a character is "in" a scene but does not speak for a long period of time, listeners will forget that she is there. Mary would purposefully "disappear" so that when she *did* speak up, her cutting remark on her blowhard husband elicited a sharp comic reaction.

14. Mercedes McCambridge, *The Quality of Mercy: An Autobiography* (New York: Times Books, 1981), 71.

15. See Harry Heuser, "Etherized Victorians: Drama, Narrative, and the American Radio Play, 1929–1954" (PhD diss., City University of New York, 2004), 302.

16. McCambridge, *The Quality of Mercy*, 69.

17. See Karin Bijsterveld, *Mechanical Sound: Technology, Culture, and Public Problems of Noise in the Twentieth Century* (Cambridge: MIT Press, 2008), 104–10.

18. Rick Altman, "The Material Heterogeneity of Recorded Sound," in Altman, *Sound Theory, Sound Practice*, 23.

19. The means of indicating distance in dramatic sound can be likened to "text painting" in music, in which ascending or descending notes color words that denote corresponding motion. See Lawrence Zbikowski, *Conceptualizing Music: Cognitive Structure, Theory and Analysis* (New York: Oxford University Press, 2002), 62–95. On ambient reverb as a property in music recording, see Albin J. Zac III, *The Poetics of Rock: Cutting Tracks, Making Records* (Berkeley: University of California Press, 2001), 76–85.

20. See Don Ihde, *Listening and Voice: Phenomenologies of Sound* (Albany: SUNY Press, 2007), 79.

21. G. E. Lessing, *Hamburg Dramaturgy*, trans. Helen Simmern (New York: Dover, 1962), 179.

22. See Erik Barnouw, *Handbook of Radio Writing: An Outline of Techniques and Markets in Radio Writing* (Boston: Little, Brown, 1939), 61.

23. Here I simplify a very complex and contested history. See Rick Altman, *Silent Film Sound* (New York: Columbia University Press, 2004); Jay Beck and Tony Grajeda, eds., *Lowering the Boom: Critical Studies in Film Sound* (Urbana: University of Illinois Press, 2008); David Bordwell and Kristin Thompson, "Fundamental Aesthetics of Sound in the Cinema," in *Film Sound: Theory and Practice*, ed. Elisabeth Weis and John Belton (New York: Columbia University Press, 1985); Evan William Cameron, *Sound and the Cinema: The Coming of Sound to American Film* (Pleasantville, NY: Redgrave, 1980); Claudia Gorbman, *Unheard Melodies: Narrative Film Music* (Bloomington: Indiana University Press, 1987); Lastra, *Sound Technology*, 92–215; and Emily Thompson, *The Soundscape of Modernity: Architectural Acoustics and the Culture of Listening in America, 1900–1933* (Cambridge: MIT Press, 2002), 276–83.

24. Lastra, *Sound Technology*, 188. See also James Lastra, "Reading, Writing and Representing Sound," in Altman, *Sound Theory, Sound Practice*, 65–86.

25. Brown is quoted in Skutch, *Five Directors*, 28; Reis is quoted in Brunel, "Radio Magic," AA5.

26. Larry Wolters, "Studio Audience Is Pet Aversion of Barrymore," *Chicago Tribune*, March 8, 1936, NW4. This is not to say that earlier broadcasts lacked paradigms of aesthetic practice. See VanCour, "The Sounds of 'Radio,'" 307–80.

27. Rudolf Arnheim, *Radio* (1936; Salem, NH: Ayer, 1986), 52–104.

28. Ibid., 78, 83–94.

29. Barnouw, *Handbook of Radio Writing*, 29, 33.

30. The Columbia Broadcasting System, *Radio Alphabet* (New York: Hastings House, 1946), 30.

31. Norman Corwin, *Thirteen by Corwin: Radio Dramas* (New York: H. Holt, 1943), 114.

32. Thompson, *Soundscape of Modernity*, 227.

33. Curiously, because ghostly voices are conventionally highly reverberant, actors playing *dead* characters delivered lines at the "live" end of the studio, while those playing *living* characters spoke at the "dead" end.

34. See Alec Nisbett, *The Technique of the Sound Studio* (New York: Hastings House, 1962), 1–53, 150–90, and Ruth Carmen, *Radio Dramatics: Instruction Lectures* (New York: John C. Yorson, 1937), 55–57. For more on techniques for "microphoning," see Susan Schmidt Horning, "Engineering the Performance: Recording Engineers, Tacit Knowledge and the Art of Controlling Sound," *Social Studies of Science* 34, no. 5 (October 2004): 709–14.

35. Earl McGill, *Radio Directing*, 130; Earle McGill, "Requirements for Good Radio Acting," *Variety*, January 5, 1944, 111. Historian Alexander Russo has argued that the standardization of the velocity-sensitive ribbon microphone around this time might be the most important technological development in studio work for dramatic voice because this device gave engineers a technology that would both perform like human hearing while giving them greater control over sound input. See Alexander Russo, "Roots of Radio's Rebirth: Audiences, Aesthetics, Economics and Technologies of American Broadcasting, 1926–1951" (PhD diss., Brown University, 2004), 229. For more on mike technique, see Jacob Smith, *Vocal Tracks: Performance and Sound Media* (Berkeley: University of California Press, 2008), 81–114.

36. See Barnouw, *Handbook of Radio Writing*, 99–118, and Robert J. Landry, *This Fascinating Radio Business* (Indianapolis: Bobbs-Merrill, 1946), 70–72.

37. From my telephone interview with Norman Corwin, conducted October 29, 2009.

38. Francis Bacon, "The New Atlantis," in *Three Early Modern Utopias*, ed. Susan Brice (New York: Oxford University Press, 1999), 182.

39. On the other hand, *Lux* is responsible for developments in vocal acting and mike technique. See Smith, *Vocal Tracks*, 103.

40. Other sound-effects artists came from silent film palaces, an area that has been studied at length by Rick Altman. See in particular his passage on Lyman Howe and early effects machines: Altman, *Silent Film Sound*, 144–55. On radio sound effects in the 1920s, see VanCour, "The Sounds of 'Radio,'" 358–65.

41. Robert L. Mott, *Sound Effects: Radio, TV, and Film* (Boston: Focal Press, 1990), 17–19; "Ora Nichols Only Woman Expert on Sound Effects in Broadcasting," *Variety*, January 12, 1938, 29; and Kahn, *Noise, Water, Meat*, 101.

42. "The Canned Sound Industry," *Variety*, January 29, 1936, 35.

43. Hilmes, *Radio Voices*, 6.

44. Lucretius, *On the Nature of Things*, trans. Martin Ferguson Smith (Indianapolis: Hackett, 2001), 115.

45. Jonathan Sterne, *The Audible Past: Cultural Origins of Sound Reproduction* (Durham, NC: Duke University Press, 2003), 214–86.

46. Remy Brunel, "Broadcasters Employ Complex Tricks to Put Over Radio Crime Thrillers," *Washington Post*, August 30, 1936, AA7.

47. See Mott, *Sound Effects*, 88–111; and Charles Tranberg, *I Love the Illusion: The Life and Career of Agnes Moorehead* (Balsburg, PA: Bear Manor, 2005), 56.

48. John K. Hutchens, "Crime Pays—on the Radio," *New York Times*, March 19, 1944, 34; Howard Blue, *Words at War: World War II Era Radio Drama and the Postwar Broadcasting Industry Blacklist* (Lanham, MD: Scarecrow, 2002), 190; Mott, *Sound Effects*, 57. See also Chion, *Audio-Vision*, 109–12.

49. Mott, *Sound Effects*, 60.

50. Laske is quoted in Schafer, *The Soundscape*, 273. See also Barry Truax, *Acoustic Communication* (Norwood, NJ: Ablex, 1984), 49–50.

51. Wylie, *Radio Writing*, 39. See also Tim Crook, *Radio Drama: Theory and Practice* (New York: Routledge, 1999), 71; VanCour, "The Sounds of 'Radio,'" 358–65.

52. Nisbett, *Technique of the Sound Studio*, 171.

53. This term is Michael Warner's. See Michael Warner, *Publics and Counterpublics* (New York: Zone Books, 2002), 114.

54. John K. Hutchens, "The Secrets of a Good Radio Voice," *New York Times*, December 6, 1942, 26–27; Robert D. Sherwood, *Roosevelt and Hopkins: An Intimate History* (New York: Harper, 1948), 217; "The Talk of the Town," *New Yorker*, June 19, 1943, 15–16.

55. J. P. McEvoy, "He Snoops to Conquer," *Saturday Evening Post*, August 13, 1938, 47.

56. Chion makes a similar point about film sound. See Chion, *Film, a Sound Art*, 225.

57. John Hutchens, "Footnotes on Radio Drama," *New York Times*, August 16, 1936, X10. See also Andrew Crisell, *Understanding Radio* (New York: Routledge, 1994), 146.

58. MacLeish is quoted in William Stott, *Documentary Expression and Thirties America* (New York: Oxford University Press, 1973), 83.

59. Friedrich von Schiller, "On the Use of the Chorus in Tragedy" in *The Bride of Messina, William Tell, Demetrius*, trans. Charles E. Passage (New York: Frederick Ungar, 1962), 7–8.

60. Gérard Genette, *Narrative Discourse: An Essay in Method*, trans. Jane E. Lewin (Ithaca: Cornell University Press, 1980), 259–60.

61. Béla Balázs, "Theory of Film: Sound," in *Film Sound: Theory and Practice*, ed. Elisabeth Weis and John Belton (New York: Columbia University Press, 1985), 123.

62. James W. Carey, *Communication as Culture: Essays on Media and Society* (Boston: Unwin Hyman, 1989), 80. See also John Durham Peters, "Satan and Savior: Mass Communication in Progressive Thought," *Critical Studies in Mass Communication* 6, no. 3 (September 1989): 247–64. Here I refer to what Carey has famously called the "Lippmann-Dewey Debate," a disagreement between writers Walter Lippmann and John Dewey on how elites ought to use the mass media in democracy. Of course, this debate is less about Lippmann or Dewey in particular, and more about problems inherent in progressive thought. See Michael Schudson, "The 'Lippmann-Dewey Debate' and the Invention of Walter Lippmann as an Anti-Democrat," *International Journal of Communication* 2 (2008): 1031–42.

63. Miriam Bratu Hansen, "The Mass Production of the Senses: Classical Cinema as Vernacular Modernism," *Modernism/Modernity* 6, no. 2 (April 1999): 69–70.

CHAPTER THREE

1. Dorothy Thompson, "The Great War of Words," *Saturday Evening Post*, December 1, 1934, 69.

2. See Michael Kammen, *The Mystic Chords of Memory: The Transformation of Tradition in American Culture* (New York: Vintage, 1991), 460–64; Richard H. Pells, *Radical Visions and American Dreams: Culture and Social Thought in the Depression Years* (New York: Harper and Row, 1973), 194–241; William Stott, *Documentary Expression and Thirties America* (New York: Oxford University Press, 1973).

3. James Agee and Walker Evans, *Let Us Now Praise Famous Men* (New York: Houghton Mifflin, 2001), 5.

4. On *The Shadow*, see Kathleen Battles, *Calling All Cars: Radio Dragnets and the Technology of Policing* (Minneapolis: University of Minnesota Press, 2010), 203–22; Tim DeForest, *Radio by the Book: Adaptations of Literature and Fiction on the Airwaves* (Jefferson, NC: McFarland, 2008), 124–31; Walter B. Gibson and Anthony Tollin, *The Shadow Scrapbook* (New York: Harcourt Brace, 1979); and Jason Loviglio, *Radio's Intimate Public: Network Broadcasting and Mass-Mediated Democracy* (Minneapolis: University of Minnesota Press 2005), 102–22.

5. Battles, *Calling All Cars*, 149.

6. After 1938, the program was titled the *Campbell's Soup Playhouse*, after the company came on as sponsor. In 1946, Welles did a revival entitled *The Mercury Summer Theater*. Here I use *Mercury Theater* in all instances.

7. "Radio Reviews," *Variety*, July 28, 1937, 42.

8. Richard B. O'Brien, "The Shadow Talks," *New York Times*, August 14, 1938, 136. Here I am indebted to a paper by Jay Beck, "The Narrator Has Your Ear': Orson Welles' First Person Singular, Voice-Over, and the Cinematic Influence of Radio Aesthetics" (paper presented at the Chicago Film Seminar, Chicago, Illinois, April 3, 2008).

9. Of course, there are other conventions for this purpose. In some dramas, for instance, an underscore signals to the listener a transition between a segment in which a character is speaking inside the fiction and a segment in which that character speaks directly to the listener in a mode of narration.

10. Gérard Genette, *Narrative Discourse: An Essay in Method*, trans. Jane E. Lewin (Ithaca: Cornell University Press, 1980), 215.

11. Barry Truax, *Acoustic Communication* (Norwood, NJ: Ablex, 1984), 90.

12. Pells, *Radical Visions*, 197–99.

13. Daniel J. Czitrom, *Media and the American Mind, from Morse to McLuhan* (Chapel Hill: University of North Carolina Press, 1982), 88.

14. See Joseph Julian, *This Was Radio: A Personal Memoir* (New York: Viking, 1975), 80–106.

15. Gertrude Stein, *Writings and Lectures, 1911–1945* (London: Peter Owen, 1967), 58–60.

16. Stott, *Documentary Expression*, 90. See also James L. Baughman, *The Republic of Mass Culture: Journalism, Filmmaking and Broadcasting in America since 1941* (Baltimore: Johns Hopkins University Press, 2006), 3, and Joan Shelley Rubin, *The Making of Middlebrow Culture* (Chapel Hill: University of North Carolina Press, 1992), 269. Jason Loviglio has developed a

sophisticated idea on this subject, arguing that cultural productions like the Fireside Chats created an "intimate public" that worked by oscillating between public and private spheres. See Loviglio, *Radio's Intimate Public*, xiii–37.

17. Carolyn Marvin, *When Old Technologies Were New: Thinking about Electric Communication in the Late Nineteenth Century* (New York: Oxford University Press, 1988), 154.

18. The legend goes that the play's adapter, Howard Koch, drafted his scenes using a map of New Jersey purchased at a gas station, choosing settings by closing his eyes and pointing at random. See Paul Heyer, *The Medium and the Magician: Orson Welles, the Radio Years, 1934–1952* (Lanham, MD: Rowman and Littlefield, 2005), 79.

19. For a first-hand description of the process behind *Mercury*, see John Houseman, *Unfinished Business: Memoirs, 1902–1988* (New York: Applause Theater Books, 1989), 190–200.

20. See Hadley Cantril, *The Invasion from Mars: A Study in the Psychology of Panic* (New York: Harper and Row, 1966), and Malcolm Goldstein, *The Political Stage: American Drama and Theater of the Great Depression* (New York: Oxford University Press, 1974), 296.

21. See Susan Douglas, *Listening In: Radio and the American Imagination* (Minneapolis: University of Minnesota Press, 1999), 161–98; David Holbrook Culbert, *News for Everyman: Radio and Foreign Affairs in Thirties America* (Westport, CT: Greenwood, 1976); Gwyneth Jackaway, *Media at War: Radio's Challenge to the Newspapers 1924–1939* (Westport: Praeger, 1994); and Michael Stamm, *Sound Business: Newspapers, Radio and the Politics of New Media* (Philadelphia: University of Pennsylvania Press, 2011), 62–81. The so-called Biltmore Agreement was designed to limit radio news coverage to two brief segments a day, and to channel all radio news through a new Radio Press Bureau. The agreement never worked very well. During the 1930s, H. V. Kaltenborn broadcast live from Irun on the front of the Spanish Civil War; stations aired updates on the Bruno Hauptmann trial in 1936; and emergency coverage aired during both the 1937 Ohio river flood and the Hindenburg disaster.

22. "Radio Reports," *Variety*, October 2, 1935, 41. The suggestion of the analogy between *March of Time* and "War" is also noted in Heyer, *The Medium and the Magician*, 89.

23. Remy Brunel, "'March of Time' Scenarists and Actors Trained to Swift Changes to Keep Up with the News," *Washington Post*, September 13, 1936, AA5.

24. Charles Tranberg, *I Love the Illusion: The Life and Career of Agnes Moorehead* (Balsburg, PA: Bear Manor, 2005), 49.

25. Julian, *This Was Radio*, 10.

26. On this, see Paddy Scannell, "Radio Times: The Temporal Arrangements of Broadcasting in the Modern World," in *Television and its Audience*, ed. Philip Drummond and Richard Paterson (London: BFI, 1988), 23.

27. MacLeish is quoted in Stott, *Documentary Expression*, 83.

28. Richard Lingeman, *Don't You Know There's a War On? The American Home Front, 1941–1945* (New York: Nation, 2003), 230; Walter Benjamin, "Theater and Radio," in *Walter Benjamin: Selected Writings*, vol. 2, *1927–1934*, trans. Rodney Livingstone et al., ed. Michael W. Jennings, Howard Eiland, and Gary Smith (Cambridge: Harvard University Press, 1999), 584; Robert J. Landry, "The Improbability of Radio Criticism," *Hollywood Quarterly* 2 (October 1946): 70; Harry Heuser, "Etherized Victorians: Drama, Narrative, and the American Radio Play, 1929–1954" (PhD diss., City University of New York, 2004), 403; Julian, *This Was Radio*, 112; Erik Barnouw, *Handbook of Radio Writing: An Outline of Techniques and Markets in*

Radio Writing (Boston: Little, Brown, 1939), 113; William Matthews, "Radio Plays as Literature," *Hollywood Quarterly* 1 (October 1945): 43.

29. Bruce Lenthall, *Radio's America: The Great Depression and the Rise of Modern Mass Media* (Chicago: University of Chicago Press, 2007), 198–201. See also Culbert, *News for Everyman*, 107.

30. From my telephone interview with Corwin, conducted on August 23, 2008.

31. Stott, *Documentary Expression*, 53.

32. R. LeRoy Bannerman, *Norman Corwin and Radio* (Tuscaloosa: University of Alabama Press, 1986), 141.

33. Jeffrey Sconce, *Haunted Media: Electronic Presence from Telegraphy to Television* (Durham, NC: Duke University Press, 2000), 114. See also Lenthall, *Radio's America*, 1–5.

34. On this aspect of "War," see Martin Spinelli, "'Masters of Sacred Ceremonies': Welles, Corwin and a Radiogenic Modernist Literature," in *Broadcasting Modernism*, ed. Debra Rae Cohen, Michael Coyle, and Jane Lewty (Gainesville: University Press of Florida, 2009), 76.

CHAPTER FOUR

1. Michael Denning, *The Cultural Front: The Laboring of American Culture in the Twentieth Century* (New York: Verso, 1998), xix–xx, 117–18. See also Chris Vials, *Realism for the Masses: Aesthetics, Popular Front Pluralism, and U.S. Culture, 1935–1947* (Jackson: University Press of Mississippi, 2009). Here I reference a series of the attitudes of inclusion emphasized in historical literature on the period: Lizabeth Cohen, *Making a New Deal: Industrial Workers in Chicago, 1919–1939* (New York: Cambridge University Press, 1990), 323–360; Lary May, *The Big Tomorrow: Hollywood and the Politics of the American Way* (Chicago: University of Chicago Press, 2000), 3–4; Richard H. Pells, *Radical Visions and American Dreams: Culture and Social Thought in the Depression Years* (New York: Harper and Row, 1973), 111–17.

2. On Corwin, see R. LeRoy Bannerman, *Norman Corwin and Radio* (Tuscaloosa: University of Alabama Press, 1986), and Harry Heuser, "Etherized Victorians: Drama, Narrative, and the American Radio Play, 1929–1954" (PhD diss., City University of New York, 2004), 317–420.

3. Erskine Caldwell and Margaret Bourke-White, *You Have Seen Their Faces* (New York: Modern Age, 1937), 33, 17.

4. Alfred Kazin, *On Native Grounds* (New York: Harcourt, Brace, 1970), 496–97.

5. Andrew Crisell, *Understanding Radio* (New York: Routledge, 1994), 159.

6. By my count, this set of broadcasts had sixty-one kaleidosonic works, twenty-nine intimate works, and ten music shows.

7. *Cavalcade* was part of a public relations campaign by sponsor DuPont to combat the bad reputation it earned for growing wealthy selling munitions during the Great War. The show also conspicuously incorporated woman scientists, singers, and pioneers, as a way to reach out to housewife consumers.

8. On this play, see Howard Blue, *Words at War: World War II Era Radio Drama and the Postwar Broadcasting Industry Blacklist* (Lanham, MD: Scarecrow, 2002), 105–6; Bannerman, *Norman Corwin and Radio*, 73–88; and John Dunning, *On the Air: The Encyclopedia of Old-Time Radio* (New York: Oxford University Press, 1998), 165–66.

9. See: Michael Kammen, *The Mystic Chords of Memory: The Transformation of Tradition in American Culture* (New York: Vintage 1991), 510–11.

10. Ibid., 464.

11. All but the final set of ellipses reflect the original script.

12. György Lukács, "The Sociology of Modern Drama," in *The Theory of the Modern Stage*, trans. Lee Bachandall, ed. Eric Bentley (New York: Penguin, 1984), 434–35.

13. "Radio Reviews" *Variety*, December 17, 1941, 44; "American Charter," *Time*, December 29, 1941, 36; "Prizes for Corwin," *Time*, March 4, 1946, 62.

14. "CBS Builds 'Em and Sells 'Em," *Variety*, November 3, 1943, 19. See also: Blue, *Words at War*, 341–70.

15. Lou Frankel, "In One Ear," *Nation*, March 15, 1947, 304.

16. "Rat Race?," *Time*, July 30, 1945, 66.

17. Corwin is quoted in Sally Bedell Smith, *In All His Glory: The Life of William S. Paley, the Legendary Tycoon and His Brilliant Circle* (New York: Simon and Schuster, 1990), 268. Another account of the same conversation is in Erik Barnouw, *The Golden Web: A History of Broadcasting in the United States, 1933–1953* (New York: Oxford University Press, 1968), 241.

18. Smith, *In All His Glory*, 301.

19. See Philip Hamburger, "Norman Corwin," *New Yorker*, April 5, 1947, 36–49. Hamburger wrote this profile in the form of a radio play.

20. See Glenn Gould and Tim Page, "Glenn Gould in Conversation with Tim Page," in *The Glenn Gould Reader*, ed. Tim Page (New York: Vintage, 1990), 457.

21. Benedict Anderson, *Imagined Communities: Reflections on the Origin and Spread of Nationalism* (London: Verso, 1983), 7.

22. Eric Bentley, *The Brecht Commentaries, 1943–1980* (New York: Grove/Atlantic, 1980) 22.

23. "Communications—the Fourth Front," *Fortune*, November 1939, 90; Chester T. Crowell, "Dogfight on the Airwaves," *Saturday Evening Post*, May 21, 1938, 23.

24. "By the Ears," *Time*, January 26, 1942, 64. Zeesen refers to a 230-foot-tall radio mast in Berlin. The Nazis used the mast to broadcast disinformation programs overseas by shortwave.

25. Sherman Dryer, *Radio in Wartime* (New York: Greenberg, 1942), 205; William Bennett Lewis, "Radio Propaganda Must Be Painless," *Variety*, January 6, 1943, 97; Robert J. Landry, "Radio: Key to National Unity," *Atlantic*, April 1942, 503.

26. Asa Briggs, *The History of Broadcasting in the United Kingdom*, vol. 3, *The War of Words* (New York: Oxford University Press, 1995), 40; John B. Whitton, "Radio after the War," *Foreign Affairs*, January 1944, 311.

27. Arch Oboler, *Oboler Omnibus: Radio Plays and Personalities* (New York: Duell, Sloan and Pearce, 1945), 191–93, and "3 Provocative Pieces Touch on Radio," *Variety*, June 3, 1942, 32; MacLeish is quoted in Gerd Horten, *Radio Goes to War: The Cultural Politics of Propaganda during World War II* (Berkeley: University of California Press, 2002), 52. See also Lary May, "Making the American Consensus: The Narrative of Conversion and Subversion in World War II Films," in *The War in American Culture: Society and Consciousness during World War II*, ed. Lewis Erenberg and Susan E. Hirsch (Chicago: University of Chicago Press, 1996), 71.

28. Landry, "Radio," 508.

29. Ibid., 506; Michael S. Sweeney, *Secrets of Victory: The Office of Censorship and the American Press and Radio in World War II* (Chapel Hill: University of North Carolina Press, 2001), 9.

30. Rubicam is quoted in Frank W. Fox, *Madison Avenue Goes to War: The Strange Military Career of American Advertising* (Provo, UT: Brigham Young University Press, 1975), 28.

31. See Emily Rosenberg, *A Date Which Will Live: Pearl Harbor in American Memory* (Durham, NC: Duke University Press, 2003), 24; Erik Barnouw, *Radio Drama in Action: Twenty-Five Plays of a Changing World* (New York: Rinehart, 1945), 112.

32. "U.S. Propaganda," *Time*, October 12, 1942, 32; "War Messages as Part of Plot," *Variety*, October 14, 1942, 33.

33. Fox, *Madison Avenue Goes to War*, 54.

34. "OWI's Kudos to Radio Industry," *Variety*, October 13, 1943, 28. The CBS figures (likely an exaggeration) come from an advertisement in *Variety*, January 6, 1943, 113.

35. See: Blue, *Words at War*, 135–268; Michele Hilmes, *Radio Voices: American Broadcasting, 1922–1952* (Minneapolis: University of Minnesota Press, 1999), 236–44.

36. Robert K. Merton, with Marjorie Fiske and Alberta Curtis, *Mass Persuasion: The Social Psychology of a War Bond Drive* (New York: Harper and Brothers, 1946); Herta Herzog, "Radio—the First Post-War Year," *Public Opinion Quarterly* 10 (Autumn 1946): 299.

37. John Hutchens, "Tracy, Superman, et al. Go to War," *New York Times*, November 21, 1943, SM15.

38. Archibald MacLeish, *The American Story: Ten Broadcasts* (New York: Duell, Sloan and Pearce, 1944), xi.

39. See Hilmes, *Radio Voices*, 244.

40. Theodore Adorno, *The Psychological Technique in Martin Luther Thomas' Radio Addresses* (Stanford, CA: Stanford University Press, 2000), 31.

CHAPTER FIVE

A version of this chapter was published as Neil Verma, "Honeymoon Shocker: Lucille Fletcher's 'Psychological' Sound Effects and Wartime Radio Drama," *Journal of American Studies* 44, no. 1 (February 2010): 137–53.

1. Lucille Fletcher, "Squeaks, Slams, Echoes and Shots," *New Yorker*, April 13, 1940, 81–96.

2. Ibid., 86.

3. See Jim Harmon, *Radio Mystery and Adventure and Its Appearances in Film, Television and Other Media* (Jefferson, NC: McFarland, 1992), 154–55. For a more detailed look at *The Shadow* and hypnosis, see Jason Loviglio, *Radio's Intimate Public: Network Broadcasting and Mass-Mediated Democracy* (Minneapolis: University of Minnesota Press, 2005), 102–22.

4. On stream-of-consciousness see: Richard J. Hand, *Terror on the Air! Horror Radio in America, 1931–1952* (Jefferson NC: McFarland & Co, 2006), 89.

5. Arthur Miller, "Notes to 'The Guardsman,'" in *The Theater Guild on the Air*, ed. William Fitelson (New York: Rinehart, 1947), 68.

6. Robert B. Heilman, *Tragedy and Melodrama: Versions of Experience* (Seattle: University of Washington Press, 1968), 79.

7. John W. Pacey, "Radio Nerve Center," *Wall Street Journal*, May 18, 1942, 16. See also Emily Rosenberg, *A Date Which Will Live: Pearl Harbor in American Memory* (Durham, NC: Duke University Press, 2003), 24.

8. Here I follow the definition of "propaganda" offered by Garth Jowett and Victoria O'Donnell: "The deliberate, systematic attempt to shape perceptions, manipulate cognitions and direct behavior to achieve a response that furthers the desired intent of the propagandist." See Garth S. Jowett and Victoria O'Donnell, *Propaganda and Persuasion* (Thousand Oaks, CA: Sage Publications, 1999), 6.

9. See William C. Ackerman, "The Dimensions of American Broadcasting," *Public Opinion Quarterly* 9 (Spring 1945): 3; Hadley Cantril, *The Invasion from Mars: A Study in the Psychology of Panic* (New York: Harper and Row, 1966), xii; "The Fortune Survey," *Fortune* (January 1938): 62; "87% of Nation Has Radios," *Variety*, March 11, 1942, 32. See also James L. Baughman, *The Republic of Mass Culture: Journalism, Filmmaking and Broadcasting in America since 1941* (Baltimore: Johns Hopkins University Press, 2006), 16–21. In 1938, *Fortune*'s survey found that 88.1 percent of American families had a radio, but the 1940 census cited in Ackerman put ownership at 82.8 percent. That same year, researcher Hadley Cantril estimated that radios were in the homes of around 27.5 million of the 32 million families in the United States. In March of 1942, *Variety* quoted industry figures to arrive at 87 percent total households.

10. "Radio's Billion $ Scope," *Variety*, January 21, 1942, 3.

11. Paul Lazarsfeld et al., *Radio and the Printed Page* (New York: Duell, Sloan and Pearce, 1940), 259; Paul Lazarsfeld et al., *The People Look at Radio: A Report on a Survey* (Chapel Hill: University of North Carolina Press, 1946), 6; John K. Hutchens, "The Secrets of a Good Radio Voice," *New York Times*, December 6, 1942, SM26; "600,000 Words or More Daily Pass through Propaganda Analysis Mill of FCC," *Variety* December 31, 1941, 33. The tendency to pit newspapers and radio against one another was common in popular discourse in the period, but the distinction between these mass media was certainly overblown. As Michael Stamm has recently shown, not only were newspapers major owners of hundreds of stations, but they helped shape the content and structure of broadcasting profoundly. See Michael Stamm, *Sound Business: Newspapers, Radio and the Politics of New Media* (Philadelphia: University of Pennsylvania Press, 2011).

12. George Raynor Thompson et al., *The Signal Corps: The Test (December 1941–July 1943)*, in *The United States Army in World War II: The Technical Services* (Washington, DC: Center of Military History, 1957), 200; George Raynor Thompson et al., *The Signal Corps: The Outcome (Mid-1943 through 1945)*, in *The United States Army in World War II: The Technical Services* (Washington, DC: Center of Military History, 1966), 351.

13. Thompson et al., *The Test*, 218, 607.

14. Susan J. Douglas, *Listening In: Radio and the American Imagination* (Minneapolis: University of Minnesota Press, 1999), 152; Michael C. C. Adams, *The Best War Ever: America and World War II* (Baltimore: Johns Hopkins University Press, 1994), 10. See also John Morton Blum, *V Was for Victory: Politics and American Culture during World War II* (New York: Harcourt and Brace, 1976); Philip D. Beidler, *The Good War's Greatest Hits: World War II and American Remembering* (Athens: University of Georgia Press, 1998); John W. Dower, *War without Mercy: Race and Power in the Pacific War* (New York: Pantheon, 1986); Richard Lingeman, *Don't You Know There's a War On? The American Home Front, 1941–1945* (New York: Nation,

2003), 168–233; and George H. Roeder Jr., *The Censored War: American Visual Experience during World War Two* (New Haven, CT: Yale University Press, 1993).

15. "The Fortune Survey," *Fortune*, August, 1938, 49.

16. "Films 400,000,000 a Year," *Variety*, January 3, 1945, 1; "Four Webs Hit 190,605,078 in 1944," *Variety*, January 10, 1945, 25; Lingeman, *Don't You Know There's a War On?*, 182; "603 War Shows on Web," *Variety*, July 29, 1942, 25.

17. See, e.g., Alan Brinkley, "World War II and American Liberalism," in *The War in American Culture: Society and Consciousness during World War II*, ed. Lewis Erenberg and Susan E. Hirsch (Chicago: University of Chicago Press, 1996), 313–30; James Kimble, *Mobilizing the Home Front: War Bonds and Domestic Propaganda* (College Station: Texas A&M University Press, 2006); David F. Krugler, *The Voice of America and the Domestic Propaganda Battles, 1945–53* (Columbia: University of Missouri Press, 2000); Daniel L. Lykins, *From Total War to Total Diplomacy: The Advertising Council and the Construction of the Cold War Consensus* (Westport, CT: Praeger, 2003); Lawrence C. Soley, *Radio Warfare: OSS and CIA Subversive Propaganda* (New York: Praeger, 1989); Michael S. Sweeney, *Secrets of Victory: The Office of Censorship and the American Press and Radio in World War II* (Chapel Hill: University of North Carolina Press, 2001); and Allan M. Winkler, *The Politics of Propaganda: The Office of War Information, 1942–1945* (New Haven, CT: Yale University Press, 1978).

18. See, e.g., Benjamin L. Alpers, *Dictators, Democracy, and American Public Culture: Envisioning the Totalitarian Enemy, 1920s–1950s* (Chapel Hill: University of North Carolina Press, 2003), 91–127; Baughman, *Republic of Mass Culture*, 1–29; Howard Blue, *Words at War: World War II Era Radio Drama and the Postwar Broadcasting Industry Blacklist* (Lanham, MD: Scarecrow, 2002), 135–268; Steven Casey, *Cautious Crusade: Franklin D. Roosevelt, American Public Opinion, and the War against Nazi Germany* (New York: Oxford University Press, 2001); Michele Hilmes, *Radio Voices: American Broadcasting, 1922–1952* (Minneapolis: University of Minnesota Press, 1999), 230–70; Gerd Horten, *Radio Goes to War: The Cultural Politics of Propaganda during World War II* (Berkeley: University of California Press, 2002); Kathy M. Newman, *Radio Active: Advertising and Consumer Activism, 1935–1947* (Berkeley: University of California Press, 2004), 109–92; Ann Elizabeth Pfau and David Hochfelder, "Her Voice a Bullet: Imaginary Propaganda and the Legendary Broadcasters of World War II," in *Sound in the Age of Mechanical Reproduction*, ed. David Suisman and Susan Strasser (Philadelphia: University of Pennsylvania Press, 2010), 47–68; Barbara Dianne Savage, *Broadcasting Freedom: Radio, War, and the Politics of Race, 1938–1948* (Chapel Hill: University of North Carolina Press, 1999), 106–53; and Holly Cowan Shulman, *Voice of America: Propaganda and Democracy, 1941–1945* (Madison: University of Wisconsin Press, 1990), 54–74.

19. Fletcher relates this story in Martin Grams Jr., *Suspense: Twenty Years of Thrills and Chills* (Kearney, NE: Morris, 1997), 13.

20. See Joseph Liss, *Radio's Best Plays* (New York: Greenberg, 1947), 324; Hand, *Terror on the Air!*, 127.

21. James Naremore, *More Than Night: Film Noir in Its Contexts* (Berkeley: University of California Press, 1998), 279; Hand, *Terror on the Air!*, 21; Allison McCracken, "Scary Women and Scarred Men: Suspense, Gender Trouble and Postwar Change, 1942–1950," in *The Radio Reader: Essays in the Cultural History of Radio*, ed. Michele Hilmes and Jason Loviglio (New York: Routledge, 2002), 183–208. See also Harry Heuser, "Etherized Victorians: Drama,

Narrative, and the American Radio Play, 1929-1954 (PhD diss., City University of New York, 2004), 230-78.

22. Douglas Gomery, *A History of Broadcasting in the United States* (Malden, MA: Blackwell, 2008), 90.

23. In 1904 Kipling famously fantasized about similarities between wireless technology and paranormal phenomena. See Rudyard Kipling, "Wireless," in *Traffics and Discoveries* (London: Macmillan, 1904), 211-39.

24. Douglas, *Listening In*, 40-82; Douglas Kahn and Gregory Whitehead, *Wireless Imagination: Sound, Radio and the Avant Garde*, (Cambridge: MIT Press, 1992); Jeffrey Sconce, *Haunted Media: Electronic Presence from Telegraphy to Television* (Durham, NC: Duke University Press, 2000), 59-91; Allan Weiss, *Phantasmatic Radio* (Durham, NC: Duke University Press, 1995).

25. Sconce, *Haunted Media*, 64; Jonathan Sterne, *The Audible Past: Cultural Origins of Sound Reproduction* (Durham, NC: Duke University Press, 2003), 290; and John Durham Peters, *Speaking into the Air: A History of the Idea of Communication* (Chicago: University of Chicago Press, 1999), 150-51.

26. Richard Thruelsen, "Men at Work: Radio Emcee," *Saturday Evening Post*, August 2, 1947, 33.

27. "Cycle of Ghost Stories Hits Radio," *Variety*, March 4, 1936, 49.

28. "Radio Reports—the Witch's Tale," *Variety*, October 8, 1936, 37. See also Alonzo Deen Cole, *The Witch's Tale: Stories of Gothic Horror from the Golden Age of Radio* (Yorktown Heights, NY: Dunwich, 1998).

29. Hand, *Terror on the Air!*, 82.

30. "Letter from the National League for Decency in Radio," *Variety*, May 15, 1935, 38-39.

31. McCracken, "Scary Women," 183; Ira Skutch, *Five Directors: The Golden Years of Radio* (Los Angeles: Directors Guild of America, 1998), 25.

32. John K. Hutchens, "The Shockers," *New York Times*, November 8, 1942, X12.

33. See the Peabody Awards website, http://peabody.uga.edu (accessed April 1, 2007); Hand, *Terror on the Air!*, 19-20.

34. See Elaine Tyler May, *Homeward Bound: American Families in the Cold War Era* (New York: Basic Books, 1988), 67-91.

35. Tzvetan Todorov, *The Fantastic: A Structural Approach to a Literary Genre*, trans. Richard Howard (Ithaca: Cornell University Press, 1975), 158-66.

36. Hutchens, "The Shockers," X12.

37. Earle McGill, *Radio Directing* (McGraw-Hill: New York, 1940), 39.

38. Elissa Guralnick, *Sight Unseen: Beckett, Pinter, Stoppard, and Other Contemporary Dramatists on Radio* (Athens: Ohio University Press, 1995), 29.

39. R. Murray Schafer, *The Soundscape: Our Sonic Environment and the Tuning of the World* (Rochester, VT: Destiny Books, 1992), 175; Barry Truax, *Acoustic Communication* (Norwood, NJ: Ablex, 1984). This usage evokes but does not quite mimic the distinction between "signal" and "noise" often discussed in communication theory, in which the former designates information from a source and the latter conveys superfluous information interfering with signal. Andrew Crisell has a similar idea about sound effects like these, which feature what he calls "extended signification." See his *Understanding Radio* (New York: Routledge, 1994), 44-48.

40. Pierre Schaeffer, "Acousmatics," in *Audio Culture: Readings in Modern Music*, ed. Christoph Cox and Daniel Warner, (New York: Continuum, 2004), 76–81. See also Michael Chion, *Audio-Vision: Sound on Screen*, trans. Claudia Gorbman (New York: Columbia University Press, 1994), 203–215, and Crisell, *Understanding Radio*, 46.

41. On "attention" in listening, see David Goodman, "Distracted Listening: On Not Making Sound Choices in the 1930s," in *Sound in the Age of Mechanical Reproduction*, ed. David Suisman and Susan Strasser (Philadelphia: University of Pennsylvania Press, 2010), 15–46; Don Ihde, *Listening and Voice: Phenomenologies of Sound*, 2nd ed. (Albany: SUNY Press, 2007), 203–15; and Ingrid Monson, "Hearing, Seeing, and Perceptual Agency," *Critical Inquiry* 34, ser. 2 (Winter 2008): S36–S58.

42. Ernie Pyle, *Here Is Your War* (New York: Lancer Books, 1943), 157.

43. Ibid., 158.

44. Pyle's passage recalls a famous moment in Rabelais's epic *Pantagruel*, when the band of heroes comes across the "bright red words" of a battle that turned into icy solid objects in the Arctic Sea. See François Rabelais, *Gargantua and Pantagruel*, trans. Burton Raffel (New York: Norton, 1990), 494–97.

45. This play has been broadcast more than once. This reading is based on a two-part version that aired on *Suspense* (05/18/44 and 05/25/44).

46. Fred Allen, *Treadmill to Oblivion: My Days in Radio* (Rockville, MD: Wildside, 1954), 57.

CHAPTER SIX

1. Douglas Gomery, *A History of Broadcasting in the United States* (Malden, MA: Blackwell, 2008), 346–52.

2. See, e.g., J. P. Telotte, *Voices in the Dark: The Narrative Patterns of Film Noir* (Urbana: University of Illinois Press, 1989), 74–87; Allison McCracken, "Scary Women and Scarred Men: Suspense, Gender Trouble and Postwar Change, 1942–1950," in *The Radio Reader: Essays in the Cultural History of Radio*, ed. Michele Hilmes and Jason Loviglio (New York: Routledge, 2002), 188–92.

3. This last line was famously flubbed during the first performance.

4. Charles Tranberg, *I Love the Illusion: The Life and Career of Agnes Moorehead* (Balsburg, PA: Bear Manor, 2005), 101. See also Martin Grams Jr., *Suspense: Twenty Years of Thrills and Chills* (Kearney, NE: Morris, 1997), 22. Fletcher got the idea for this play from the irritating voice of a woman ahead of her in a grocery store queue.

5. This chapter ought to be read alongside Michele Hilmes's discussion of what she calls the "disciplined audience." See Michele Hilmes, *Radio Voices: American Broadcasting, 1922–1952* (Minneapolis: University of Minnesota Press, 1999), 183–229.

6. See Kathy M. Newman, *Radio Active: Advertising and Consumer Activism, 1935–1947* (Berkeley: University of California Press, 2004), and Elena Razlogova, *The Voice of the Listener: Americans and the Radio Industry, 1920–1950* (PhD diss., George Mason University, 2003).

7. For some background, see Daniel J. Czitrom, *Media and the American Mind, from Morse to McLuhan* (Chapel Hill: University of North Carolina Press, 1982), 122–46; Timothy Glander, *Origins of Mass Communications Research during the American Cold War: Educational Effects*

and Contemporary Implications (Mahwah, NJ: Lawrence Erlbaum, 2000); Ellen Herman, *The Romance of American Psychology: Political Culture in the Age of Experts* (Berkeley: University of California Press, 1995), 1–123; Donald Hurwitz, "Market Research and the Study of the US Radio Audience," *Communication* 10 (Spring 1988): 223–41; and Bruce Lenthall, *Radio's America: The Great Depression and the Rise of Modern Mass Media* (Chicago: University of Chicago Press, 2007), 143–74.

8. Glander, *Origins of Mass Communications Research*, 1.

9. "Radio and Ideas," *Nation*, September 28, 1940, 278.

10. John Durham Peters, *Speaking into the Air: A History of the Idea of Communication* (Chicago: University of Chicago Press, 1999), 22; Steven Chaffee et al., *The Contributions of Wilbur Schramm to Mass Communication Research*, Journalism Monographs, no. 36 (October 1974), 1–44; Glander, *Origins of Mass Communications Research*, 46; and Herman, *Romance of American Psychology*, 55.

11. Susan Douglas, *Listening In: Radio and the American Imagination* (Minneapolis: University of Minnesota Press, 1999), 130.

12. "Pioneers of Audience Research," *Variety*, December 2, 1936, 41.

13. "NBC Drops Rates," *Advertising Age*, January 12, 1935, 17.

14. Writers fiercely debated how to measure radio "circulation" in industry and popular publications. See, e.g., Matthew Chappell, "Expected Variations in Radio Programs and Ratings," *Printer's Ink*, October 2, 1942, 58–62; C. E. Hooper, "Looking behind the Radio Dial," *Printer's Ink*, May 15, 1942, 20; C. E. Hooper, "How Adequate are 'Telephone Homes' as a Cross-Section of Radio Audience?," *Printer's Ink*, July 24, 1942, 43. See also John J. Karol, "Analyzing the Radio Market," *Journal of Marketing*, April, 1938, 309; Collie Small, "The Biggest Man in Radio," *Saturday Evening Post*, November 22, 1947, 25–144; "Crossley and Hooper's Radio Ratings Clash," *Advertising Age*, May 9, 1938, 33; "Exit Crossley," *Time*, Sept. 30, 1946, 96–98.

15. Robert Tallman, "Radio and the War: Totems and Taboos," *New York Times*, September 27, 1942, X10; Richard Manville, "How to Find Best Radio Station and Time to Advertise Your Product," *Printer's Ink*, April 24, 1942, 19.

16. "Electric Light Test for Radio Programs," *Advertising Age*, January 25, 1937, 40; Maurice Zolotow, "Washboard Weepers," *Saturday Evening Post*, May 29, 1943, 16–52.

17. Tor Hollonquist and Edward A. Suchman, "Listening to the Listener," in *Radio Research, 1942–43*, ed. Paul Lazarsfeld and Frank Stanton (New York: Duell, Sloan and Pearce, 1944), 265–334.

18. John W. Pacey, "Building Radio Microscopes," *Wall Street Journal*, June 4, 1942, 23; "What Do They Like?," *Time*, June 29, 1942, 52.

19. Paul Lazarsfeld et al., *Radio and the Printed Page* (New York: Duell, Sloan and Pearce, 1940), 135–36; Paul Lazarsfeld and Patricia L. Kendall, *Radio Listening in America* (New York: Prentice-Hall, 1948), 18–42; Paul Lazarsfeld, *The People's Choice*, 2nd ed. (New York: Columbia University Press, 1948), 131.

20. Jacques Ellul, *Propaganda: The Formation of Men's Attitudes* (New York: Knopf, 1965), 4–5.

21. David Jenemann, *Adorno in America* (Minneapolis: University of Minnesota Press, 2007), 9; T. J. Jackson Lears, *Fables of Abundance: A Cultural History of Advertising in America* (New York: Basic Books, 1994), 245. See also Kenneth Cmiel, "On Cynicism, Evil and the Discovery

of Communication in the 1940's," *Journal of Communication* 46 (Summer 1996): 88–108, and Todd Gitlin, "Media Sociology: The Dominant Paradigm," *Theory and Society* 6 (September 1978): 205–53.

22. Christopher Simpson, *The Science of Coercion: Communications Research and Psychological Warfare, 1945–1960* (New York: Oxford University Press, 1996), 62.

23. Lazarsfeld and Kendall, *Radio Listening*, 113.

24. James W. Carey, *Communication as Culture: Essays on Media and Society* (Boston: Unwin Hyman, 1989), 147.

25. Paul Lazarsfeld and Elihu Katz, *Personal Influence: The Part Played by the People in the Flow of Mass Communications* (Glencoe, IL: Free Press, 1955), 1.

26. Edward L. Bernays, *Propaganda* (New York: Liveright, 1928), 37.

27. Harold D. Lasswell, *Propaganda Technique in World War I* (Cambridge: MIT Press, 1971), 9; Harold D. Lasswell, *Democracy through Public Opinion* (Menasha, WI: George Banta, 1941), 25, 35.

28. Hadley Cantril, *The Invasion from Mars: A Study in the Psychology of Panic* (New York: Harper and Row, 1966), 130. See also Hadley Cantril, "Causes and Control of Riot and Panic," *Public Opinion Quarterly* 7 (Winter 1943): 669–79.

29. See the following works by Theodor Adorno: "An Analytical Study of the NBC *Music Appreciation Hour*," *Musical Quarterly* 78 (Summer 1994): 325–77; "On the Fetish Character in Music and the Regression of Listening," in *The Culture Industry: Selected Essays on Mass Culture*, ed. J. M. Bernstein (New York: Routledge, 2003), 29–60; *The Psychological Technique in Martin Luther Thomas' Radio Addresses* (Stanford, CA: Stanford University Press, 2000); "Radio Physiognomics," in *Theodore Adorno: Current of Music, Elements of a Radio Theory*, ed. Robert Hullot-Kentor (Frankfurt: Suhrkamp, 2006), 175–213; "A Social Critique of Radio Music," *Kenyon Review* 18 (Summer 1996): 229–36; and, with Max Horkheimer, "The Culture Industry: Enlightenment as Mass Deception," in *Dialectic of Enlightenment* (New York: Continuum, 1994), 120–67. For the classic account of Adorno's work in this period, see Martin Jay, *The Dialectical Imagination: A History of the Frankfurt School and the Institute of Social Research, 1923–1950 (Boston: Little, Brown and Company, 1973)*, 188–83, 219–24. See also Theodor Adorno, "Scientific Experiences of a European Scholar in America," in *The Intellectual Migration: Europe and America, 1930–1960*, ed. Donald Fleming and Bernard Bailyn (Cambridge: Harvard University Press, 1969); Jenemann, *Adorno in America*, 1–46; Robert Witkin, *Adorno on Popular Culture* (New York: Routledge, 2003).

30. Jay, *The Dialectical Imagination*, 190; Adorno, "Analytical Study," 331; Adorno, "On the Fetish Character," 51; Adorno, *Martin Luther Thomas*, 84.

31. Adorno, "Music Fetishism," 51; Adorno, "Social Critique," 232; Theodor Adorno, *Minima Moralia*, trans. E. F. N. Jephcott (New York: Verso, 2005), 201; Adorno, "Radio Physiognomics," 183–200.

32. Theodor Adorno, "The Culture Industry Reconsidered," in *The Culture Industry: Selected Essays on Mass Culture*, ed. J. M. Bernstein (New York: Routledge, 2003), 106; Horkheimer and Adorno, "The Culture Industry," 140; Adorno, *Minima Moralia*, 109. See, also Theodor Adorno et al., *The Authoritarian Personality* (New York: Harper and Brothers, 1950), 759–62.

33. "NBC Is Host as General Mills Changes Nets," *Advertising Age*, May 30, 1938, 27.

34. Lazarsfeld and Katz, *Personal Influence*, 32.

35. Walter Lippmann, *Public Opinion* (New York: Free Press, 1997), 158.

36. Horkheimer and Adorno, "The Culture Industry," 159.

37. See Richard J. Hand, *Terror on the Air! Horror Radio in America, 1931–1952* (Jefferson, NC: McFarland, 2006), 58.

38. Peters, *Speaking into the Air*, 6, 104.

39. For more on radio drama and influence of opinion, see Jason Loviglio, *Radio's Intimate Public: Network Broadcasting and Mass-Mediated Democracy* (Minneapolis: University of Minnesota Press, 2005).

40. See Terry Eagleton, *Ideology: An Introduction* (London: Verso, 1991), 89–91.

41. Daniel Foster raised the notion of characters behaving like radio announcers in his paper "'I See a Voice': Radio and the Need to Narrativize" (paper presented to the annual meeting of the Modern Language Association, Chicago, Illinois, 2007).

42. Homer, *The Odyssey*, trans. Robert Fagles (New York: Penguin, 1996), 276–77.

43. Briankle G. Chang, *Deconstructing Communication: Representation, Subject and Economies of Exchange* (Minneapolis: University of Minnesota Press, 1996) 48, 171–220. See also Jacques Derrida, *The Post Card: From Freud to Socrates and Beyond*, trans. Alan Bass (Chicago: University of Chicago Press, 1987). Chang touches on a debate between Derrida and analytical philosopher John Searle over J. L. Austin's philosophy of language. For details, see Gordon C. F. Bearn, "Derrida Dry: Iterating Iterability Analytically," *Diacritics* 25, no. 3 (Fall 1995): 3–25.

CHAPTER SEVEN

1. See Ann Gaylin, *Eavesdropping in the Novel from Austen to Proust* (New York: Cambridge University Press, 2002), 7–18.

2. Dziga Vertov, "Kinoks: A Revolution," in *Kino-Eye: The Writings of Dziga Vertov*, ed. Michelson, trans. Kevin O'Brien (Berkeley: University of California Press 1984), 18. See also Lucy Fischer, "*Enthusiasm*: From Kino-Eye to Radio-Eye," in *Film Sound: Theory and Practice*, ed. Elisabeth Weis and John Belton (New York: Columbia University Press, 1985), 247–64.

3. Carl Van Doren, introduction to Norman Corwin, *Thirteen by Corwin: Radio Dramas* (New York: H. Holt, 1943), vii.

4. This model of communication is also sometimes called a "hypodermic needle" theory, "vertical theory," or "magic bullet" model.

5. John Durham Peters, *Speaking into the Air: A History of the Idea of Communication* (Chicago: University of Chicago Press, 1999), 29, 62, 268.

6. For more on the aesthetics of sound and the use of filters, see Jean-François Augoyard and Henry Torgue, *Sonic Experience: A Guide to Everyday Sounds* (Kingston, Canada: McGill-Queens University Press, 2006), 48–58.

7. Erik Barnouw, *Handbook of Radio Writing: An Outline of Techniques and Markets in Radio Writing* (Boston: Little, Brown, 1939), 99.

8. Jacques Attali, *Noise: The Political Economy of Music*, trans. Brian Massumi (Minneapolis: University of Minnesota Press, 1985), 7.

9. See Steven Connor, *Dumbstruck: A Cultural History of Ventriloquism* (New York: Oxford University Press, 2000), 366–402; Gerald Nachman, *Raised on Radio* (Berkeley: University of

California Press, 1998), 133–38; Valentine Vox, *I Can See Your Lips Moving: The History and Art of Ventriloquism* (Kinswood, UK: Kay and Ward, 1981), 106.

10. Connor, *Dumbstruck*, 374; Max Horkheimer and Theodor Adorno, "The Culture Industry: Enlightenment as Mass Deception," in *Dialectic of Enlightenment* (New York: Continuum, 1994), 159.

11. See Pierre Schaeffer, "Acousmatics," in *Audio Culture: Readings in Modern Music*, ed. Christoph Cox and Daniel Warner (New York: Continuum, 2004), 76–81.

12. Christoph Riedweg, *Pythagoras: His Life, Teaching and Influence*, trans. Steven Rendall (Ithaca: Cornell University Press, 2002), 1–41.

13. Michael Chion, *The Voice in Cinema,* ed. and trans. Claudia Gorbman (New York: Columbia University Press, 1999), 16–29.

14. Barry Truax, *Acoustic Communication* (Norwood, NJ: Ablex, 1984), 156.

15. Benny was famous for making this rhetorical stretch as long and awkward as possible, which made the product endorsement both absurd and potent—perhaps the earliest effective use of the ironic style in broadcast advertising.

16. The "insincerity" of radio speech may be one reason why researchers found that "sincerity" was the most important quality that listeners cited in surveys about aural appeals. See Robert K. Merton, with Marjorie Fiske and Alberta Curtis, *Mass Persuasion: The Social Psychology of a War Bond Drive* (New York: Harper and Brothers, 1946), 71–106 and 144.

17. M. M. Bakhtin, *The Dialogic Imagination: Four Essays*, ed. Michael Holquist, trans. Caryl Emerson and Michael Holquist (Austin: University of Texas Press, 1981), 337.

18. See Marvin Carlson, "Theater and Dialogism," in *Critical Theory and Performance*, ed. Janelle G. Reinhelt and Joseph R. Roach (Ann Arbor: University of Michigan Press, 1992), 313–23.

19. Ibid., 341.

20. David M. Kennedy, *Freedom from Fear: The American People in World War II* (New York: Oxford University Press, 1999), 351–54; Michael C. C. Adams, *The Best War Ever: America and World War II* (Baltimore: Johns Hopkins University Press, 1994), 80.

21. George Raynor Thompson et al., *The Signal Corps: The Outcome (Mid-1943 through 1945)*, in *The United States Army in World War II: The Technical Services* (Washington, DC: Center of Military History, 1966), 580.

22. Harold Lavine and James Wechsler, *War Propaganda and the United States* (New Haven, CT: Yale University Press, 1940), 153, 165.

23. Asa Briggs, *The History of Broadcasting in the United Kingdom*, vol. 3, *The War of Words* (New York: Oxford University Press, 1995), 605.

24. Tubbs and Bernstein are quoted in Library of America, *Reporting World War II*, part 2, *American Journalism, 1944–1946* (New York: Library of America, 1995), 13, 24; Ernie Pyle, *Here Is Your War* (New York: Lancer Books, 1943), 86.

25. "What They See in the Papers" *Time*, April 19, 1943.

26. See William Saroyan, *The Human Comedy* (New York: Dell, 1943), 111.

27. Kathleen Battles has gone so far as to say that the Shadow's use of a portable radio prefigures cell phone use. See Kathleen Battles, *Calling All Cars: Radio Dragnets and the Technology of Policing* (Minneapolis: University of Minnesota Press, 2010), 220.

28. On the radio show, Clark Kent was always portrayed as an ace reporter, not the bumbling character in other forms of the Superman story.

29. Similar scenes can be found in several contemporaneous film thrillers. See, e.g., *The House on 92nd Street* (1945), *The Naked City* (1948), *Journey into Fear* (1943), *The Asphalt Jungle* (1950), and *He Walked by Night* (1948).

30. On this play, see also Harry Heuser, "Etherized Victorians: Drama, Narrative, and the American Radio Play, 1929–1954" (PhD diss., City University of New York, 2004), 345.

31. Virgil, *The Aeneid*, trans. W. F. Jackson Knight (New York: Penguin, 1964), 102–3.

32. Cooper's radio work deserves a book-length treatment. For insight, see Richard J. Hand, *Terror on the Air! Horror Radio in America, 1931–1952* (Jefferson, NC: McFarland, 2006), 83–87, 145–61; Arch Oboler, *Oboler Omnibus: Radio Plays and Personalities* (New York, Duell, Sloan and Pearce, 1945), 20.

33. This tale was among the most frequently adapted. Other versions of it appear on programs including *Suspense*, *The Columbia Workshop*, *Nightfall*, *Weird Circle*, and *Beyond Midnight*.

34. Augoyard and Torgue, *Sonic Experience*, 84.

35. Paul Ricoeur, *Interpretation Theory: Discourse and the Surplus of Meaning* (Fort Worth: Texas Christian University Press, 1976), 15.

36. Ibid., 16.

37. Aristotle, *De Anima*, trans. R. D. Hicks (Amherst, NY: Prometheus Books, 1991), 58.

CHAPTER EIGHT

1. On Oboler, see Erik Barnouw, *The Golden Web: A History of Broadcasting in the United States, 1933–1953* (New York: Oxford University Press, 1968), 71–73; John Dunning, *The Encyclopedia of Old Time Radio* (New York: Oxford University Press, 1998), 37–39; Bruce Lenthall, *Radio's America: The Great Depression and the Rise of Modern Mass Media* (Chicago: University of Chicago Press, 2007), 197–205; Gerald Nachman, *Raised on Radio* (Berkeley: University of California Press, 1998), 311–15; Arch Oboler, *Oboler Omnibus: Radio Plays and Personalities* (New York: Duell, Sloan and Pearce, 1945); Richard J. Hand, *Terror on the Air! Horror Radio in America, 1931–1952* (Jefferson, NC: McFarland, 2006), 83–105; Ira Skutch, *Five Directors: The Golden Years of Radio* (Los Angeles: Directors Guild of America, 1998), 52–82; and Christopher H. Sterling, *The Biographical Encyclopedia of American Radio* (New York: Routledge, 2011) 278–79.

2. For details, see Oboler, *Oboler Omnibus*, 44–46.

3. *Radio Life* is quoted in Dunning, *Encyclopedia of Old Time Radio*, 38.

4. Erik Barnouw, *Radio Drama in Action: Twenty-Five Plays of a Changing World* (New York: Rinehart, 1945), 386; John K. Hutchens, "Drama on the Air: Some Notes, Suggested by an Anthology, on Radio Writing as a Serious Art," *New York Times*, November 2, 1941, X12; William Matthews, "Radio Plays as Literature," *Hollywood Quarterly* 1 (October 1945): 40–50; Oboler, *Oboler Omnibus*, 153.

5. Orrin E. Dunlap, "When 13 Is Lucky," *New York Times*, July 23, 1939, X10.

6. Hand, *Terror on the Air!*, 85; Dunning, *Encyclopedia of Old Time Radio*, 39.

7. Arch Oboler, "I Submit to the Radio Networks the Following Proposition," *Variety*, May 3, 1944, 39.

8. Mildred Adams, "The Poor Young Art of Radio," *Nation*, May 4, 1946, 545; Robert J. Landry, "The Improbability of Radio Criticism," *Hollywood Quarterly* 2 (October 1946), 66.

9. See Jim Cox, *Say Goodnight, Gracie: The Last Years of Network Radio* (Jefferson, NC: McFarland, 2000), and Mickie Edwardson, "James Lawrence Fly's *Report on Chain Broadcasting* (1941) and the Regulation of Monopoly in America" *Historical Journal of Film, Radio, and Television* 22 (October 2002): 397–423.

10. Kenneth Bilby, *The General: David Sarnoff and the Rise of the Communications Industry* (New York: Harper and Row, 1986), 248.

11. See David L. Morton, *Sound Recording: The Life Story of a Technology* (Baltimore: Johns Hopkins University Press, 2004), 111–24. For an exploration of the use of transcription prior to tape, see Alexander Russo, *Points on the Dial: Golden Age Radio beyond the Networks* (Durham, NC: Duke University Press, 2010).

12. Friedrich Kittler, *Gramophone, Film, Typewriter*, trans. by Geoffrey Winthrop-Young and Michael Wutz (Stanford, CA: Stanford University Press, 1986), 107.

13. Arch Oboler, "Arch Oboler Sounds Requiem for Radio," *Variety*, January 3, 1945, 71. Oboler gave a similar speech at the University of Oklahoma one year later.

14. With apologies to two standard works on the period: Richard Hofstadter, *The Paranoid Style in American Politics and Other Essays* (New York: Knopf, 1965); Richard Pells, *The Liberal Mind in a Conservative Age: American Intellectuals in the 1940s and 1950s* (Cambridge: Harper and Row, 1985).

15. Nachman, *Raised on Radio*, 486; James L. Baughman, *The Republic of Mass Culture: Journalism, Filmmaking and Broadcasting in America since 1941* (Baltimore: Johns Hopkins University Press, 2006), 65; Barnouw, *Radio Drama in Action*, 228.

16. "FCC grants 13 More FM applications," *Broadcasting*, January 21, 1946, 71.

17. Susan J. Douglas, *Listening In: Radio and the American Imagination* (Minneapolis: University of Minnesota Press, 1999), 225; Christopher H. Sterling and John M. Kittross, *Stay Tuned: A Concise History of American Broadcasting* (Belmont, CA: Wadsworth, 1978), 246–309. See also Lynn Spigel, *Make Room for TV: Television and the Family Ideal in Postwar America* (Chicago: University of Chicago Press, 1992).

18. "Nielsen Study: Radio Group up since '49," *Broadcasting*, December 15, 1952, 26; Nicholas J. Cull, *The Cold War and the United States Information Agency: American Propaganda and Public Diplomacy, 1945–1989* (New York: Cambridge University Press, 2008), 492. See also Alan Heil Jr., *Voice of America: A History* (New York: Columbia University Press, 2003), and Holly Cowan Shulman, *Voice of America: Propaganda and Democracy, 1941–1945* (Madison: University of Wisconsin Press, 1990).

19. Michele Hilmes, *Radio Voices: American Broadcasting, 1922–1952* (Minneapolis: University of Minnesota Press, 1999), 271–77.

20. Fred Allen, *Treadmill to Oblivion: My Days in Radio* (Rockville, MD: Wildside, 1954), 239.

21. Douglas Gomery, *A History of Broadcasting in the United States* (Malden, MA: Blackwell, 2008), 142–64; Asa Briggs and Peter Burke, *A Social History of the Media: From Gutenberg to the Internet* (Malden, MA: Polity Press, 2009), 208–9.

22. Kathy M. Newman, *Radio Active: Advertising and Consumer Activism, 1935–1947* (Berkeley: University of California Press, 2004), 166–92. See also Howard Blue, *Words at War: World War II Era Radio Drama and the Postwar Broadcasting Industry Blacklist* (Lanham,

MD: Scarecrow, 2002), 227–36; Robert L. Hilliard and Michael C. Keith, *The Quieted Voice: The Rise and Demise of Localism in American Radio* (Carbondale: Southern Illinois University Press, 2005), 52–54; Gerd Horten, *Radio Goes to War: The Cultural Politics of Propaganda during World War II* (Berkeley: University of California Press, 2002), 177–83; and Susan Smulyan, *Selling Radio: The Commercialization of American Broadcasting, 1920–1934* (Washington, DC: Smithsonian Institution Press, 1994), 58–59.

23. Newman, *Radio Active*, 184. See also the Federal Communications Commission, "The Blue Book [excerpts]," in *Documents of American Broadcasting*, ed. Frank J. Kahn (New York: Appleton-Century-Crofts, 1968).

24. Baughman, *Republic of Mass Culture*, 32; Hilliard and Keith, *The Quieted Voice*, 53. See also Cox, *Say Goodnight, Gracie*, 120–50.

25. See Lizabeth Cohen, *A Consumer's Republic: The Politics of Mass Consumption in Postwar America* (New York: Vintage, 2003), 292–344.

26. For the classic account, see Barnouw, *The Golden Web*, 253–303.

27. David Everitt, *A Shadow of Red: Communism and the Blacklist in Radio and Television* (Chicago: Ivan R. Dee, 2007). See also Blue, *Words at War*, 339–42.

28. Joseph Julian, *This Was Radio: A Personal Memoir* (New York: Viking, 1975), 172–200.

29. On *Red Channels*, see Thomas Doherty, *Cold War, Cool Medium: Television, McCarthyism and American Culture* (New York: Columbia University Press, 2003), 24–59; Michael Kackman, *Citizen Spy: Television, Espionage and Cold War Culture* (Minneapolis: University of Minnesota Press, 1998), xxv; Ted Morgan, *Reds: McCarthyism in Twentieth-Century America* (New York: Random House, 2003), 256; and Ellen Schrecker, *Many Are the Crimes: McCarthyism in America* (New York: Little, Brown, 1998), 218.

30. Doherty, *Cold War, Cool Medium*, 36.

31. See chapter 7. I used the following reference works for data on program schedules and their content: Dunning, *Encyclopedia of Old Time Radio*; J. David Goldin, *Radio Yesteryear Presents the Golden Age of Radio* (New York: Yesteryear, 1998); and Mitchell E. Shapiro, *Radio Network Prime Time Programming, 1926–1967* (Jefferson, NC: McFarland, 2002).

32. Exceptions include Cox, *Say Goodnight, Gracie*; Dunning, *Encyclopedia of Old Time Radio*; J. Fred MacDonald, *Don't Touch That Dial! Radio Programming in American Life from 1920 to 1960* (Chicago: Nelson-Hall, 1991); Nachman, *Raised on Radio*; and Jennifer Hyland Wang, "The Case of the Radioactive Housewife: Relocating Radio in the Age of Television," in *The Radio Reader: Essays in the Cultural History of Radio*, ed. Michele Hilmes and Jason Loviglio (New York: Routledge, 2002), 343–66.

33. Paul Edwards, *The Closed World: Computers and the Politics of Discourse in Cold War America* (Cambridge: MIT Press, 1996), 13.

34. See Alan Nadel, *Containment Culture: American Narratives, Postmodernism, and the Atomic Age* (Durham, NC: Duke University Press, 1995), 1–37; George Kennan, "The Long Telegram," *Containment: Documents of American Policy and Strategy, 1945–1950*, ed. Thomas H. Etzold and John Lewis Gaddis (New York: Columbia University Press, 1978), 50–63. The secondary literature on Kennan's containment theory is vast. See, e.g., John Lewis Gaddis, *The United States and the Origins of the Cold War, 1941–47* (New York: Columbia University Press, 1972), and Anders Stephenson, *Kennan and the Art of Foreign Policy* (Cambridge: Harvard University Press, 1989).

35. See Tom Gunning, *The Films of Fritz Lang: Allegories of Vision and Modernity* (London: BFI, 2000), 9–11, 475–80.

36. I borrow this term from Kenneth D. Rose, *One Nation Underground: A History of the Fallout Shelter in American Culture* (New York: New York University Press, 2001), 38–77. See also Paul Boyer, *When Time Shall Be No More: Prophecy Belief in Modern American Culture* (Cambridge, MA: Belknap, 1995), 117–25; and Allan Winkler, *Life under a Cloud: American Anxiety about the Atom* (New York: Oxford University Press, 1993).

37. Paul Boyer, *By the Bomb's Early Light: American Thought and Culture at the Dawn of the Atomic Age* (New York: Pantheon, 1985), 64.

38. Hofstadter, *The Paranoid Style*, 29–30.

39. Pells, *The Liberal Mind*, 345.

CHAPTER NINE

1. Kathleen Battles, *Calling All Cars: Radio Dragnets and the Technology of Policing* (Minneapolis: University of Minnesota Press, 2010); Jason Loviglio, *Radio's Intimate Public: Network Broadcasting and Mass-Mediated Democracy* (Minneapolis: University of Minnesota Press, 2005), 102–22; and Elena Razlogova, "True Crime Radio and Listener Disenchantment with Network Broadcasting, 1935–1946," *American Quarterly* 58 (March 2006): 137–58. See also Jim Cox, *Radio Crime Fighters* (Jefferson, NC: McFarland, 2002); and Gerald Nachman, *Raised on Radio* (Berkeley: University of California Press, 1998), 298–311.

2. Battles, *Calling All Cars*, 154.

3. See, e.g., David Cochran, *American Noir: Underground Writers and Filmmakers of the Postwar Era* (Washington, DC: Smithsonian Institution Press, 2000); Carl Malmgren, *Anatomy of Murder: Mystery, Detective and Crime Fiction* (Bowling Green: Bowling Green State University, 2001); William Marling, *The American Roman Noir: Hammett, Cain and Chandler* (Athens: University of Georgia Press, 1995); Sean McCann, *Gumshoe America: Hard-Boiled Crime Fiction and the Rise and Fall of New Deal Liberalism* (Durham, NC: Duke University Press, 2000); and Erin Smith, *Hard-Boiled: Working-Class Readers and Pulp Magazines* (Philadelphia: Temple University Press, 2000).

4. James Naremore, *More than Night: Film Noir in Its Contexts* (Berkeley: University of California Press, 1998). See also Alain Silver, introduction to *Film Noir Reader*, ed. Alain Silver and James Ursini (New York: Limelight, 1996).

5. Tom Gunning, *The Films of Fritz Lang: Allegories of Vision and Modernity* (London: BFI, 2000), 300.

6. Raymond Chandler, "The Simple Art of Murder," in *Later Novels and Other Writings* (New York: Library of America, 1995), 990. See also Marling, *The American Roman Noir*, 201–63; and McCann, *Gumshoe America*, 145–50.

7. Nachman, *Raised on Radio*, 299.

8. Karen Hollinger, "Film Noir, Voice Over and the *Femme Fatale*," in *Film Noir Reader*, ed. Alain Silver and James Ursini (New York: Limelight, 1996), 243–60. See also Sarah Kosloff, *Invisible Storytellers: Voice Over Narration in American Fiction Film* (Berkeley: University of California Press, 1988).

9. Jay Beck made this point in a recent presentation: Jay Beck, "The Narrator Has Your Ear:

Orson Welles' First Person Singular, Voice-Over, and the Cinematic Influence of Radio Aesthetics" (paper presented at the Chicago Film Seminar, Chicago, Illinois, April 3, 2008).

10. See, e.g., Paul Young, *The Cinema Dreams Its Rivals: Media Fantasy Films from Radio to the Internet* (Minneapolis: University of Minnesota Press, 2006), 73–135. For touchstone works in the film/radio field, see Rick Altman, "Deep Focus Sound: Citizen Kane and the Radio Aesthetic," in *Perspectives on Citizen Kane*, ed. Ronald Gottesman (New York: G. K. Hall, 1996), 94–121; Michel Chion, *Film, a Sound Art*, trans. Claudia Gorbman (New York: Columbia University Press, 2009); and Claudia Gorbman, *Unheard Melodies: Narrative Film Music* (Bloomington: Indiana University Press, 1987).

11. Paul Schrader, "Notes on Film Noir," in *Film Noir Reader*, ed. Alain Silver and James Ursini (New York: Limelight, 1996), 53–64.

12. Cox, *Radio Crime Fighters*, 22.

13. See Mieke Bal, *Narratology: Introduction to the Theory of Narrative*, 2nd ed. (Toronto: University of Toronto Press, 1999), 89–90.

14. Chandler, "The Simple Art of Murder," 991.

15. Warren I. Susman, *Culture as History: The Transformation of American Society in the Twentieth Century* (Washington, DC: Smithsonian Institution Press, 2003), 204. See also Malmgren, *Anatomy of Murder*, 71–109; R. Austin Freeman, "The Art of the Detective Story," in *The Art of the Mystery Story*, ed. Howard Haycraft (New York: Biblio and Tannen, 1976), 7–17.

16. Edward Dimendberg, *Film Noir and the Spaces of Modernity* (Cambridge: Harvard University Press, 2004), 13–14, 171.

17. McCann, *Gumshoe America*, 141, 159–60. See also David Lehman, *The Perfect Murder: A Study in Detection* (New York: Free Press, 1989), 159–167.

18. See Dimendberg, *Film Noir*, 21–85.

19. For more on this type of policing and the media, see Battles, *Calling All Cars*; Claire Bond Potter, *War on Crime: Bandits, G-Men, and the Politics of Mass Culture*, (New Brunswick, NJ: Rutgers University Press, 1998); and Thomas Schatz, *Boom and Bust: American Cinema in the 1940's* (Berkeley: University of California Press, 1999), 378–86.

20. Hofstadter, *The Paranoid Style in American Politics and Other Essays* (New York: Knopf, 1965), 39; Mark Fenster, *Conspiracy Theories: Secrecy and Power in American Culture* (Minneapolis: University of Minnesota Press, 1999), 18–19. See also Timothy Melley, *Empire of Conspiracy: The Culture of Paranoia in Postwar America* (Ithaca: Cornell University Press, 2000).

21. See Ellen Schrecker, *Many Are the Crimes: McCarthyism in America* (New York: Little, Brown, 1998), 230.

22. Richard Gid Powers, *Secrecy and Power: The Life of J. Edgar Hoover* (New York: Free Press, 1987), 201. See also Battles, *Calling All Cars*, 33–146; Richard Gid Powers, "One G-Man's Family: Popular Entertainment Formulas and J. Edgar Hoover's F.B.I.," *American Quarterly* 30 (Autumn 1978): 471–92.

23. J. Fred MacDonald, *Don't Touch That Dial! Radio Programming in American Life from 1920 to 1960* (Chicago: Nelson-Hall, 1991), 169–170. See also Martin Priestman, *Crime Fiction from Poe to the Present* (Estover, UK: Northcote, 1998), and Ronald R. Thomas, *Detective Fiction and the Rise of Forensic Science* (New York: Cambridge University Press, 1999).

24. Nachman, *Raised on Radio*, 306.

25. Ibid., 473.

26. Cox, *Radio Crime Fighters*, 254–55; Ron Lackman, *The Encyclopedia of American Radio* (New York: Checkmark Books, 2000), 272.

27. Battles, *Calling All Cars*, 95.

28. Schrecker, *Many Are the Crimes*, 119–53; Melley, *Empire of Conspiracy*, 2, 9.

29. See Wayne Booth, *The Rhetoric of Fiction*, 2nd ed. (Chicago: University of Chicago Press, 1983), 175.

30. See Alan Nadel, *Containment Culture: American Narratives, Postmodernism, and the Atomic Age* (Durham, NC: Duke University Press, 1995), 17–89.

31. "Real Thriller," *Time*, May 15, 1950, 59.

32. John Dunning, *The Encyclopedia of Old Time Radio* (New York: Oxford University Press, 1998), 209; Nachman, *Raised on Radio*, 304.

33. Jacob Smith, *Vocal Tracks: Performance and Sound Media* (Berkeley: University of California Press, 2008), 109–12.

34. William H. Whyte Jr., *The Organization Man* (New York: Doubleday and Company, 1956), 3, 46, 171; Archibald MacLeish, "The Conquest of America," *Atlantic*, August 1949, 82; and David Riesman, *The Lonely Crowd: A Study of the Changing American Character* (New Haven, CT: Yale University Press, 1950).

35. Richard Pells, *The Liberal Mind in a Conservative Age: American Intellectuals in the 1940s and 1950s* (Cambridge: Harper and Row, 1985), 247.

CHAPTER TEN

1. See United States Congress, House Committee on Un-American Activities, *Exposé of the Communist Party of Western Pennsylvania Based upon Testimony of Matthew Cvetic, Undercover Agent: Hearings before the Committee on Un-American Activities, House of Representatives, Eighty-first Congress, Second Session* (Washington, D.C.: GPO, 1950).

2. William S. White, "Democrats Seek Spy Charge Study," *New York Times*, February 22, 1950, 9.

3. "Undercover Agent Lists Many as Reds," *Washington Post*, Feb 27, 1950, 2.

4. For more on Cvetic, see Daniel J. Leab, *I Was a Communist for the FBI: The Unhappy Life and Times of Matt Cvetic* (University Park: Pennsylvania State University Press, 2000).

5. "Ex-Agent 'Mislaid' Lost Inquiry File," *New York Times*, March 26, 1950, 25.

6. Leab, *I Was a Communist*, 70, 116.

7. Pete Martin and Matthew Cvetic, "I Posed as a Communist for the FBI," *Saturday Evening Post*, July 15, 22, and 29, 1950.

8. See Stephen J. Whitfield, *The Culture of the Cold War* (Baltimore: Johns Hopkins University Press, 1991), 132–34.

9. Thomas Doherty, *Cold War, Cool Medium: Television, McCarthyism and American Culture* (New York: Columbia University Press, 2003), 23.

10. Jim Cox, *Say Goodnight, Gracie: The Last Years of Network Radio* (Jefferson, NC: McFarland, 2000), 139–40.

11. Ellen Schrecker, *Many Are the Crimes: McCarthyism in America* (New York: Little,

Brown, 1998), 119–54, 152; Powers is quoted in Stephen J. Whitfield, *The Culture of the Cold War* (Baltimore: John's Hopkins University Press, 1991), 135; David Everitt, *A Shadow of Red: Communism and the Blacklist in Radio and Television* (Chicago: Ivan R. Dee, 2007), 191.

12. See Michael Kackman, *Citizen Spy: Television, Espionage and Cold War Culture* (Minneapolis: University of Minnesota Press, 1998), 45; Doherty, *Cold War, Cool Medium*, 140–53; Whitfield, *Culture of the Cold War*; and Michael Kazin, *The Populist Persuasion: An American History* (New York: Basic Books, 1995), 165–93.

13. Alan Nadel, *Containment Culture: American Narratives, Postmodernism, and the Atomic Age* (Durham, NC: Duke University Press, 1995), 86–87. See also Catherine Lutz, "Epistemology of the Bunker: The Brainwashed and Other New Subjects of Permanent War," in *Inventing the Psychological: Toward a Cultural History of Emotional Life in America*, ed. Joe Pfister and Nancy Schnog (New Haven, CT: Yale University Press, 1997), 245–69; and Christopher Simpson, *The Science of Coercion: Communications Research and Psychological Warfare, 1945–1960* (New York: Oxford University Press, 1996), 63–106.

14. Lutz, "Epistemology of the Bunker," 254. See also James H. Capshew, *Psychologists on the March: Science, Practice and Professional Identity in America, 1929–1969* (New York: Cambridge University Press, 1999), 159–86; Nathan G. Hale Jr., *The Rise of Psychoanalysis in the United States: Freud and the Americans, 1917–1985* (New York: Oxford University Press, 1995); and Ellen Herman, *The Romance of American Psychology: Political Culture in the Age of Experts* (Berkeley: University of California Press, 1995), 124–52.

15. See Hale, *The Rise of Psychoanalysis*, 299.

16. Ibid., 283.

17. Elaine Tyler May, *Homeward Bound: American Families in the Cold War Era* (New York: Basic Books, 1988), 191.

18. See Lutz, "Epistemology of the Bunker," 253–69, and David Seed, *Brainwashing: The Fictions of Mind Control, a Study of Novels and Films since World War II* (Kent, OH: Kent State University Press, 2005), 27–49.

19. Roland Barthes, "Diderot, Brecht, Eisenstein," in *Image-Music-Text*, trans. Stephen Heath (New York: Noonday, 1992), 70.

20. Warren Susman, *Culture as History: The Transformation of American Society in the Twentieth Century* (Washington, DC: Smithsonian Institution Press, 2003), 158–60.

CODA

1. *On a Note of Triumph: The Golden Age of Norman Corwin*, dir. Eric Simonson (Nona Films, 2005).

2. From my telephone interview with Norman Corwin, conducted September 20, 2008.

3. I have in mind Schaeffer's notion of the "sonorous object," as considered through the prism of Bill Brown's work on "Thing Theory." See Pierre Schaeffer, "Acousmatics," in *Audio Culture: Readings in Modern Music*, ed. Christoph Cox and Daniel Warner (New York: Continuum, 2004), 76–81, and Bill Brown, "Thing Theory," *Critical Inquiry* 28 (Autumn 2001): 1–22.

4. Friedrich Kittler, *Gramophone, Film, Typewriter*, trans. Geoffrey Winthrop-Young and Michael Wutz (Stanford, CA: Stanford University Press, 1999), xl.

5. On "media archaeology" as an approach, see Erkki Huhtamo and Jussi Parikka, eds., *Media Archaeology* (Berkeley: University of California Press, 2011); Wendy Hui Kyong Chun and Thomas Keenan, eds., *New Media, Old Media* (New York: Routledge, 2006); and Siegfried Zielinski, *Deep Time of the Media: Toward an Archaeology of Hearing and Seeing by Technical Means*, trans. Gloria Custance (Cambridge: MIT Press, 2006).

6. Miriam Bratu Hansen, "Why Media Aesthetics?," *Critical Inquiry* 30 (Winter 2004): 394.

INDEX

The letter *f* denotes figures.

ABC (American Broadcast Company), 164; programming on, 169, 186; Red Scare and blacklisting at, 186
Academy Award Theater, 184
Ackerman, William C., 260n9
acousmatic media, 146–48
acoustics, 34, 48. *See also* sound; sound effects
action, 5
actor unions, 21
"actual" sound, 8–11
Addams, Charles, 40f
Adler, Alfred, 213
Adorno, Theodor: on the passive listener, 122–23, 126; on radio demagoguery, 87; on the split consciousness of listeners, 11
adventure programs, 13, 59–60
Adventures by Morse, 35–36
Adventures of Phillip Marlowe, The, 184, 187–90
Adventures of Sam Spade, The, 186
advertising: brand sponsorship in, 27–28; persuasion of, 126–27; in product-based programming, 81–82; radio production of, 21, 27; as secondary messages in radio dramas, 147–48, 267nn15–16; social science research on, 118–23, 125, 131–32; spending on, 1, 14; for suspense dramas, 101; on television, 14; World War II role of, 84
aesthetic experience, 7–12; "actual" sound in, 8–11; of Benjamin's *aisthesis*, 10, 245n32; of Corwin's People's Radio, 11, 73–82, 225–26; of dramatic conventions of radio, 9–10; impermanence of radio in, 74–78; process of craft and reception in, 10–11; scholarship on, 8–9; space and time in, 13, 17–19, 25–31; structure of feeling in, 11–12
aesthetic ideologies, 73–74
afterlife of radio dramas, 14, 227–28
Agee, James, 59, 74
"Air Raid" (MacLeish), 72
Alcott, Louisa May, 76
Allen, Fred, 113, 167
Allen, Gracie, 86
Allen, Woody, 2
Allport, Gordon, 26, 118, 119
Altman, Rick, 8, 37, 253n40
ambient resonance, 42
Ameche, Don, 163
American Exodus, An (Lange and Taylor), 74

American in England, An, 64, 74–75, 168
Americans All Immigrants All, 76
American School of the Air, The, 21
America's Town Meeting of the Air, 27
Ames, Walter, 1
Amos 'n' Andy, 24, 29, 56, 124
amplitude, 34; dramatic uses of, 37, 42; measurement in decibels of, 37
Anderson, Benedict, 26, 82, 249n51
Anderson, Sherwood, 59, 77
Andrews, Dana, 184, 204, 205
announcers and hosts, 50–52
"Anonymous" (Cooper), 130
anthology programming, 170–71
anticommunism, 165, 168–70. *See also* Red Scare
antitrust law, 23
apocalypse dramas, 176–80
appel du vide effect, 156–57
architectural acoustics, 48
Arch Oboler's Plays, 92
Ardrey, Richard, 101
Aristotle, 5, 158, 245n32
Army Command and Administrative Network (ACAN), 95, 96f
Arnheim, Rudolf, 8, 10, 25, 39–40
atomic age, 172, 176–80, 203
audience, 8; audioposition of, 35–38, 86–88; aural competence of, 225–26; creation of, 6; listening choices of, 8; opinion leaders of, 123–25; passive listener tropes of, 122–27, 136, 263n5; World War II–era research on, 117–23. *See also* listeners
audioposition, 35–38, 86–88; in bounded narratives of the Cold War era, 173–76; Corwin's hybrid perspectives in, 69–72, 77–82, 87–88; creative mike techniques for, 38–45, 48–49; in "The Fall of the City," 49–56, 88; in "The Hitch-Hiker," 98–99; intimate style in, 56–65, 69–73; kaleidosonic style in, 57–58, 63–72, 256n18; listening for, 49–50; political contexts of, 58–60, 66–67, 73–74; *The Shadow*'s thematizing of, 60; standardized sound effects in, 45–48

audiotape, 164
aural depictions of action, 5

Bacon, Francis, 41
Bakhtin, Mikhail, 148, 149
Balázs, Béla, 54
Ball, Lucille, 99
Bandstand USA, 167
Bankhead, Tallulah, 20
Barnouw, Erik, 40–41, 163
Barry, Walter, 123
Barrymore, John, 20
Barrymore, Lionel, 39, 77–78, 128
Barthes, Roland, 5, 34, 222
Battles, Kathleen, 60, 182, 194, 267n27
Battle Stations, 152
BBC (British Broadcasting Corporation), 83–84
behavioral psychology, 211–12
Being and Time (Heidegger), 25
Bell for Adano, A (Hersey), 148
Benét, Stephen Vincent, 2; death of, 82; scripts of, 69, 74, 77, 101
Benjamin, Walter, 10, 245n32
Benny, Jack, 2; advertising sponsors of, 24, 27, 147, 267n15; *The Jack Benny Show* of, 86, 164, 167; move to television of, 167, 169; satire of, 115
Bentley, Elizabeth, 203
Bentley, Eric, 20, 82
Berelson, Bernard, 118
Berg, Gertrude, 250n59
Berg, Louis, 120
Bergen, Edgar, 86, 146
Berlin blockade, 203
Bernays, Edward, 118, 119, 122
Bernstein, Walter, 151
Best Years of Our Lives, The (Wyler), 107–8
"Between Americans" (Corwin), 69
bidirectional polarity, 43f
Big Clock, The (Farrow), 179
"Big Knife, The" (*Dragnet*), 199–200
"Big Parrot, The" (*Dragnet*), 200
"Big Phone Call, The" (*Dragnet*), 200

"Big Tear, The" (*Dragnet*), 201–2
Big Town, 154
Bill of Rights, 77–80
Biltmore Agreement, 66, 256n21
"Black Cat, The" (Poe), 100, 109
Blackett-Sample-Hummert, 21
blacklists, 82, 102, 165, 186, 204
blocking, 39
Blue Book (1946 FCC report), 167
Boston Blackie, 186, 199
Bourke-White, Margaret, 74
Boyd, James, 77
Boyer, Paul, 179
Bracken, Eddie, 212–13
Bradbury, Ray, 146, 149, 217
brainwashing, 217–23
Brecht, Bertolt, 34, 251n8
"Broadway Evening" (Proser), 30
Broadway Is My Beat: crime dramas on, 182, 190–93; models of the psyche on, 212; ventriloquism on, 146
Bromo-Seltzer, 102
Brown, Bill, 274n3
Brown, Himan, 33–34, 39, 100, 102
Buck Rogers, 172
Bulletin of the Atomic Scientists, 179
Burns, George, 86
Burns, Ken, 2

Cage, John, 22
Cain, James M., 184
Caldwell, Erskine, 74
California studios, 27–28
Calling All Cars, 63, 100, 141, 152
Campbell's Soup Playhouse, 255n6
Cantor, Eddie, 24
Cantril, Hadley, 26, 118, 119, 122, 260n9
Captain Midnight, 139–40
carbon microphones, 42
cardioid polarity, 43f
"Careless Talk Got There First," 142f
Carey, James, 55, 121, 254n62
Carleton, Dean, 194, 196, 197
Carr, John Dickson, 125

Carter's Little Liver Pills, 102
"Cartwheel" (*Columbia Workshop*), 30
Casey, Crime Photographer, 186
Cavalcade of America, 20, 76–78; morale-boosting dramas on, 109–10; political contexts of, 77–81; quasi-kaleidosonic style on, 76–77, 256nn6–7; signalmen of, 152; sponsors of, 256n7; World War II–era appeals on, 85; writers for, 77
CBS (Columbia Broadcasting System), 20–27; competition with NBC of, 21–22, 164; crime serials on, 186; Hollywood studios of, 27; local programming on, 167; network system of, 20–27, 167; Paley raids of, 164; prestige of, 21, 22; radio dramas on, 1, 13, 17, 23–24, 81–85, 100–101, 169, 186, 217; Red Scare and blacklisting at, 186; sound effects on, 105; television broadcasts of, 166; World War II role of, 119, 145. See also *Columbia Workshop, The*
Ceiling Unlimited, 144–45
Chambers, Whittaker, 203
Chandler, Raymond, 184, 185, 187–89
Chandu, the Magician, 139
Chang, Briankle, 134, 266n43
Chase and Sanborn Hour, The, 163
Chion, Michel, 35, 147
chorus (Greek), 50–51
Clock, The, 178
closed world narratives, 175–76
close readings, 10–11
Cohen, Lizabeth, 167
Cold War era (1945–55), 12–13, 159, 163–71; anthology programming of, 170–71; anxiety and paranoia of, 165, 179–81, 192–93, 204; apocalyptic thrillers on, 176–80; atomic technology of, 172, 176, 203; brainwashing dramas of, 217–23; closed world narratives of, 173–76; conformity and repression of, 202, 214, 219; decline of radio drama of, 168–71, 179–80, 222–23; domestic radio use in, 166–67; FBI dramas of, 193–97, 204–12; growth of television in, 164–68; implosion of aural scene in, 222–23; models of the

Cold War era (1945–55) (*cont.*)
psyche of, 211–23; propaganda of, 166; psychological ordeals of, 207–11; Red Scare culture of, 165, 168–70, 180, 181–82, 186, 203–12; Soviet containment policies of, 175; technological developments of, 172–73; time-out-of-time tropes in, 176–80. *See also* crime dramas

Cole, Alonzo Deen, 101

Columbia Workshop, The, 1–2, 19–25; audioposition in, 37–38; Fletcher's work for, 91; impressions of locale in, 30–31; intimate styles in, 58–59; perspective in, 13; political contexts of, 58–60; psychological dramas of, 92; Reis's vision for, 21–25, 30; signalmen of, 153; sound on, 91; staff of, 23–24; transmitting persuasion in, 127. *See also* "Fall of the City, The"

comedy shows, 170f, 171f

commercial radio, 81–88; criticism of, 167; secondary messages of, 147–48, 267nn15–16

Communist Party, 203–4

Congress of Industrial Organizations (CIO), 203

Conrad, William, 184

conventions, 6

cool media, 9, 245n28

Cooper, Wyllis: apocalyptic thrillers of, 178; signalmen dramas of, 154–55, 268n32; stream of consciousness method of, 102, 163

"Correction" (*Radio City Playhouse*), 153–54

Corwin, Norman, 13, 19, 73–82, 170, 225–29; on audioposition, 41; *Cavalcade of America* of, 76–78; depictions of space by, 28–29; later works of, 225; People's Radio of, 69–72, 77–82, 87–88, 225–26; radical politics and blacklisting of, 82, 165; "We Hold These Truths" of, 77–81, 87, 91, 225; World War II dramas of, 64, 69–72, 101, 129

Cotton, Joseph, 99, 184

Counterspy. See *David Harding, Counterspy*

Count of Monte Cristo, The (Dumas), 35

Craig, Edward Gordon, 20

Crews, Albert, 156–58

Crime Cases of Warden Lewis, The, 186

crime dramas, 13, 165–66, 170, 182–89; authors and actors of, 184; brainwashing on, 217–23; direct address in, 193–97; disordered underworlds of, 188–89, 197; evolution of, 182, 184; FBI's political commentary in, 193–97; hard-boiled language of, 185; links to film noir of, 184–89; lonewolf detectives in, 185–89, 192; police procedurals in, 187–93; psychotic disintegration in, 191–92; Red Scare, spies, and the FBI in, 204–11; testimony and confessions in, 197–202; themes of meaninglessness in, 200–202; underplaying in, 199. *See also* Cold War era (1945–55)

criminology, 211–12

Crisell, Andrew, 8, 74, 262n39

Crossley service, 119–20, 122

cultural contexts, 3

Cvetic, Matt: Congressional testimony of, 203–4; dramatization of, 204–11

Czitrom, Daniel, 64

Dangerous Assignment, 172, 179, 186, 206

Darrow, Whitney, Jr., 183f

David Harding, Counterspy, 124, 128, 138, 186; Cold War programming on, 172–76; new technology on, 172; signalmen of, 152; undercover stories on, 206

Davis, Elmer, 85

Day, Doris, 179

"Daybreak" (Corwin), 69–70

dead booths, 41

dead walls, 41, 253n33

"Death Talks Out of Turn" (*Molle Mystery Theater*), 141–44

decibels, 37

"Defenders, The" (Dick), 176

Defense Attorney, 186

della-Cioppa, Guy, 1–2, 5–6, 7, 12, 14

DeMille, Cecil B., 24
Denning, Michael, 22, 73
Depression era (1937–45), 12–13, 86–88, 222; acoustic perspective in, 19, 33–56; brand sponsorship in, 27–28; construction of radio rhythms of daily life in, 28; Corwin's aesthetic ideology in, 19, 73–82, 225–26; development of dramaturgy in, 19–25; documentary realism in, 74–78; homogenization of culture in, 19, 26–31; impermanence in, 74–78; political contexts of, 17–18, 19, 34, 50, 55–56, 58–60, 66–67, 73–74; role of celebrity in, 27–28; shaping of space and time in, 13, 17–19, 25–31; social and historical topics in, 76–78; studios for drama during, 41–42, 44–45*f*; supernatural radio of, 99–101; the thirteen-week season in, 28. *See also* dramas of space and time
Derrida, Jacques, 133–34
Destiny-machines, 177–78, 218–20
Devine, Jerry, 194, 198
Dewey, John, 254n62
Diary of Fate, The: apocalyptic thrillers on, 178; crime dramas on, 215; eavesdropping on, 139, 140; transmission thrillers on, 127
Dick, Phillip K., 176
Dickens, Charles, 22, 46, 156–58, 268n33
Dick Tracy, 20, 138
Diderot, Denis, 5
Dietz, John, 99, 106–7, 190
Dimendberg, Edward, 188–89
Dimension X, 172, 217–18; closed worlds on, 176; temporal dislocation in, 177. See also *X Minus One*
dioptric arts, 222
direct address: in "The Fall of the City," 51–52; in "The Hitch-Hiker," 103; on *I Was a Communist for the FBI*, 210; on *Mercury Theater*, 60–61; in "Nine Prisoners," 58; in postwar crime dramas, 193–97; in "The Tell-Tale Heart," 130
disappearing character gags, 252n13
distant readings, 10–11

documentary realism: of Depression-era books, 74–78; sympathy-empathy structures in, 59
Doherty, Thomas, 168, 204
"Donovan's Brain" (*Suspense*), 111–12, 207, 263n45
"Don't Talk" campaign, 141–45
doomsday clock, 179
Dorsey, Tommy, 179
Dos Passos, John, 25
Douglas, Susan, 7–8, 119, 166
Dracula (Stoker), 60, 101
Dragnet, 6, 100, 182–84, 186; apocalyptic thrillers on, 179; models of the psyche on, 212; new technology on, 172; police realism on, 193, 198–99; radio-television tie-in of, 166; tagline of, 181; testimony and confessions in, 199–202; undercover stories on, 205–6
drama criticism, 11
drama in Aristotelian theory, 5
dramas of space and time, 13, 17–19, 86–88; Corwin's hybrid perspectives in, 69–82, 87–88, 225–26; creating perspective in, 19, 33–56; discrepancies of detail in, 34; dissolution of spatial boundaries in, 25–28; impermanence of, 74–78; intimate styles of, 13, 19, 57–65, 69–72, 86–87; introductions of, 29; kaleidosonic styles of, 13, 19, 57–58, 63–72, 87, 256n18; political contexts of, 17–18, 19, 34, 50, 55–56, 58–60, 66–67, 73–74, 83; pseudopsychology in, 92; replacement by commercial radio of, 81–88; scriptwriting for, 19–25; social and historical topics in, 76–78; spatial textures in, 28–31; studio effects in, 17, 18–19, 41–42, 44–45*f*; temporal dislocation in, 177. *See also* Depression era (1937–45); "Fall of the City, The"; sound effects
Dreiser, Theodore, 59
Dryer, Sherman, 82
Dumas, Alexandre, 35
Dunning, John, 164
DuPont, 256n7

early radio era: drama in, 20; establishment of narration in, 28; first stations of, 24; paranormal qualities of, 100; reception problems in, 100, 262n23; sound effects in, 45
Easy Aces, 29, 85
eavesdropping, 94, 136–45, 155; active listening in, 139–40; devices for, 139–45
echo chambers, 41
Edgar Bergen Show, The, 164
Edwards, Paul, 175
ekphrasis, 33
Eliot, T. S., 22
Ellis, Anthony, 217–18
Ellul, Jacques, 121, 242n20
"Elwood" (*Suspense*), 212–13, 214
Empire Builders, The, 76
Empire of the Air (Burns), 2
Enemy of the People, An (Ibsen), 128
"Enjoy Yourself, It's Later than You Think," 179
Ericson, Eric, 143f
Escape, 102; actors on, 184; brainwashing dramas on, 218; closed worlds on, 176; models of the psyche on, 215; new technology on, 172; programming on, 172
espionage dramas: of the Cold War, 203–12; of World War II, 138–45
Evans, Walker, 74
Everitt, David, 165
excavational listening, 228–29
exteriority, 19. *See also* dramas of space and time

Fadiman, Clifton, 20
Falcon, The, 186
"Fall of the City, The," 17–19, 88, 221–22; actors on, 24; armory studio of, 17, 18–19, 49; artistic aims of, 19; audioposition and creation of perspective in, 25, 34, 49–56; critical response to, 17–18; diagram of, 54f; opening announcer of, 50–52; political context of, 17–18, 19, 34, 50, 55–56, 58; production staff of, 23–24, 49; script for, 19–20; temporal dislocation in, 177; Welles's role in, 51–53
false consciousness, 132–33
Fat Man, The, 186
FBI (Federal Bureau of Investigation): dramatic depictions of, 193–97, 204–12; HUAC testimonies of, 203–5; scientific crime fighting of, 193. *See also This Is Your FBI*
FBI in Peace and War, The, 152, 186
Federal Communications Commission (FCC), 164; Blue Book of, 167; FCC Act of 1934, 5–6, 169; licensing policies of, 166; public service requirements of, 167
Federal Theater, 21
Fenster, Mark, 192–93
Fibber McGee (character), 124
film noir, 179, 184–89; disordered underworlds of, 188–89; three stages of, 185–86, 187f. *See also* crime dramas
filters, 42
"Finger of God, The" (Reis), 38–39
fins, 41
Fireside Chats, 25–26, 101
First Nighter, The, 28, 148
Five for the Fourth, 163
Fletcher, Lucille, 91, 94; "The Hitch-Hiker" of, 97–99, 103–7, 112; signal-based themes of, 110–13, 163, 262n39; "Sorry, Wrong Number" of, 115–18, 128–29, 263nn3–4
Fly, James, 164
FM radio, 166
footsteps, 29–30
Ford Theater, The, 92, 129–30
Frankel, Lou, 81
Freedom's People, 76
Freud/Freudianism, 7, 211–12, 213
Front Page Drama, 109
Fuchs, Klaus, 203
Fussell, Paul, 9, 245n28

Gallup, George, 119
Galsworthy, John, 92

Gangbusters, 186
Geiger, Milton, 187, 188–89
General Mills, 118
Genette, Gérard, 51, 61
Gilman, Charlotte Perkins, 214
Gitelman, Lisa, 14
Globe Theater, The, 101, 139
Gomery, Douglas, 100
Good Neighbors, 102
gothic thrillers. *See* supernatural radio
Gould, Glenn, 82
Gouzenko, Igor, 203, 204
Granberry, Edwin, 58
Grand Hotel, 163
Great Plays, 20
Greek tragedy, 5, 50–51
"Green Light" (Cooper), 155, 207
Griffith, D. W., 31
Gulliver's Travels, 30
"Gumpert" (Corwin), 129, 132
Gun Crazy (Lewis), 185
Gunning, Tom, 177, 185, 250n74
Gunsmoke, 184
Guralnick, Elissa, 8, 105

Habermas, Jürgen, 26, 249n51
habituation syndrome, 63
Hammett, Dashiell, 184, 186
Hand, Richard, 99, 101, 164
Hansen, Miriam, 10, 55–56, 229
Hawkins, Sherman, 175
Hayworth, Rita, 184
"Headless Horseman of Sleepy Hollow, The" (Irving), 101
Hearst, William Randolph, 23
Heidegger, Martin, 25
Heilman, Robert, 92
Helfin, Van, 184
Hemingway, Ernest, 20
Here Is Your War (Pyle), 107
Hermit's Cave, 126, 127
Herrmann, Bernard, 78, 91, 97, 106
Hersey, John, 148

Herzog, Herta, 118, 119
Hilmes, Michele, 8; on the disciplined audience, 263n5; on the growth of television, 166–67; on imagined community, 97; on sound effects, 46
Himmler, Heinrich, 144–45
Hiss, Alger, 203
historical listening, 228–29
"Hit and Run" (*Suspense*), 214–15
Hitchcock, Alfred, 214
"Hitch-Hiker, The" (Fletcher), 97–99; atmospheric richness of, 102–3; score of, 104–5, 106; signal-based sound effects of, 103–7, 109, 112, 262n39; single audioposition of, 98–99; temporal dislocation in, 177
Hofstadter, Richard, 179, 192–93
Hollywood film of World War II, 95
Hollywood Hotel, 27
Honzl, Jindřich, 5
Hooper ratings, 119–20, 122
Hoover, J. Edgar, 193–94, 204
Hope, Bob, 86
Hop Harrigan, 172
Horkheimer, Max, 126
"Horla, The" (Maupassant), 126–27, 207
hot media, 245n28
Houseman, John, 60, 62, 71
House on 92nd Street, The (Hathaway), 185
House Un-American Activities Committee (HUAC), 203–5
Hucksters, The (Wakeman), 167
Hugo, Victor, 24
Human Comedy, The (Saroyan), 151
Hume, David, 3
Hurston, Zora Neale, 21
Hutchens, John, 86, 101, 102
hypercardioid polarity, 43f
hypodermic needle theory, 137, 266n4

Ibsen, Henrik, 21, 128
I Deal in Crime, 186
I Led 3 Lives, 204
Illustrated Man, The (Bradbury), 217

I Love a Mystery, 35–36, 186
imagination. *See* pictures in the mind
imagined community, 26, 97, 249n51
impermanence of radio dramas, 74–78
independent stations, 167
Information, Please!, 20
Inner Sanctum Mysteries, 27, 100–102, 141, 186; actors on, 184; "The Horla" on, 126; introductions in, 29; models of the psyche on, 212; supernatural sounds on, 124, 127; "The Tell-Tale Heart" on, 109, 130
Insull, Sam, 23
interiority, 6–7, 13, 135, 158; in Cold War-era psychological ordeals, 208, 222–23; in doubt, 149–50; of the late radio period, 165; shaping by technology of, 223; sound effects of, 91, 103–7. *See also* psychological radio
intimate styles, 13, 19, 56–65, 69–72, 86–87; in adventure programs, 59–60; confusion with radio medium of, 65, 255n16; in Corwin's "This Is War," 69–71; focus on places in, 73; impermanence of, 75–76; political use of, 58–60, 73–74; social purpose of, 65; Welles's use of, 60–65, 72
Irene Rich Dramas, 163
Iron Curtain, The (Wellman), 204
Ironized Yeast, 102
Irving, Washington, 101
"I Saw Myself Running" (*Suspense*), 213–14
"It's a Gift" (*Ford Theater*), 129–30
"It's later than you think" tagline, 164, 176–77, 178
I Was a Communist for the FBI, 184, 186, 204–11

Jack Benny Show, The, 164, 167, 169
Jay, Martin, 122
Jenemann, David, 121
"Joe Blair" (*Broadway Is My Beat*), 191–92
Joe Friday (character). See *Dragnet*
journalism dramas, 150–55. *See also* signalmen
Julian, Joseph, 2, 168

"Julie Dixon Case, The" (*Broadway Is My Beat*), 190
Julius Caesar, 60

Kahn, Douglas, 25
kaleidosonic styles, 13, 19, 57–58, 63–72, 87; announcer introductions in, 67–68; in Corwin's "This Is War," 69–71; decline of, 81–88; focus on events in, 73; impermanence of, 75–76; in *The March of Time*, 66–69, 83; political contexts of, 66–67, 73–74; two-dimensionality of, 75; in "The War of the Worlds," 65–66, 71–72, 256n18
Kammen, Michael, 77
Karloff, Boris, 109, 130
Kate Smith, 27
Katz, Elihu, 123
Kazin, Alfred, 74, 75–76
Keats, John, 22
Kempton, Murray, 63
Kennan, George, 175
"Kenny Purdue" (*Broadway Is My Beat*), 192
Kesten, Paul, 22, 25
keynote sounds, 33, 105
"Kill" (Oboler), 126–27
"Killer, Come Back to Me" (Bradbury), 149
Kipling, Rudyard, 262n23
Kirkpatrick, Bill, 27
Kittler, Friedrich, 164, 228
Kittross, John, 166
Knight, Vic, 30
Koch, Howard, 62, 256n18
Korean War, 203, 217
Kyser, Kay, 86

Landry, Bob, 82
Lange, Dorothea, 74
Laske, Otto, 47
Lasswell, Harold, 118, 119, 122
Lastra, James, 39
late radio age. *See* Cold War era (1945–55)
Laughton, Charles, 129
Laura (Preminger), 185

Lawrence, Charlotte, 213
Lazarsfeld, Paul: mass psychology theories of, 118–22, 125, 131–32, 209, 210; on who says what to whom, 135
Lears, T. J. Jackson, 121
Lenthall, Bruce, 26, 67–68
Lessing, G. E., 5, 38, 242n14
Let Us Now Praise Famous Men (Agee and Evans), 59, 74
Lewis, Elliot, 37–38, 173–76, 190
Lewis, William Bennett, 23, 82
Lights Out!, 4, 85, 100; psychological drama on, 92, 102; signalmen on, 156–58; sponsors of, 102; stream of consciousness on, 102, 154; tagline of, 164, 176–77, 178
linking of mind and medium, 6–7, 14, 228–29, 242n20
Lippmann, Walter, 125, 254n62
Lippmann-Dewey Debate, 254n62
listeners: aural competence of, 225–26; as characters of radio dramas, 124–28; eavesdropping archetypes of, 136–45, 155; excavational aspects of, 228–29; mobility of, 152, 268n27; reciprocity of, 226–28; social science paradigms of, 121–23, 125, 131–32; solitary state of, 158. *See also* audience
listening imagination. *See* pictures in the mind
"Listen to the People" (Benét), 69
live walls, 41, 253n33
Livingston, Mary, 252n13
localism: creation of impressions of, 30–31, 33; detachment from, 26–28
local networks, 23
Lombardo, Guy, 179
Lone Ranger, The, 44f, 186
lone-wolf detectives, 185–87. *See also* crime dramas
Lord, Phillips H., 46
Lorentz, Pare, 63
Lorre, Peter, 126
Lovejoy, Frank, 204
Loviglio, Jason, 182, 255n16
loyalty programs, 82, 182, 194

Lucretius, 46
Lukács, György, 80
Lutz, Catherine, 207
Lux Radio Theater, The, 4, 21, 24, 253n39; crime thriller authors for, 184; espionage dramas in, 138; gothic thrillers on, 101; proxy vocalizations, 148; psychological drama on, 92; "Sorry, Wrong Number" on, 115; sound quality of, 42; supernatural sounds on, 124
Lynd, Robert and Helen, 26

MacDonald, J. Fred, 193
Macdonnell, Norman, 37–38
MacLeish, Archibald, 6; "Air Raid" of, 72; on announcer introductions, 67–68; on development of radio, 86, 87; pictures of America of, 69; on public opinion, 84; radio dramas of, 19–21, 34, 171; World War II work of, 82, 86, 119. *See also* "Fall of the City, The"
MacMurray, Fred, 184
Mail Call, 115
Major Bowes, 27
Maltese Falcon, The, 185
Man Behind the Gun, The, 24, 92, 102, 152
Man Called X, The, 186
"Man Who Liked Dickens, The" (Waugh), 130–31
March, William, 58
March of Time, The, 24, 66–69, 83
Marinetti, F. T., 25
Marlowe. See Adventures of Phillip Marlowe, The
Martian Chronicles, The (Bradbury), 217
Marvin, Carolyn, 65, 242n18
Maryland Mysteries, 27
mass communication: of Cold War–era testimonies, 181–82; of elites vs. the masses, 54–56, 254n62; FCC policies on, 5–6, 164, 166–67, 169; signal-based communication in, 13, 94–97, 107–9, 263n44; World War II–era focus on, 150–51, 156, 158, 181

Mauldin, Bill, 108f
Maupassant, Guy de, 126
May, Elaine Tyler, 214
Mayor of the Town, The, 128, 129, 139
McCambridge, Mercedes, 36
McCarthy, Joseph, 168–69, 180, 203–4
McChesney, Robert, 7–8, 21–22
McCracken, Allison, 99, 101
McGill, Earle, 23, 30; creation of sound perspective by, 34, 41, 105; union duties of, 82
McLuhan, Marshall, 9, 228, 242n20, 245n28
media archeology, 228
media effects, 94
media fantasy, 16–17, 223
media history, 7–12, 228–29
Meet the Author, 21
melodramas, 171f
Menefee, Selden, 64
Mercury Theater on the Air, The, 13, 19, 91, 255n6; crime drama authors for, 184; first-person voice-overs on, 185; intimate styles on, 60–65; psychological drama on, 92; supernatural drama on, 101. *See also* "War of the Worlds, The" (Welles)
"Meridian 7-1212" (Reis), 153, 156
Merton, Robert K., 118, 119
metalistening. *See* eavesdropping
Michael Shayne, 186
microphones: architectural acoustics in, 48; blocking of, 39; common polar patterns of, 43f; creating perspective with, 38–45, 48–49; parabolic apparatus in, 38–39, 42f; types of, 41, 42, 43f, 253n35; velocity sensitivity in, 30, 253n35
Milland, Ray, 184
Millay, Edna St. Vincent, 21
Miller, Arthur, 20, 92
mind: linking with radio medium of, 6–7, 14, 228–29, 242n20; as a stage, 2–3
Misérables, Les (Hugo), 24, 35, 60
Mitchell, W. J. T., 2, 3, 33
Mitchum, Robert, 184
mobility, 152, 268n27

Mohr, Gerald, 187
Molière, 20
Molle Mystery Theater, The, 101, 109; apocalyptic thrillers on, 178; eavesdropping on, 141–44; transmitting persuasion on, 128; ventriloquism on, 149; writers for, 184, 217
Moorehead, Agnes, 67, 115–16
morale programs, 92; signal-based sound in, 109–10; signalmen of, 153–54; transmitting persuasion in, 128, 135–36
Morgan, Brewster, 23, 58, 82
Morse, Carleton E., 35–37, 60, 127
Mott, Robert, 45, 47
Mr. District Attorney, 128, 199; crime dramas on, 182; eavesdropping on, 139; models of the psyche on, 212–13; new technology on, 172
Mrs. Stevenson. *See* "Sorry, Wrong Number" (Fletcher)
Mueller, Ken, 2
Mumford, Lewis, 25
Murrow, Edward R., 31
musical perspective, 33
music shows, 24, 167; acoustic perspective on, 41; programming of, 170f; studio surfaces for, 41
Mutual Network, 23, 154, 164, 169; blacklisting at, 186; crime serials on, 186; local programming on, 167
"My Client Curly" (Fletcher), 91
Mysterious Traveler, 172
Mystery in the Air, 126, 146
mystery serials, 170. *See also* crime dramas

Nachman, Gerald, 185, 193
Nadel, Alan, 207
Naremore, James, 99, 184–85
narration, 28, 52, 192, 194; by external narrators, 60, 92, 127; to a fictional second person, 52, 192, 194; by first-person characters, 28, 60–61, 66, 156, 173; intimate and kaleidosonic styles of, 70; retrospective forms of, 102, 185; in shocker anthologies,

127; in time-out-of-time tropes, 177; using present tense, 61–62; in "Zero Hour," 228–30. *See also* direct address
narratology, 11
narrowcasting, 167
NBC (National Broadcasting Company): advertising fees of, 119; competition with CBS of, 21–22, 164; Hollywood studios of, 27; network system of, 20–27; radio dramas on, 45, 84, 100–101, 152, 163, 169; sound effects department of, 45; variety programs on, 146; World War II role of, 95
NBC Symphony Orchestra, 20
Nero Wolfe, 186
nerve centers, 152–53
network systems, 20–27, 226–27; Blue Book challenges to, 167; centralization in, 23, 27–28; emergence of television in, 166–67; financial scandals in, 164; local content requirements of, 167; staff of, 23–24
New Deal, 55–56; conflicting ideology of, 34; cultures of unity in, 19, 73, 87; post–World War II rejection of, 192–93; WPA Federal Theater of, 21
Newman, Kathy, 167
Newman, Robert, 130
newscasts, 28; Biltmore Agreement on, 66, 256n21; programming of, 170f; sense of locale in, 31; up-to-the-minute feel of, 66–67
Ney, Richard, 131
Nichols, Ora, 45, 47
"Night Tide" (*The Adventures of Phillip Marlowe*), 187–89
noir. *See* film noir
nostalgia for radio, 226–28
NPR (National Public Radio), 225
numbers of radio plays, 4, 241n9

Oboler, Arch, 63, 70, 82; directing career of, 163; *Lights Out!* series of, 92, 163–64, 176–77, 178; "Night Flight" of, 178; radio-of-ideas proposal of, 164, 269n13; Requiem for Radio of, 164, 169; thrillers of, 100, 102; transmission dramas of, 126, 127; on war propaganda, 84; writing career of, 163–64, 170
Office of Radio Research (ORR), 119–23, 124
Office of War Information (OWI), 84–86, 141–45
old-time radio revivals, 226–28
omnidirectional polarity, 43f
"On a Note of Triumph" (Corwin), 81, 101
On a Note of Triumph (Simonson), 2
O'Neill, Eugene, 20, 92
On Stage, 178
Open Letter on Race Hatred, An, 24
opinion leaders, 123
opinion research, 119–23
Out of the Deep, 110
Out of This World, 146

Paley, Bill, 22, 82, 164
Pantagruel (Rabelais), 263n44
parabolic microphones, 38–39, 42f
paranoia movies, 204
paranormal radio. *See* supernatural radio
Parker, Dorothy, 22
"Passenger to Bali, A" (Welles), 62–63
passive listener tropes, 122–27, 136, 263n5
"Pass to Berlin" (*Escape*), 215
Pat Novak, for Hire, 186
Pearl Harbor attack, 77–78, 83–84
Peck, Gregory, 99
Pells, Richard, 180, 202
People's Radio, 69–72, 77–82, 87–88, 225–26
perdition, 156–57
performance theory, 11–12
Personal Influence (Lazarsfeld and Katz), 123
perspective, 13, 19, 33–56; audioposition for, 35–38, 86–88; Corwin's synthesis of, 69–82; diagrams of, 81f; differentiation of characters in, 36–37, 252n13; discrepancies in detail in, 34; in "The Fall of the City," 49–56, 88; intimate styles of, 13, 19, 57–65, 69–73, 86–87; kaleidosonic styles of, 13, 19, 57–58, 63–73, 81–88, 87, 256n18; microphone

perspective (cont.)
use for, 38–45, 48–49; musical creation of, 33; sound effects in, 45–47
Peters, John Durham, 118–19, 128, 137
Philbrick, Herbert, 204
Phillip Marlowe. See *Adventures of Phillip Marlowe, The*
Phillip Morris Playhouse, The, 101
Philo Vance, 186
Pickford, Mary, 24
pictures in the mind, 1–2; linking of mind and medium in, 6–7, 14, 228–29, 242n20; listeners as central in, 9–10; pragmatic understandings of, 3
"Pied Piper of Hamelin, The" (Reis), 127
Pierson, Walter, 33
"Pigeon in a Cage" (*Suspense*), 173–76
"Pit and the Pendulum, The" (Poe), 125–27
Plays for Americans, 163
pluralized perspectives, 69–72
Poe, Edgar Allan, 100, 109, 125, 130
point of audition, 35
political commentary, 193–97
political contexts, 3; of audioposition, 58–60, 66–67, 73–74; of the Cold War's Red Scare, 165, 168–70, 180, 181–82, 186, 203–12; of Depression-era dramas, 17–18, 19, 34, 50, 55–56, 58–60, 66–67, 73–74, 78–81; U.S. Constitution and Bill of Rights in, 77–80
postal principle of intercommunication, 134
post–World War II era. *See* Cold War era (1945–55)
Powell, Dick, 184
Powers, Richard Gid, 193, 205
Preminger, Otto, 185
present-tense narration, 61–62
President's Research Committee on Social Trends, 26
Price, Vincent, 176, 184
propaganda: of the Cold War era, 166; definition of, 260n8; hypodermic needle theory of, 137, 266n4; of the Office of War Information, 84–86, 141–45; shortwave radio for, 83, 85, 258n24; social science research on, 94, 121–23, 125, 133; of World War II, 58, 92–94
Proser, Leopold, 30
proxy vocalizations, 148–50
"Psalm for a Dark Year" (Corwin), 69
pseudopsychology, 92
Psycho (Hitchcock), 214
psychoanalysis, 211–12
psychological radio, 6–7, 13, 19, 83, 91–94, 155–59; atmospheric richness of, 102–3; brainwashing dramas of, 217–23; breakdown of message in, 156–58; development of, 91–94; dream sequences in, 92; eavesdropping in, 94, 136–45, 155; fast-talking dialogue on, 124–25; first-person voice-overs of, 185; gothic thrillers of, 99–103; listener experiences of metaphysical jeopardy in, 94; persuasion and instruction in, 94, 126–33, 135–36; in Red Scare spy dramas, 207–17; signal-based communication in, 91, 94, 97–99, 103–13, 135–36, 262n39; signalmen and journalists in, 94, 136, 150–59; single-voice scenes in, 92; sponsorship in, 102; temporal dislocation in, 177; transmission (sender/receiver) dramas of, 94, 115–18, 123–29, 207–8, 266n41; ventriloquism in, 94, 137, 146–50, 155. *See also* "Hitch-Hiker, The" (Fletcher); supernatural radio; World War II era (1941–50)
Psychology of Radio, The (Cantril and Allport), 26
Public Opinion Quarterly, 119
public service broadcasting, 167
public sphere, 26, 249n51
Pyle, Ernie, 64, 107–8, 151, 262n44, 263n44
Pythagorean acousmatics, 147, 267nn15–16

Quiet, Please, 130; apocalyptic thrillers on, 178, 179; confessions in, 198; new technology on, 172; signalmen dramas on, 154–55
quiz shows, 170f

Rabelais, 263n44
Radio (Arnheim), 39–40

Radio Act of 1927, 23, 100
Radio City Playhouse, 109; apocalyptic thrillers on, 178; models of the psyche on, 212, 214–15; signalmen of, 153–54; writers for, 217
Radio Days (Allen), 2
Radio Director's Guild, 82
radio dramas/plays: diminished role of, 13–14; ideology and values in, 11; necessity of scenes in, 5, 242n14; as positive imaginary endeavor, 9–11; process of listening in, 10–11; programming changes over time of, 168–71; scholarship on, 4–7; as term, 4–5
Radio Life, 163
Radio Writing (Barnouw), 40–41
Razlogova, Elena, 182
receiver positions, 94, 207–8; eavesdroppers in, 136–45, 155; passive listener tropes of, 122–27, 136, 263n5. *See also* transmission (sender/receiver) dramas
reciprocity, 226–28
recordings, 11–12, 227–28
Red Channels, 168–69
Red Menace, The (Springsteen), 204
Red Scare, 165, 168–70, 180, 181–82, 203–12; blacklists of, 82, 102, 165, 186, 204; brainwashing in, 217–23; criminalizing of Communism in, 205; FBI spy dramas of, 193–97, 204–11; HUAC testimonies on, 203–5; psychological ordeal dramas on, 207–17. *See also* Cold War era (1945–55)
Red smears. *See* blacklists
Red Snow (Petroff), 204
regional networks, 23
Reis, Irving, 6; creative vision of, 21–25, 30, 81; "The Fall of the City" of, 49, 50, 53–55, 58; film work of, 82; instructions to actors of, 39; "Meridian 7-1212" of, 153, 156
Requiem for Radio, 164, 169
research methodology, 4–7; "actual" sound in, 8, 10–12; close and distant readings in, 10; historical connectedness in, 6; inner life in, 6–7, 242n20; study of conventions in, 6; terminology of, 4–5; time frame of, 5–6, 12–14

reverberation, 37, 252n19
revivals of old-time radio, 1, 226–28
rhetoric, 39, 48–49, 57, 70, 180
"Ria Bouchinska" (Bradbury), 146–47, 149
ribbon microphones, 42, 253n35
Rice, Rosemary, 24
Richard Diamond, Private Detective, 186
Ricoeur, Paul, 157–58
Riesman, David, 202
Robinson, Edward G., 184
Robson, William N., 23–24; audioposition work of, 37–38, 42, 63; blacklisting of, 168; supernatural thrillers of, 102; World War II work of, 82
Romance of the Rancho, The, 76
Ronald Adams (character). *See* "Hitch-Hiker, The" (Fletcher)
Roosevelt, Eleanor, 67
Roosevelt, Franklin D.: on Corwin's scripts, 60; Fireside Chats of, 25–26, 101, 255n16; MacLeish's work for, 86; Pearl Harbor speech of, 48; public opinion on, 120; use of microphones by, 48
Rosenberg, Julius, 203
Rosten, Norman, 21, 81–82
Rowan, Roy, 188
Rubicam, Raymond, 84
Russo, Alexander, 9, 27, 253n35
"Ruth Larson" (*Broadway Is My Beat*), 190–91

Saint, The, 186
Sam Spade, 186
Sanctum. See Inner Sanctum Mysteries
Sandburg, Carl, 63, 74
Sarnoff, David, 25
Saroyan, William, 22, 77, 151, 171
scenes, 5
Schaeffer, Pierre, 106, 147, 274n3
Schafer, R. Murray, 33, 106–7, 262n39
Schiller, Friedrich von, 51
scholarship on radio plays, 4–7
Schrader, Paul, 185
Schramm, Wilbur, 118, 119

Schrecker, Ellen, 205
science fiction, 13, 172
Sconce, Jeffrey, 6–7, 71
Scotland Yard's Inspector Burke, 186
Screen Director's Playhouse, The, 184
Screen Guild Theater, 184
Sealed Book, The, 100, 124, 127
"Secret Agent 23" (Welles), 144–45
segues, 42
Seldes, Gilbert, 59
semantic tasks, 94
semantic tasks of signal-based sound effects, 103–13, 135–36, 163, 262n39
senders. *See* transmission (sender/receiver) dramas
Sergeant Preston of the Yukon, 186
Shadow, The, 24; actors on, 184; intimate styles on, 59–60; perspective on, 19; portable radio of, 152, 268n27; pseudopsychology on, 92; sponsors of, 148; studio layout for, 45f; thematization of audioposition on, 60
Shakespeare, William, 20, 22, 60, 225
Shaw, Artie, 21, 127
Shaw, George Bernard, 20
Shaw, Irwin, 20
Sheriff, The, 186
Sherwood, Robert, 48, 77, 85
Shirer, William L., 168
shockers. *See* supernatural radio
shortwave radio, 30, 83, 85, 250n69, 258n24
shunting, 140–41
"Signalman, The" (Dickens), 156–58, 268n33
signalmen, 94, 136, 150–59; breakdown of messages of, 156–58; information society of, 151–52; nerve centers and networks of, 152–55; responsibilities to the public of, 153–54
signals, 13, 227; Pavlovian impact of, 107–9; in psychological dramas, 94, 103–13, 135–36, 163, 262n39
Simonson, Eric, 2
"Simple Art of Murder" (Chandler), 185

Simpson, Christopher, 121
Siodmak, Curt, 111
situation comedies, 171f
Smith, Jacob, 199
Smith, Kate, 85
Smith, Smilin' Jack, 2
Smulyan, Susan, 7–8, 23
soap operas, 28, 171f
social science research, 118–23, 125, 131–32
Soldiers of the Press, 152–53
Solitude Series (Gould), 82
song/singing dialogue, 148
sonological competence, 47
sonorous object, the (*l'objet sonore*), 106, 274n3. *See also* signals
"Sorry, Wrong Number" (Fletcher), 115–18, 221, 263nn3–4; active/passive binary in, 124; transmission of persuasion in, 128–29
sotto voce, 92
sound: "actual" sound and, 8–11; amplitude of, 34, 37, 42; aural rules of, 28; creating perspective with, 33–56; discrepancies in detail in, 34; engineer jargon for, 41; measurement in decibels of, 37; prioritizing of narrative in, 39, 44f; props for, 41; proxy vocalizations in, 148–50; in psychological dramas, 92; studio environments for, 17, 18–19, 41–42, 44–45f; technological development of, 164. *See also* microphones
sound caves, 41, 44f
sound effects, 34, 45–48; for affective states, 91, 97–99, 103–7; creation of spatial tiers with, 174–76; for eavesdropping, 139–45; footsteps, 29–30; keynote coding of scenes in, 33, 105; perdition and the *appel du vide* in, 156–57; post–World War II use of, 164; recordings of, 45–46; segues, 30, 42; shared vocabularies of, 45–48, 253n40; shunting of voices as, 140–41; as signals, 94, 103–13, 135–36, 163, 226–27, 262n39; studios for, 41, 44f; in supernatural thrillers, 103–13
sound studies, 11
sound tents, 41, 44f

"Sound That Kills, The" 143f
space-binding, 25–31; homogenized content in, 25–28; introductions in, 29; textured spaces in, 28–31. *See also* dramas of space and time
Spier, William, 67, 82, 99, 106
spy dramas: of the Cold War, 203–12; of World War II, 138–45
Stamm, Michael, 260n11
Standard Brands, 27, 28, 87
Stanton, Frank, 119
Stanwyck, Barbara, 184
Starch, Daniel, 119
Steele, Wilbur Daniel, 92
Stein, Gertrude, 64–65
Sterling, Christopher, 166
Sterne, Jonathan, 12, 46
Stewart, James, 78–81, 99
St. Joseph, Ellis, 62
Stoker, Bram, 101
Stoops, Herbert Morton, 142f
Stott, William, 69
Strange Dr. Weird, The, 100, 124–25, 127, 138
Strange Interlude (O'Neill), 92
Stranger, The, 179
Stravinsky, Igor, 22
structure of feeling, 11–12
Studio One, 92, 101, 128
"Subway" (*Suspense*), 215–17
Superman, 138; Atom Man villain of, 172; eavesdropping in, 139; journalism on, 152, 268n28
supernatural radio, 92, 99–103; atmospheric richness of, 102–3; of the early radio era, 100; opposition to, 101; retrospective narrators in, 102; signal-based sound effects of, 103–13, 135–36, 163, 262n39; sponsorship in, 102; stream of consciousness in, 102; *Suspense* anthology in, 99–100; transmission dramas of, 115–18, 123–29, 266n41. *See also* "Hitch-Hiker, The" (Fletcher); psychological radio
survey-based research, 119–23

Susman, Warren, 188, 222
Suspense, 6, 99–100; audioposition on, 37–38; brainwashing dramas on, 217–23; casts on, 184; character confessions on, 198; Cold War programming on, 172; crime dramas on, 182; "Donovan's Brain" on, 111–12, 263n45; espionage dramas on, 139; first-person voice-overs on, 185; Fletcher's plays on, 97–99, 103–7, 110–13; "The Hitch-Hiker" on, 97–99, 103–7, 109, 112, 262n39; models of the psyche on, 212–21; Peabody Award of, 101; proxy vocalizations on, 148; radio-television tie-in of, 166; retrospective narrators on, 102; "Ria Bouchinska" on, 146–47; role of listeners on, 124; scriptwriters for, 184; signalmen on, 152; "Sorry, Wrong Number" on, 115–18, 128–29, 263nn3–4; temporal dislocation on, 177; transmitting persuasion on, 127–28, 130–31; undercover stories on, 206; "Zero Hour" on, 165
switchboard operators, 153
sympathy-empathy structures, 59

Tale of Two Cities, A (Dickens), 46
Tales of the Texas Rangers, 186
talk shows, 170f
Tallman, Robert, 63–64, 120
Taylor, Paul S., 74
technology, 3, 242n11; of the atomic age, 172, 176–80; FM radio, 166; post–World War II developments in, 164, 172–73; of television, 14, 164–68; of transmission, 242n17
television, 14, 164–68; critical scholarship on, 170; tie-ins with radio of, 166–67
"Tell-Tale Heart, The" (Poe), 100, 109, 130
temporal dislocation, 164, 177–80
tense (in grammar), 61, 70
Terry and the Pirates, 102
testimony, 197–202
Theater Guild on the Air, The, 148
theater of the mind (as term), 1–3, 9–10, 12, 223

theories of mind, 3, 209
Thin Man, The, 186
This Gun for Hire, 185
"This Is War" (Corwin), 69–72
This Is Your FBI, 60; crime dramas on, 182; eavesdropping on, 140–41; ideological underworld on, 197; as police procedural, 193; political commentary on, 194–97, 212; positive role of truth on, 200; testimony on, 197–98
Thompson, Dorothy, 58
Thompson, Emily, 41
Thompson, George, 95
Thor, Larry, 190
"Three Skeleton Key" (*Escape*), 176
time: Destiny-machines in, 177–78, 218–20; radio-based rhythms of, 28, 250n60; time-out-of-time tropes of, 164, 177–80; in wired networks, 30, 250n69. *See also* dramas of space and time
"Time to Reap, A" (Benét), 69
Todorov, Tzvetan, 101–2
Toffler, Alvin, 242n20
Tolstoy, Leo, 20
Toscanini, Arturo, 20
To the President, 102, 163
"Towards the Century of the Common Man" (Benét), 69
tragedy in Aristotelian theory, 5
transcriptions, 227–28
transmission (sender/receiver) dramas, 13, 94, 115–18, 123–33, 207–8; compulsive listeners in, 124–25; eavesdroppers in, 136–45; false consciousness hypothesis of, 132–33; passive listener tropes in, 122–27, 136; pathological talkers in, 124–25; persuasion and transformation in, 127–33, 135–36, 266n41; ventriloquism in, 137, 146–50, 155
travel stories, 59
"Trip to Czardis, A" (Granberry), 58
Truax, Barry, 63, 106–7, 147, 262n39
Truman, Harry, 179
Tubbs, Vincent, 151

Twain, Mark, 76
"12 to 5" (Corwin), 154–55
26 by Corwin, 63
2000 Plus, 172

underplaying, 199
United Fruit Company, 84
"Unity Fair" (Corwin), 75
University of Chicago Round Table, The, 27
U.S.A. (Dos Passos), 25
US Army Signal Corps, 94–97, 150
US Constitution and Bill of Rights, 77–80

Valentino, Tom, 45–46
VanCour, Shawn, 28
Van Dine, S. S., 186
Van Doren, Carl, 137
van Voorhis, C. Westbrook ("The Voice of Time"), 67–68
variety shows, 170f
velocity microphones, 30
ventriloquism, 94, 137, 146–50, 155; acousmatic media of, 146–48; proxy vocalization as, 148–49; unified voices in, 146
verse literature, 20
Vertov, Dziga, 137
Voice of America, 166
voices: differentiation of characters using, 36–37, 252n13; disappearing character gags using, 252n13; in first-person voice-overs, 185; shunting of, 140–41; in single-voice scenes, 92; in ventriloquism, 94, 137, 146–50, 155

Wakeman, Frederick, 167
War Advertising Council (WAC), 84–85
War Bonds, 85
"War of the Worlds, The" (Welles): Cantril's study of, 122; intimate style of, 72; kaleidosonic style of, 65–66, 71–72, 256n18; political context of, 66–67; sound on, 91
Waugh, Evelyn, 130–31
"Weather Ahead" (*Radio City Playhouse*), 153

Webb, Jack, 166, 199
"We Hold These Truths" (Corwin), 77–82, 87; diagram of, 81f; hybridized perspective in, 78–81; musical score of, 78, 91; NPR's remake of, 225; political context of, 77–82
Weird Circle, The, 92, 126
Welles, Orson, 24, 81; in "Donovan's Brain," 111–12, 263n45; in "The Fall of the City," 51–53; film work of, 82; narration by, 28, 61–62; in "Secret Agent 23," 145; sound effects of, 46; supernatural drama of, 101; use of audioposition by, 35–36; use of intimate style by, 60–65, 72; use of kaleidoscopic style by, 65–66, 71–72, 256n18; in "We Hold These Truths," 77, 79–80
Wendell Wilkie "One World" Award, 82
West, Mae, 163
Whistler, The, 100, 124, 127–28, 141
White, E. B., 26
Whitman, Walt, 63–64
Whitton, John, 84
Whyte, William H., 202
Widmark, Richard, 184
Williams, Raymond, 6, 11–12, 242n17, 250n60
Willkie, Wendell, 120
Winchell, Walter, 28, 48
Wings to Victory, 92
wired networks, 30, 250n69
Witch's Tale, The (Cole), 101
Wolters, Larry, 27
Woollcott, Alexander, 20, 185
Woolrich, Cornell, 101, 184
Words at War, 64, 110; espionage dramas on, 138–39; proxy vocalizations on, 148; signalmen on, 152
workshop era. *See* dramas of space and time
Works Progress Administration (WPA), 21
World War II era (1941–50), 12–13, 86, 91–113, 155–59; centralization of communication in, 95, 96f, 150–51, 156, 158, 181; "Don't Talk" campaign of, 141–45; espionage dramas of, 137–45; gothic thrillers of, 101–3; government use of radio in, 82–86, 92–94, 118; journalism dramas of, 150–55; loyalty programs in, 82; morale-boosting/patriotic programs of, 92, 101, 109–10, 128, 153–54; Pearl Harbor attack of, 77–78, 83–84; product-based programming of, 81–82; propaganda of, 6, 58, 92–94, 121–23, 260n8; radio saturation of life during, 94–97, 260n9, 260n11; shortwave radio of, 83, 85, 258n24; signal-based communication in, 13, 94–97, 107–9, 263n44; social science research of, 94, 117–23, 125, 131–32; transmission (sender/receiver) dramas, 13, 94, 115–18, 123–33, 207–8; war dramas of, 85–86. *See also* propaganda; psychological radio
Wright, Frank Lloyd, 164
Writer's Guild, 82
Wurtzler, Steve, 27
Wyler, William, 107–8
Wylie, Max, 30

X Minus One: apocalyptic thrillers on, 177; closed worlds on, 176; models of the psyche on, 212–13; new technology on, 172. See also *Dimension X*

"Yellow Wallpaper, The" (Gilman), 214
You Have Seen Their Faces (Caldwell and Bourke-White), 74
Young, Paul, 26–27
Young Widder Brown, 85
Your Hollywood, 163

"Zero Hour" (Ellis), 165, 217–22
Ziv, Frederic W., 186, 204

www.ingramcontent.com/pod-product-compliance
Lightning Source LLC
Chambersburg PA
CBHW070820300426
44111CB00014B/2467